Lecture Notes in Electrical Engineering

Volume 760

The book series *Lecture Notes in Electrical Engineering* (LNEE) publishes the latest developments in Electrical Engineering - quickly, informally and in high quality. While original research reported in proceedings and monographs has traditionally formed the core of LNEE, we also encourage authors to submit books devoted to supporting student education and professional training in the various fields and applications areas of electrical engineering. The series cover classical and emerging topics concerning:

- Communication Engineering, Information Theory and Networks
- Electronics Engineering and Microelectronics
- Signal, Image and Speech Processing
- Wireless and Mobile Communication
- Circuits and Systems
- Energy Systems, Power Electronics and Electrical Machines
- Electro-optical Engineering
- Instrumentation Engineering
- Avionics Engineering
- Control Systems
- Internet-of-Things and Cybersecurity
- Biomedical Devices, MEMS and NEMS

For general information about this book series, comments or suggestions, please contact leontina.dicecco@springer.com.

To submit a proposal or request further information, please contact the Publishing Editor in your country:

China

Jasmine Dou, Editor (jasmine.dou@springer.com)

India, Japan, Rest of Asia

Swati Meherishi, Editorial Director (Swati.Meherishi@springer.com)

Southeast Asia, Australia, New Zealand

Ramesh Nath Premnath, Editor (ramesh.premnath@springernature.com)

USA, Canada:

Michael Luby, Senior Editor (michael.luby@springer.com)

All other Countries:

Leontina Di Cecco, Senior Editor (leontina.dicecco@springer.com)

**** This series is indexed by EI Compendex and Scopus databases. ****

More information about this series at http://www.springer.com/series/7818

Shelly Vadhera · Bhimrao S. Umre · Akhtar Kalam
Editors

Latest Trends in Renewable Energy Technologies

Select Proceedings of NCRESE 2020

Springer

Editors
Shelly Vadhera
School of Renewable Energy and Efficiency
National Institute of Technology
Haryana, India

Akhtar Kalam
College of Engineering and Science
Victoria University
Melbourne, VIC, Australia

Bhimrao S. Umre
Department of Electrical and Electronics Engineering
Visvesvaraya National Institute of Technology
Nagpur, India

ISSN 1876-1100 ISSN 1876-1119 (electronic)
Lecture Notes in Electrical Engineering
ISBN 978-981-16-1185-8 ISBN 978-981-16-1186-5 (eBook)
https://doi.org/10.1007/978-981-16-1186-5

This Springer imprint is published by the registered company Springer Nature Singapore Pte Ltd.
The registered company address is: 152 Beach Road, #21-01/04 Gateway East, Singapore 189721, Singapore

Committees

Patrons

Dr. Hanif Qureshi, IPS, (Director, DNRE & HAREDA)
Padma Shri Dr. Satish Kumar, (Director, NIT Kurukshetra)

Co-patrons

Prof. Lillie Dewan, (Coordinator, SREE, & Prof. EED, N.I.T. Kurukshetra)
Sh. O. D. Sharma, (Additional Director DNRE/HAREDA)
Sh. P. K. Nautiyal, (Scientific Engineer, A, DNRE/HAREDA)

Conveners

Dr. Shelly Vadhera, (Assoc. Prof. EED, N.I.T. Kurukshetra)
Mr. Sukhchain Singh, (Project Manager DNRE/HAREDA)

Secretaries

Dr. Shashi Bhushan Singh, (Asst. Prof. EED, N.I.T. Kurukshetra)
Dr. Rahul Sharma, (Asst. Prof. EED, N.I.T. Kurukshetra)

Organizing Committee

Dr. Gulshan Sachdeva, (Asst. Prof. MED, N.I.T. Kurukshetra)
Dr. Avadhesh Yadav, (Asst. Prof. MED, N.I.T. Kurukshetra)
Mr. Gaurav Sharma, (Faculty, SREE, N.I.T. Kurukshetra)
Ms. Amandeep Kaur, (Faculty, SREE, N.I.T. Kurukshetra)

Preface

National Conference on Renewable Energy and Sustainable Environment (NCRESE-2020) opened the pan-India platform for researchers and scientists to present and discuss research results in the areas of Renewable Energy and Environmental Engineering. The NCRESE-2020 conference was organized by the School of Renewable Energy and Efficiency, NIT Kurukshetra, in association with Bureau of Energy Efficiency, Government of India (GOI); Department of New and Renewable Energy, Haryana; and Haryana Renewable Energy Development Agency, Panchkula during August 28–29, 2020. The aim of this conference was to disseminate state-of-the-art information about Green Energy and Alternate Energy Resources, Wind Energy, Solar Energy, Solar Thermal, Bio-Energy Technologies, Energy Storage, while encouraging collaboration among multidisciplinary investigators across various diverse engineering disciplines. The conference wholeheartedly welcomed the contributions at the junction between theory and practice with immediate impact on applications, as well as interdisciplinary research involving cutting edge technologies. The NCRESE-2020 conference also discussed the practical challenges encountered and the solutions adopted by encapsulating the various theories, methods, applications, and tools for sustaining the environment.

NCRESE-2020 unfolded the stimulating and informative platform with a wonderful array of keynote and invited speakers in the field of Renewable Energy. The online format of this conference was conducive to foster significant interaction among attendees.

For the NCRESE-2020 conference, about 100 papers were received, out of which 41 made it to the final. We would like to take this opportunity and warmly thank all the authors who submitted their work to NCRESE-2020. Their contribution to the conference is greatly appreciated. Each paper submitted to NCRESE-2020 was peer-reviewed by at least two members of the reviewer committee (37 members, from 20 reputed institutes), who judged it for originality, significance, technical contents, application contents, and presentation style. In fact, most of the papers were reviewed by three or more reviewers who provided their input in the acceptance decision process.

Moreover, we would like to thank the developers and other professional staff of EasyChair.org, who made it possible for us to use the resources of this excellent and comprehensive conference management system, from the call of papers and inviting reviewers to handling paper submissions, communicating with the authors, and creating the online volume of the conference proceedings. In addition, we owe great thanks to the NCRESE-2020 conference organization members for their dedication and professionalism in taking care of the many aspects of conference management.

Collectively, the research discussed at the conference enhanced the knowledge and did value addition to one and all. This knowledge will lead to progress in renewable energy technologies and strategies. It will also empower all the readers to think out of the box solutions for sustainable growth of our environment.

Haryana, India
Nagpur, India
Melbourne, Australia
August 2020

Shelly Vadhera
Bhimrao S. Umre
Akhtar Kalam

About This Book

This book presents select proceedings of the National Conference on Renewable Energy and Sustainable Environment (NCRESE-2020) and examines a range of reliable energy-efficient harvesting technologies, their applications and utilization of available alternative energy resources. The topics covered include alternate energy technologies, smart grid topologies and their relevant issues, solar thermal and bioenergy systems, electric vehicles and energy storage systems and their control issues. The book also discusses various properties and performance attributes of advance renewable energy techniques and their impact on environmental sustainability. Further, importance of the key issues related to policy for renewable energy field and energy audit will lead the way to enhance the capacity of alternate energy. Low power energy sources apart from solar, wind and hydropower such as tidal power, hydrogen energy, and biomass, geothermal also highlight the importance of these emerging technologies and future challenges.

From the sustainable environment perspective, this book covers advancement in waste and resource management, green energy deployment, green buildings, integration of green energy and energy efficiency. Moreover, important topics are covered related to Bio-diversity: its past and prospects such as advancing sustainable food systems, sustainable agriculture, atmosphere, climate issues, meteorology and production through natural oils and mitigation technologies (e.g. Carbon Capture and Storage (CCS) and Carbon Capture and Utilization (CCU)).

This book encompasses the wide range of recent advancements in the field of Smart grids like microgrid topologies and their control issues; hybrid energy systems; HVDC &FACTS devices; power quality issues; fault coordination and protection of DC grids; autonomous, islanded and remote networks; smart distribution networks; AC and DC load flow analysis; Advance Metering Infrastructure (AMI); control and automation schemes; Wide Area Monitoring, Protection and Control (WAMPAC); PMUs; communication infrastructure; cyber security and IoT applications in smart grid.

Similarly, the recent developments and role of Hybrid vehicles integrated with renewable resources are discussed in this book. This throws light on the futuristic and dynamically changing automobile sector. The topics that set the future challenges in the related technologies such as energy storage, electric machine design, efficient braking system and their V2G/G2V charging schemes are also tapped upon.

Apart from these, the book encapsulates the wide range of solar thermal and bioenergy systems, their applications and benefits to the society, as well as to the environment. The book serves as a valuable reference to beginners, researchers and professionals interested in renewable energy and sustainable environment and the allied fields.

Contents

About the Editors

Dr. Shelly Vadhera is currently working as an Associate Professor in National Institute of Technology, Kurukshetra, Haryana, India. She received the B.E. degree in Electrical Engineering from Thapar Institute of Engineering & Technology, Patiala, India, the M. Tech. degree in Power Systems, and the Ph.D. degrees in Electrical Engineering both from the National Institute of Technology, Kurukshetra, India. She has more than 20 years of experience in teaching and is presently engaged in research activities circumscribing the guidance to post-graduate and Ph.D. students for their dissertation and thesis work, organizing the conferences/ workshops/ short term training programmes/ short term courses under the wide spectrum of renewable energy and power system technologies.

She is one of the authors of the book on "Power Systems and Protection", Bharat Publications 2007 and "Reference Handbook on Power, Control & Communication Systems: Recent Headways" Elixir Publications. Her Book Chapter entitled "Voltage Stability Assessment Techniques for Modern Power Systems" (Chapter "Adaptive Volterra Filtered-X Logarithmic Cost Least Mean *lp*-Norm Control for Grid-Tied PV Ultracapacitor Battery Fuel Cell System"), pp. 128-176 in Book "Novel Advancements in Electrical Power Planning and Performance", IGI Global, 2020 surfaces the issues concerning the voltage stability and also empowers with the solutions to mitigate the problems in the domain of power systems. She has more than 100 papers to her credit in reputed Journals and International Conferences.

Presently she holds expert talks on Solar Energy Scenario in Indian context and had delivered lectures in NITs, renowned government colleges and reputed private universities. She is a reviewer of many IEEE conferences and international journals. Along with the academic activities she had successfully handled the key administrative posts in various capacities whereby it gives her an opportunity to keep the young generation motivated for producing the academic work/projects that can benefit the society by upgrading their condition even though if the change is incremental and not breakthrough. Her current research area focuses on renewable energy technologies, power systems, high voltage engineering and artificial intelligence.

Dr. Bhimrao S. Umre received the B.E. degree in Electronics and Power in 1983 and M. Tech. degree in 1986 from Nagpur University and his Ph.D. degree in 2009 from RSTM, Nagpur University. In Ph.D., he worked on "Establishing an expert system for ascertaining torsional stresses in a turbine generator set on account of the electrical disturbances in an interconnected power system". He is currently Professor in the Department of Electrical Engineering, Visvesvaraya National Institute of Technology, Nagpur, India. He has almost 35 years of teaching experience and guided many Ph.D. theses and has published various research articles in reputed journals and conferences. He has filed 7 patents and had also worked on 2 externally sponsored R&D project 1) DST, Multiphase (Six Phase) Matrix Converter For Traction, Science And Engineering Research Board (SERB), New Delhi, 42 lakh. 2) DST, FIST: 2008 to 2014, Electrical Measurement/Machines/Drives. He had also published a book "Laboratory Manual For Electric Machines," I. K. International Publishing House Pvt. Ltd. NewDelhi" in 2017. He got PPSA award on the project "Diagnosis of Incipient Fault Development in Transformer by Condition." by POSOCO and IIT Delhi. Semifinalists award on the project "Design and Development of An Intelligent Fault Tracking System Optical Fiber Cable In Railway Tracks." He was a member of Board of Studies, Electrical, Nagpur University, from 1995 to 2000 and from 2010 to 2015. Currently he is a member of RRC Electrical, Gondwana University, Gadchiroli and Board of Studies, Electrical, Yeshwantrao Chavan College of Engineering (YCCE), Nagpur. He is a mentor of Dr. Babasaheb Ambedkar College of Engineering, Nagpur. His research interest mainly includes condition monitoring of electrical machines, torsional oscillations and power systems.

Dr. Akhtar Kalam received the B. Sc. degree (1966-1969) from St Xavier's College, Calcutta, India in Applied Science and the B. Sc. Engineering degree (1969-1973) from Aligarh Muslim University, Aligarh, India in Electrical Engineering. He received his MS degree in Electrical Engineering from the University of Oklahoma, Norman, USA, and the PhD degree from the University Of Bath, Bath UK in Electrical Engineering. He has been a professor with Victoria University (VU), Melbourne, VIC, Australia since 1985 and a former Deputy Dean with the Faculty Of Health, Engineering and Science. He is currently the head of engineering and the Director of externalization with the College of Engineering and Science, VU. He is also the chair of the academic board and lecturer's in the master's by coursework program in the Engineering Institute of Technology, Perth, WA, Australia. Furthermore, he has the distinguished professorship position with the University of New South Wales, Sydney, NSW, Australia; Maharaja Ranjit Singh Punjab Technical University, Bhatinda, India; Crescent University Chennai, India; Vellore Institute of Technology, Vellore, India and five Malaysian Universities. He has wide experience in educational institutions and industry across four continents. He provides consultancy for major electric utilities, manufacturers and other industry bodies in his field of expertise. He is the international editor for the journal of Electrical Engineering (Elektrika), Malaysia and also works as a guest editor for special issues of journal of Electrical and Electronics, Engineers Australia and Journal of Renewable Energy Engineering and reviewer of papers of various journals including Electrical and Electronics Engineering, Engineers Australia, IET, International Journal of Management and System etc. He is the former convener of the API/ESAA Power Engineering residential school and had pioneered the Electrical and Electronic Engineering summer school in January 1996 and in 1997. His major areas of interests are Power system analysis, Communication, Control, Protection, Renewable energy, Smart grid, IEC61850 implementation and Cogeneration systems. He is the editor-in-chief for the Australian Journal of Electrical and Electronics Engineering. He is also a fellow of EA, IET and AIE and a member of CIGRE APB5 study committee.

Modeling and Simulation of Grid-Connected Wind Power Plant for Electric Vehicle Charging Station with Solid Oxide Fuel Cell

Jakkoju Nikhilesh and Shashi Bhushan Singh

1 Introduction

Owing to the mass use of natural resources, global warming and the depletion of fossil fuels have become an increasingly acknowledged world issue in recent years. The installation of renewable energy systems that are not dependent on fossil fuels is an effective countermeasure to control these problems [1].

The implementation of electric vehicles (EV) in the transport sector is regarded as an effective and environmentally sustainable way of achieving low carbon emissions and guaranteeing energy protection. There are also many challenges to be overcome in order to effectively achieve the previously mentioned goals [2].

Initially, the use of electric vehicles (EVs) requires more effort and additional incentive to invest to increase charging infrastructure development and to expand transmission, distribution and generation capacities to support enhanced demand for charging.

Second, we need to overcome the indirect emission of the EVs, as they are sensitive to the generating source. If the coal-fired power plant dominates the power generation, the obvious advantage of EVs, that is to say less emission, cannot be achieved [3]. So renewable energy sources such as tides, sunlight, geothermal heat and wind are used for less emission [4, 5].

The wind energy is an important renewable energy resource. But due to change in seasonal and weather it is discontinuing source. So the wind power plant along with the grid is used to supply the power for the electric charging station.

J. Nikhilesh (✉)
School of Renewable Energy and Efficiency, NIT Kurukshetra, Kurukshetra, Haryana, India
e-mail: Jakkoju_31810116@nitkkr.ac.in

S. B. Singh
Department of Electrical Engineering, NIT Kurukshetra, Kurukshetra, Haryana, India
e-mail: sbsingh@nitkkr.ac.in

S. Vadhera et al. (eds.), *Latest Trends in Renewable Energy Technologies*, Lecture Notes in Electrical Engineering 760, https://doi.org/10.1007/978-981-16-1186-5_1

Fig. 1 Scheme of solid oxide fuel cell (SOFC)

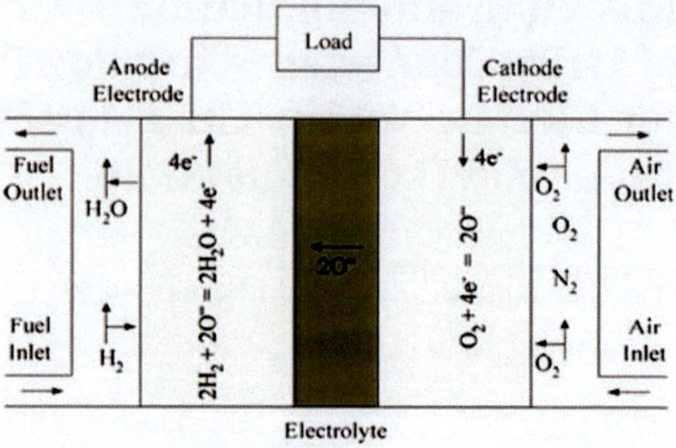

Solid oxide fuel cell (SOFC) is an electrochemical device that operates at high temperatures, directly converting a fuel's chemical energy into electrical energy. This can also transform hydrocarbon fuels directly to a hydrogen-rich gas.

by internal reforming within the stack of the fuel cell itself. This works in the range of 600–1000 °C at temperatures, making them highly efficient as well as versatile in fuel. In SOFC, the electrolyte is a dense solid that involves ceramic materials such as yttrium-stabilized zirconium dioxide, the function which is designed to prevent electrons from passing through the charged oxygen ions [6]. Scheme of solid-oxide fuel cell is shown in Fig. 1.

In SOFC the chemical reactions that occur within the cell most of which are actively engaged in electricity generation are as follows [7]:

At anode:

$$2H_2 + 2O^- \rightarrow 2H_2O + 4e^- \quad (1)$$

$$2CO + 2O^{2-} \rightarrow 2C_2O + 4e^- \quad (2)$$

At cathode:

$$O_2 + 4^{e-} \rightarrow -2O_2^- \quad (3)$$

Overall cell reaction:

$$2H_2 + O_2 \rightarrow 2H_2O \quad (4)$$

Three different fast charging levels are defined in accordance with the SAE J1772 standard which are classified as.

- DC Level-1 rating: Voltage (200/450 V), Current (80 A) and power up to (36 kW)
- DC Level-2 rating: Voltage (200/450 V), Current (200 A) and power up to (90 kW)
- DC Level-3 rating: Voltage (200/600 V), Current (400 A) and power up to (240 kW) [8].

In this paper the DC level-2 charging station is modeled. This means that these DC fast charging stations would have to be built either by the government or by the car companies in cooperation with the government.

The paper is structured as follows: System description is explained in Sect. 2. Simulation results for the system are given in Sect. 3.

2 System Description

The rating of wind farm is of 9 MW, which consists of six 1.5 MW wind turbines that are connected to a 25 kV distribution system exports power over a 30 km, 25 kV feeder, to a 120 kV grid. Block diagram of system is shown in Fig. 2.

The wind turbine with a doubly-fed induction generator (DFIG) consists of a wound rotor induction generator and an PWM converter (IGBT-based AC/DC/AC) powered on voltage sources. The generator stator winding is directly connected to the 50 Hz grid, while the AC/DC/AC converter feeds the rotor at variable frequency. The maximum wind energy to be extracted for low wind speeds by using the DFIG technology allows optimizing turbine rpm, thus during wind gusts the mechanical stress on the turbine is reduced.

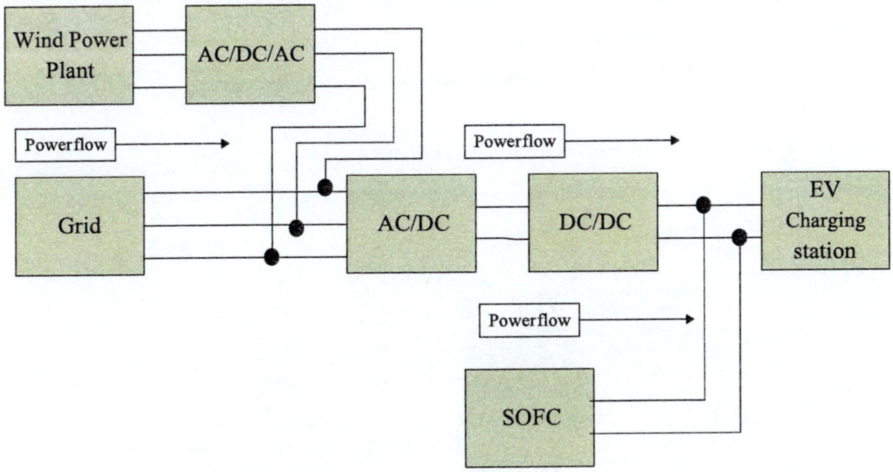

Fig. 2 Block diagram of system

2.1 Design of Wind Power Generation System

Wind power system's key components include a generator, turbine rotor, control electronics system, gearbox, and connecting transformer to grid.

The wind turbine is a spinning mechanism that the kinetic energy in the wind is converted into mechanical energy, and the wind generator transforms mechanical energy into electrical energy.

The different types of wind generators are DC generator, synchronous generator, induction generator (IG), doubly-fed induction generator (DFIG), wound-rotor induction generator (WRIG), permanent magnet synchronous generator (PMSG), squirrel cage induction generator (SCIG) and the commonly used wind generators [9].

The DFIGs are commonly used in recent wind turbines because of their power control capability; low converter costs, reduced power loss and variable speed operation are contrasted with fully equipped converter systems and fixed speed induction generators [10].

DFIG technology enables maximum wind energy to be generated by optimizing turbine rpm for low wind speeds, thereby reducing the mechanical stress on the turbine during wind gusts.

The stator of DFIG is linked to the utility grid and the rotor is connected via a slip ring that is linked into the utility grid via an insulated gate bipolar transistors (IGBT)-based four-quadrant AC-to-AC converter [11]. Block diagram of DFIG with grid is shown in Fig. 3.

The speed (ωm) and slip (S) equations are given as follows:

$$\omega_{\mathrm{m}} = \omega_1 - \omega_2 \tag{5}$$

$$\mathrm{S} = \omega_1 / \omega_2 \tag{6}$$

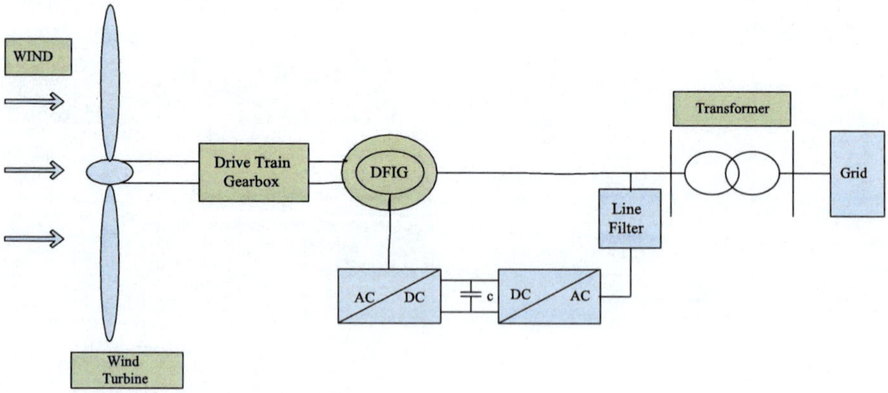

Fig. 3 Block diagram of DFIG with grid

where ω_1 is the stator frequency and ω_2 is the variable frequency.

2.2 *Design of SOFC System*

SOFC has three primary components: cathode, anode and electrolyte [12]. The electrolyte is typically a metal or porous ceramic, based on the temperature in operation. Another very general electrolyte in SOFC is yttria-stabilized zirconia (YSZ), which at these high temperatures is a conductor of oxide ions [6]. The basic representation of chemical reaction in a SOFC is shown in Fig. 4.The dynamic model proposed in Fig. 5 depends on the corresponding chemical and physical principles [13]:

(1) The electrochemical model is based on component material balance equations [14].
(2) The thermal model is based on energy balance equations [15].
(3) The ohmic losses, voltage activation and concentration is based on Nernst voltage equation [16].

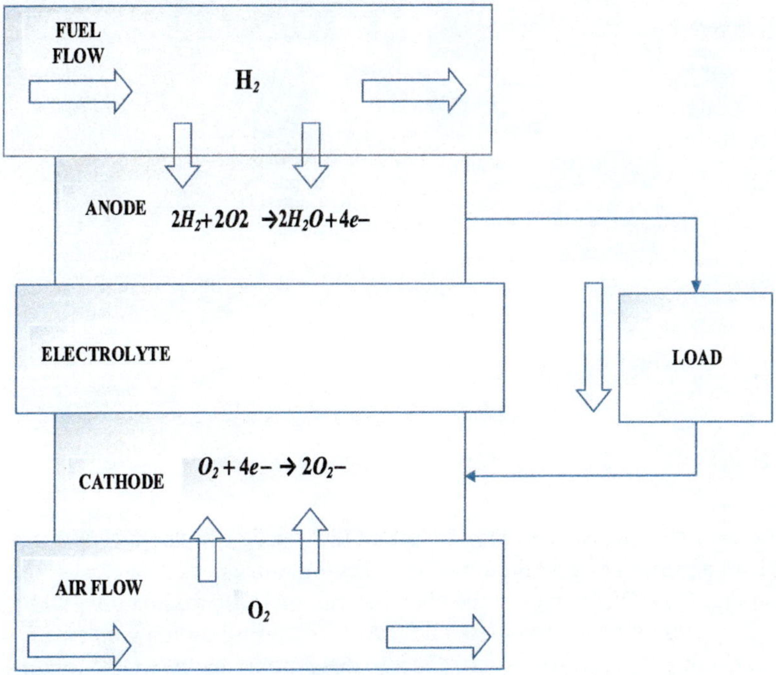

Fig. 4 The basic representation of chemical reaction in a solid oxide fuel cell (SOFC)

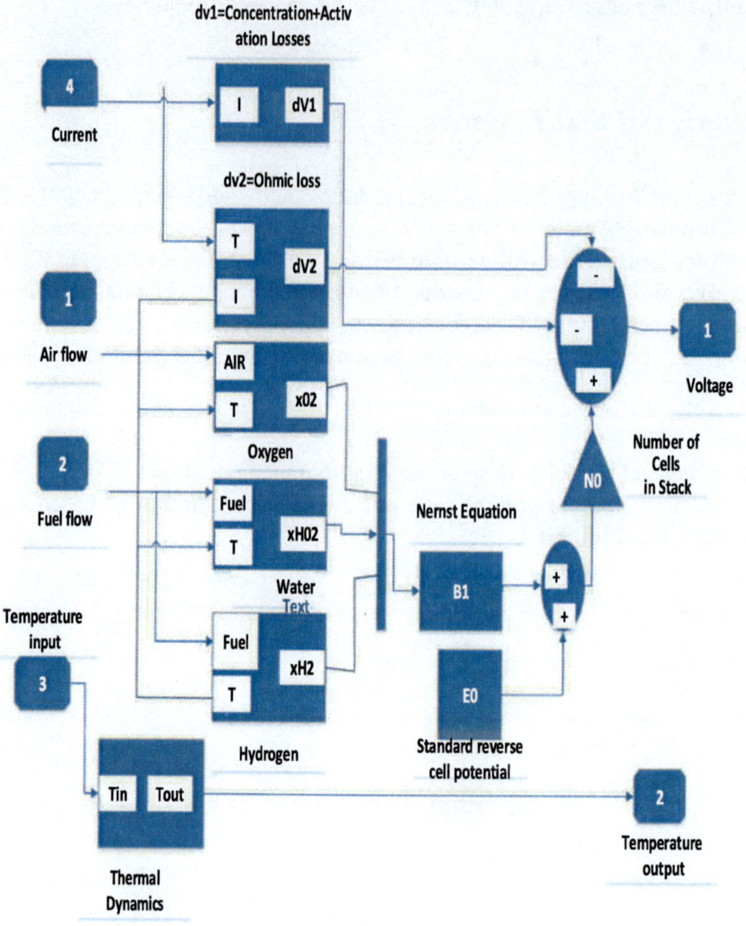

Fig. 5 Dynamic model of SOFC

3 Results

The grid power input is supported by the SOFC during the load at peak time. By giving load a portion of grid input power is smoothened.

From Fig. 6 the DC voltage at the charging station is approximately 480 V, which is equal to the voltage required for the DC level-2 charging station voltage, i.e. 450 V.

As shown in Fig. 7, at a time of 0.02 s into the simulation, the SOFC is considered to charge EV and the grid is disconnected. From there, the SOFC provides power to a load of 1 kW and a load of 10 kW at 0.020 s, which models an EV scale down link to the charging system. The operation of the system is regulated by constant DC link voltage. Figure 8 shows the load current, i.e. 200 A.

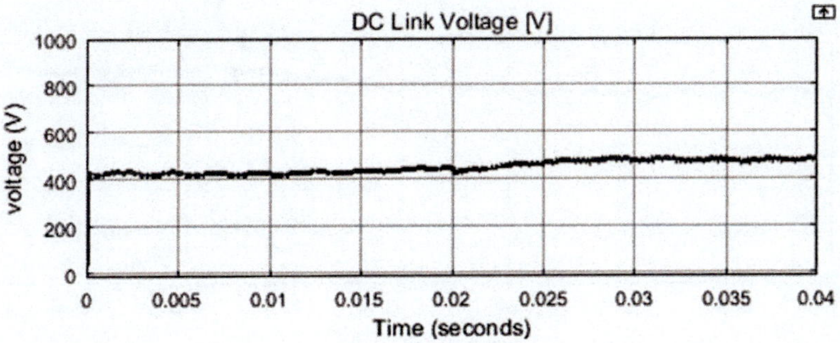

Fig. 6 C voltage

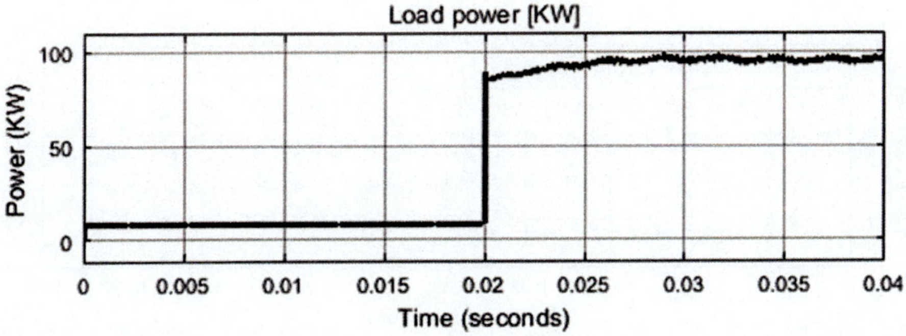

Fig. 7 Load power

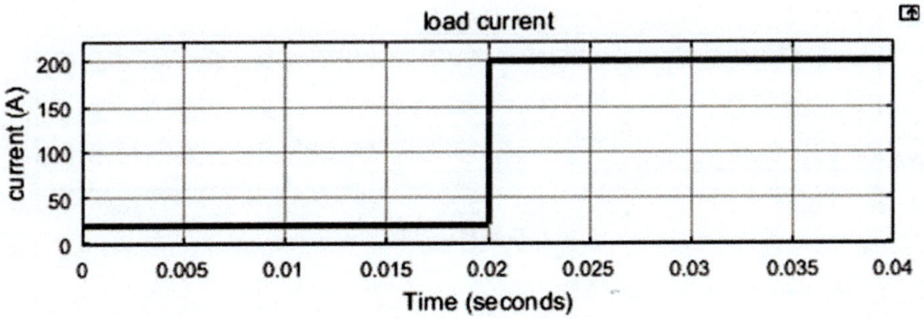

Fig. 8 Load current

Figure 9 shows the SOFC voltage which is approximately about 400 V and the voltage is step-up by using DC-to-DC boost converter to get the required DC output voltage. Figures 10 and 11 show the wind power in megawatts, i.e. 9 MW and wind reactive power in Mvar.

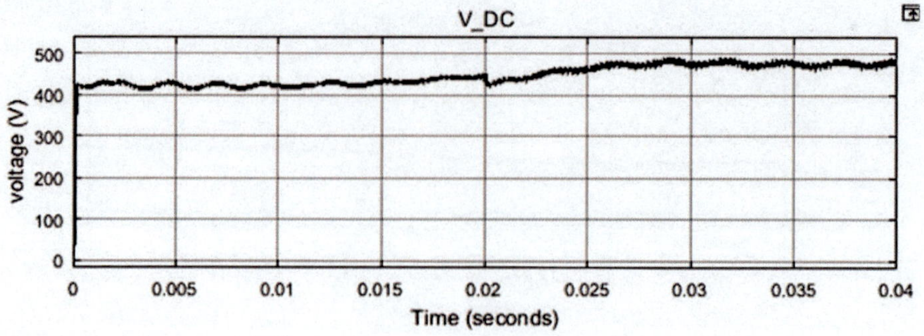

Fig. 9 SOFC voltage

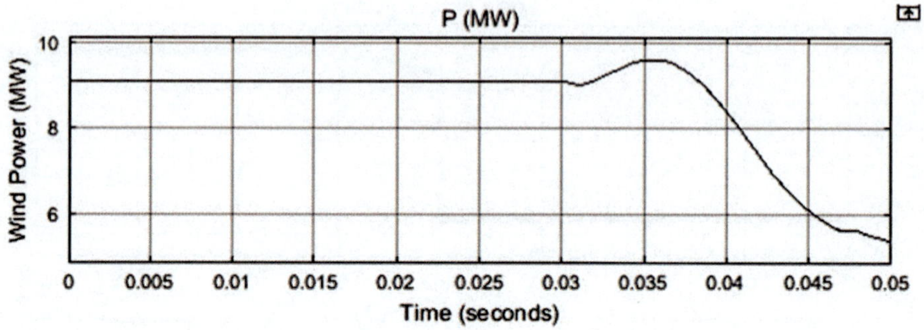

Fig. 10 Wind power in megawatt

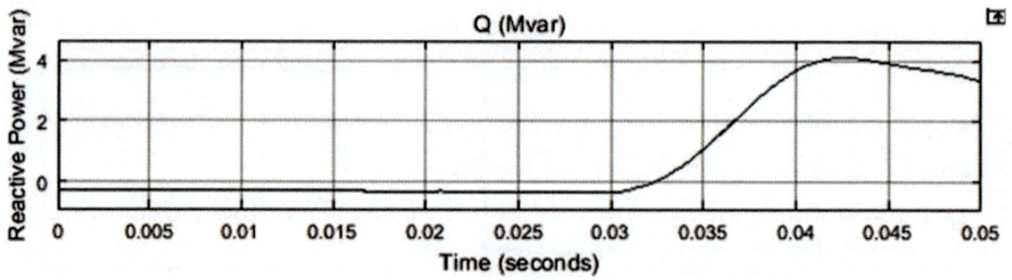

Fig. 11 Wind reactive power in Mvar

The power output of the DFIG wind farm is about 9 MW initially, as shown in Fig. 10, and the wind turbine speed is about 1.2 per unit, as shown in Fig. 13. The wind power plant DC voltage is set to 1150 V as shown in Fig. 12 and the reactive power is maintained at 0 Mvar. The voltage of the positive sequence suddenly decreases to 0.5 per unit at t = 0.03 s. This causes an oscillation of the DFIG power output and DC bus voltage. The control system attempts to regulate the DC voltage and reactive

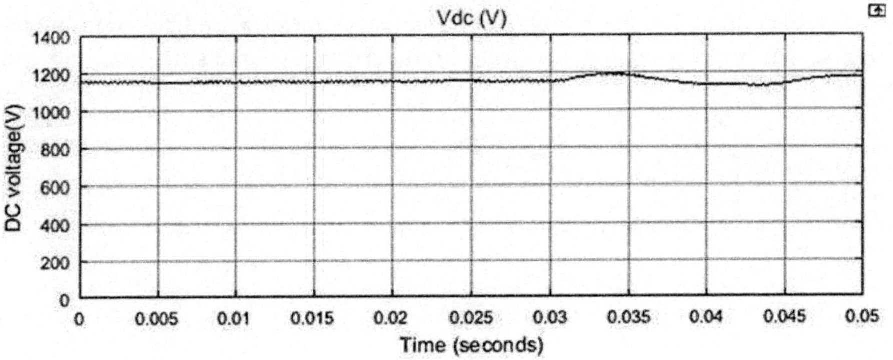

Fig. 12 DC voltage

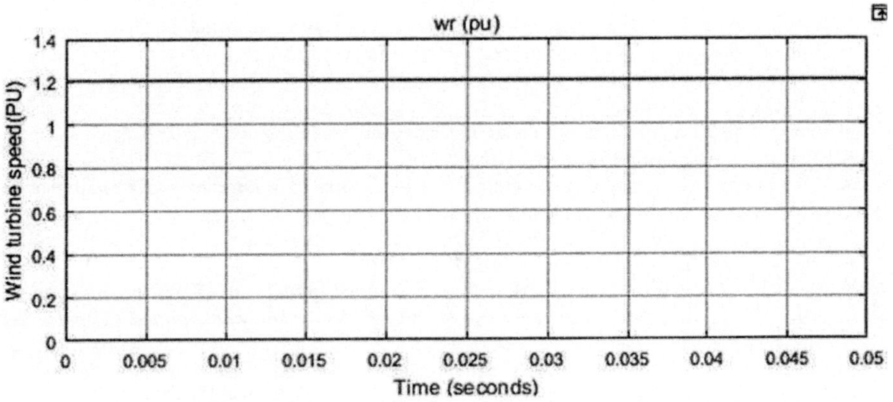

Fig. 13 Wind turbine speed in pu

power at its setpoints (1150 V, 0 Mvar) during the voltage sag. The system recovers in around four cycles.

4 Conclusion

In this paper, the model of SOFC is suggested together with wind and grid for a DC level-2 eV charging station. This study shows that the proposed system can provide sufficient quantities of electrical energy for charging EV. It can be concluded from these simulation results that the proposed SOFC system will successfully produce electricity to charge an electric vehicle while retaining high energy efficiency. The wind plant, grid and SOFC for electric vehicle charging station have been implemented. The experimental results have been verified by simulating the proposed system in MATLAB.

Advantages of the system are:

- Low carbon emission
- Reduced prices of fossil fuels
- Improving environmental quality

References

1. A. Smith Yoza, A. Yona, T. Senjyu, T. Funabashi, Optimal capacity and expansion planning methodology of PV and battery in smart house. Renew. Energy **69**, 25–33 (2014)
2. S. Memon Khan, T.P. Sattar, Analyzing Integrated renewable energy and smart-grid systems to improve voltage quality and harmonic distortion losses at electric-vehicle charging stations IEEE. Access **6**, 26404–26415 (2018)
3. P.M. De Quevedo, G. Muñoz-Delgado, J. Contreras, Impact of electric vehicles on the expansion planning of distribution systems considering renewable energy storage and charging stations. IEEE Trans. Smart Grid (2017)
4. A. Ashwin Kumar, A study on renewable sources in India, *2010 international Conference on Environmental Engineering and Applications* (2010), pp. 10–12
5. P. Nunes, R. Figueiredo, M.C. Brito, The use of parking lots to solar-charge electric vehicles: Renew. Sustain. Energy Rev. **66**, 679–693 (Dec. 2016)
6. M. Lee, G. Park, V. Radisavljevic, Modeling of solid oxide fuel cells (SOFCs): an overview, in *2013 5th International Conference on Modeling, Simulation and Applied Optimization (ICMSAO)* (2013)
7. M. Bavarian, M. Soroush, Mathematical modeling and steady-state analysis of a proton-conducting solid oxide fuel cell. J. Process Control **22**(8), 1521–1530 (2012)
8. S. Rajagopalan, A. Maitra, J. Halliwell, M. Davis, M. Duvall, Fast charging: An in- depth look at market penetration, charging characteristics, and advanced technologies, in *Proceedings of the World Electric Symposium and Exhibition (EVS27)* (Barcelona Spain, 2013), pp. 1–11
9. G. Akhilesh kumar, B. Himanshu, S. Paulson, Generator topologies with Power elec- tronics converters for a Wind energy conversion system: A review. JJSER **4**(6), 2686–2693 (2013)
10. L. Xu, D. Zhi, B. Williams, Predictive current control of doubly fed induction generators. IEEE Trans. Indus. Electron **56**(10), 4143–4153 (2009)
11. Ch. Pooja, K. Pranjali, A review of different power converter topologies for PMSGs wind turbine, in *2016 International Conference on Communication and Electronics Systems (ICCES)* (2017), pp. 1–6
12. C.M.L. Zuo, Solid oxide fuel cells, in*Sol-Gel Processing for Conventional and Alternative Energy* (2012)
13. K. Sedghisigarchi, A. Feliachi, Dynamic and transient analysis of power distribution systems with fuel cells—part I: fuel-cell dynamic model. IEEE Trans. Energy Conversion **19**(2), 423–42 (2004)
14. Fuel Cell HandBook, 5th ed., EG&G Services Parsons Inc., U.S. Department of Energy (2000)
15. Achenbach, Three-dimensional and time-dependent simulation of a planar SOFC stack. J. Power Sources **49** (1994)
16. M.D. Lukas, K.Y. Lee, H. Ghezel-Ayagh, Development of a stack simulation model for control study on direct reforming molten carbonate fuel cell power plant: IEEE Trans. Energy Conversion **14**, 1651–1657 (1999)

Parameter Computation and Current Control Loop Tuning of Non-salient PMSM Motor

Nitish Madan and Sandeep Kakran

1 Introduction

In industrial applications, motor drives are essential for automotive, power tools, robotics, and the energy industry [1–3]. For driving the motor drives, it is required to collect the information for rotor speed to achieve speed control of the motor. The measurement of rotor speed is done by sensors attached to the rotor shaft. This sends the signal to the motor drive by either analog to digital converters or by other means. However, the use of sensors for calculating rotor speed is not always practical and reliable. So, it is essential to create a scheme/working environment where we would not be using any external devices like shaft sensors, which reduce the reliability of the system. This requirement in the industry led to the research in the field of controlling the motor by sensorless methods. A scheme known as sensorless vector control or field-oriented control (FOC) of permanent magnet synchronous motor (PMSM) results is excellent torque control, smoother torque ripple, and also improves in efficiency when compared to AC control methods. A lot of methods have been explained in the past decades for PMSM control like sensing rotor position by using hall sensors [4, 5], back EMF voltage sensing technique [6, 7], and rotor position detection by use of encoders/resolvers [8].

But for controlling the rotor angle and the speed of the PMSM motor under the sensorless vector control operation, it is important to first control the current controller. Both speed and position control controllers are dependent on the current controller. In this paper, a vector control scheme using three shunt current sensing for current controlling of PMSM motor is being discussed. The sensorless vector control technique is shown in Fig. 1 [8]. Here, current controller PI gains need to be

N. Madan (✉) · S. Kakran
Department of Electrical Engineering, NIT Kurukshetra, Kurukshetra, Haryana, India
e-mail: nitishmadan1995@gmail.com

S. Kakran
e-mail: sandeepkakran@nitkkr.ac.in

S. Vadhera et al. (eds.), *Latest Trends in Renewable Energy Technologies*, Lecture Notes in Electrical Engineering 760, https://doi.org/10.1007/978-981-16-1186-5_2

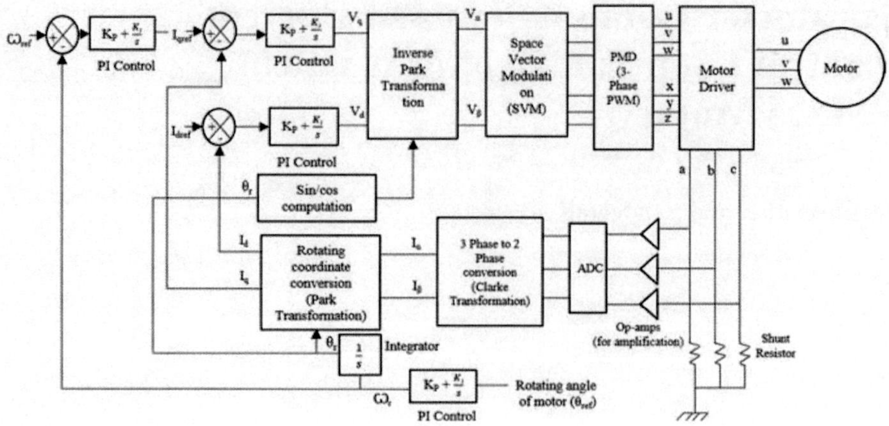

Fig. 1 Sensorless vector control technique

properly tuned for the accurate functioning of the motor. As per sensorless vector control, three phase current is being converted to d-axis (I_d) and q-axis (I_q) current components [9]. The d-axis component (I_d) is responsible for flux production in the motor windings and the q-axis component (I_q) is responsible for torque production on the motor shaft. Under starting condition, d-axis current, $I_d \neq 0$. As the motor rotor angle is calculated accurately, $I_d = 0$, but $I_q \neq 0$. This is done for producing maximum torque for PMSM motor. The current flowing in the motor windings depends on motor parameters, which also need to be calculated. In this paper, at first, the estimation of motor parameters is done, and based on the parameters obtained, tuning of PI gains is being presented.

2 System Overview

A detailed algorithm for parameter estimation and current control mechanism for PMSM motor is being provided using experimental verifications. The motor parameters as per the datasheet of the motor are given in Table 1.

3 The Proposed Model

The sensorless vector control working hardware model uses the components which are provided in Table 2. The microcontroller used is TMPM370FYFG, which is having inbuilt sensorless vector control in the hardware. The bus voltage provided is 24 V and parameter calculation is done on voltage based on the PWM duty cycle provided. The opto-couplers provide isolation between the low power circuit and

Table 1 Parameters of PMSM motor

Motor parameter	Motor parameter values
No. of poles	6
Rated output [W]	135
Rated speed [r/min]	3500
Rated voltage [V_{rms}]	24
Rated torque [N-m]	0.37
Rated current [A_{rms}]	8
The line to line stator resistance, R_S [Ω]	0.24 ± 10%
Line to line stator inductance, L_S [mH]	0.29 ± 20%

Table 2 Components used in the hardware model

Components	Name of components
Microcontroller	TMPM370FYFG
Motor	DB59S024035-A
Opto-couplers	TLP2955
MOSFETs	TK100E10N1
Pre-driver	IR-2110
Op-Amps	LM358
The resistance of the shunt resistor [Ω]	0.10 (wire-wound)

the inverter board. The working model uses current sensing by three shunt resistors, as feedback [8]. This is to be noted that the PWM frequency used for parameter calculation is 150 kHz so that there would not be any transient current ripples in the system and accuracy can be attained.

In this model, microcontroller TMPM370FYFG is being used. This microcontroller is having an in-built vector control module, i.e., this microcontroller will take care of the algorithm for Clarke, park, inverse park, space vector module in its hardware only as provided in Fig. 2. So, it is easier to build an algorithm for controlling the PMSM/BLDC motor.

For parameter estimation, three transition states are considered: T0, T1, and T2.

- T0: When all lower MOSFETs 'X', 'Y', and 'Z' are turned ON
- T1: MOSFETs U and Y turned ON (Fig. 3)
- T2: MOSFETs V and X turned ON (Fig. 5)

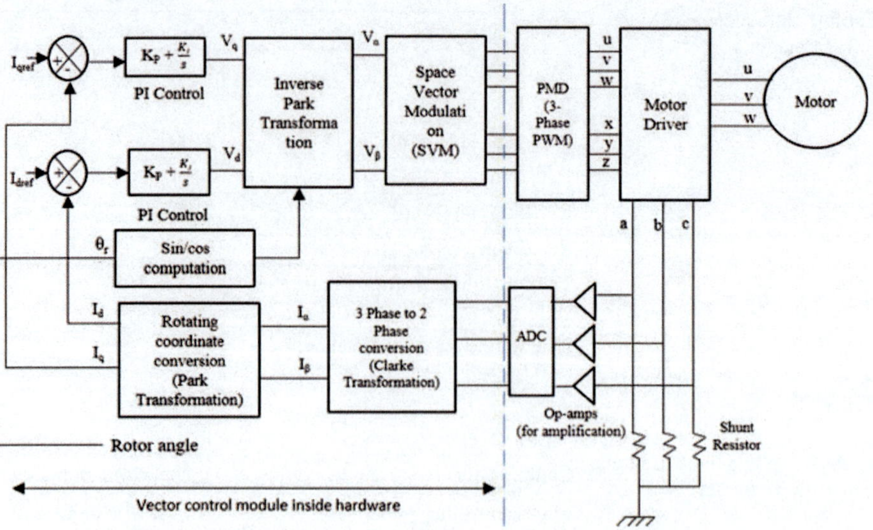

Fig. 2 The in-built field-oriented control module in TMPM370FYFG

Fig. 3 Transition state T1

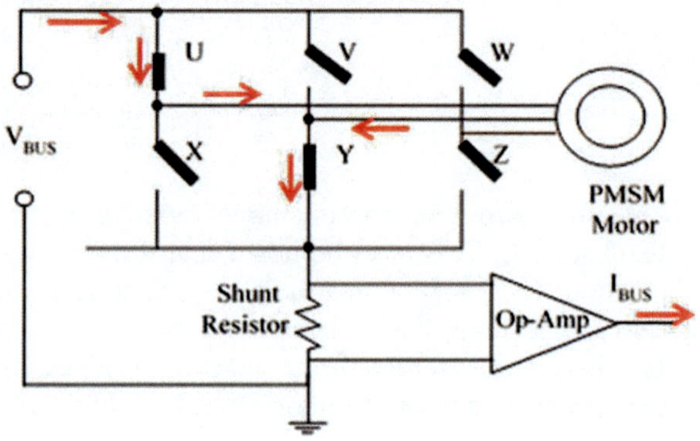

3.1 Resistance Calculation

The working hardware model involves two N-channel MOSFETs, in one leg of the inverter circuit. So to turn on the upper MOSFETs, pre-drivers are used. In pre-drivers, bootstrap capacitors need to be charged for turning ON 'U', 'V', 'W' MOSFETs. So, at first turn ON all the lower MOSFETs 'X', 'Y', and 'Z' (transition state T0) for some time to charge the bootstrap capacitors. For the resistance measurement, a 40–50% PWM duty cycle (approximately) is applied for accurate calculation. If this much voltage will be applied to the motor, the rotor will feel a jerk. To avoid that, position the rotor by applying a 5% duty cycle of PWM voltage that is 1.2–2.4 V using transition state T1.

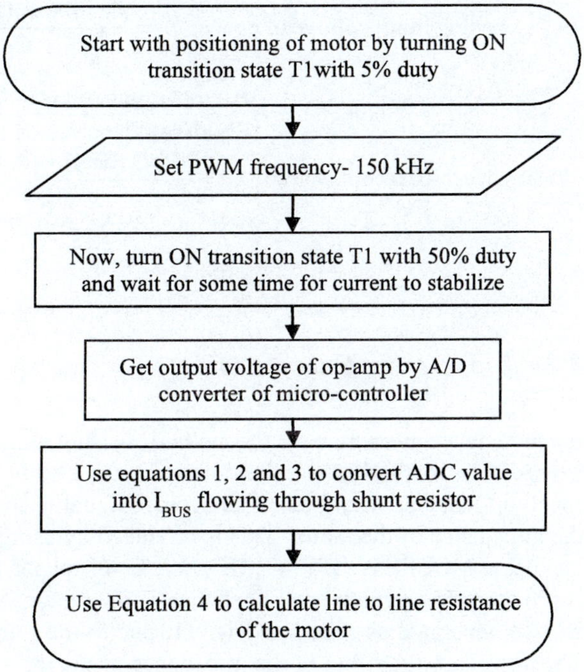

Fig. 4 Flowchart of motor resistance measurement

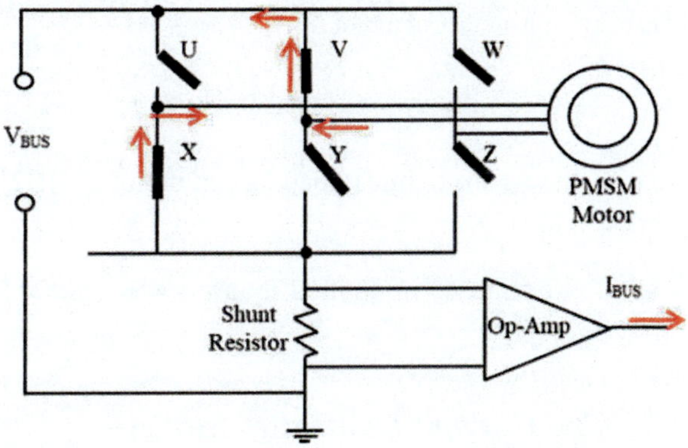

Fig. 5 Transition state T2

It is to be noted that the PWM frequency applied for parameter calculation is 150 kHz. This is done to reduce the ripple content due to inductance in the circuit. The flowchart for the resistance calculation is shown in Fig. 4.

$$\text{Analog output voltage of op - amp} = \frac{\text{Digital output of op - amp as read byADC} * 5}{\left(2^{\text{No. of bits of ADC}} - 1\right)} \tag{1}$$

$$\text{Analog input voltage of op - amp} = \frac{\text{Analog output voltage of op - amp}}{\text{Gain of op - amp}} \tag{2}$$

$$\mathrm{I_{BUS}}(A) = \frac{\text{Analog input voltage of op - amp}}{\text{Shunt resistance value } (\approx 0.1\ \Omega)} \tag{3}$$

$$\text{Line to line resistance of motor } (\mathrm{R_S}) = \frac{\text{(PWM duty)} * \text{(Rated voltage)}}{\mathrm{I_{BUS}}} - \text{Shunt resistance} \tag{4}$$

3.2 *Inductance Calculation*

For inductance measurement, the method of calculation of the impedance of the motor is used. The method of transient analysis can also be used, but for that very high-speed op-amps are required. A sinusoidal signal needs to be produced, to calculate the impedance of the motor. This is produced by using single-phase bridge inverter, i.e., using MOSFETs 'U', 'V', 'X', 'Y'. In this paper, a square wave of period 1 ms is produced to calculate the impedance of the motor. A triangular current waveform will be generated as shown in Fig. 11, due to the transient behavior of the circuit. The flowchart/algorithm of the impedance of the motor is provided in Fig. 6.

$$\text{Line to line impedance of motor } (\mathrm{Z_S})[\Omega] = \frac{\text{(PWM duty)} * \text{(Rated voltage)}}{\mathrm{I_{BUS}}} - \text{Shunt resistance} \tag{5}$$

$$\text{Line to line impedance of motor } (\mathrm{L_S})[\Omega] = \frac{\sqrt{Z_S^2 - R_S^2}}{2\pi(f)} \tag{6}$$

where f = frequency of applied square wave = 2000 Hz.

3.3 *The Current Controller Model*

After calculating motor resistance (R_s) and motor inductance (L_s), current controller gains can be estimated. Current controller gains, i.e., proportional gain (K_P) and integral gain (K_I), can be calculated by forming up a control structure of the d-axis and q-axis current. This control loop structure for I_d and I_q is shown in Fig. 7. For maximum torque production under steady stage, $I_{dref} = 0$, $I_{qref} \neq 0$. I_d is responsible for producing flux in the motor windings and I_q is responsible for providing torque to the motor shaft. The controller gain tuning is done under frequency response. So, the method involved in gain tuning is by using desired bandwidth of the controller and setting the phase margin of the current controller.

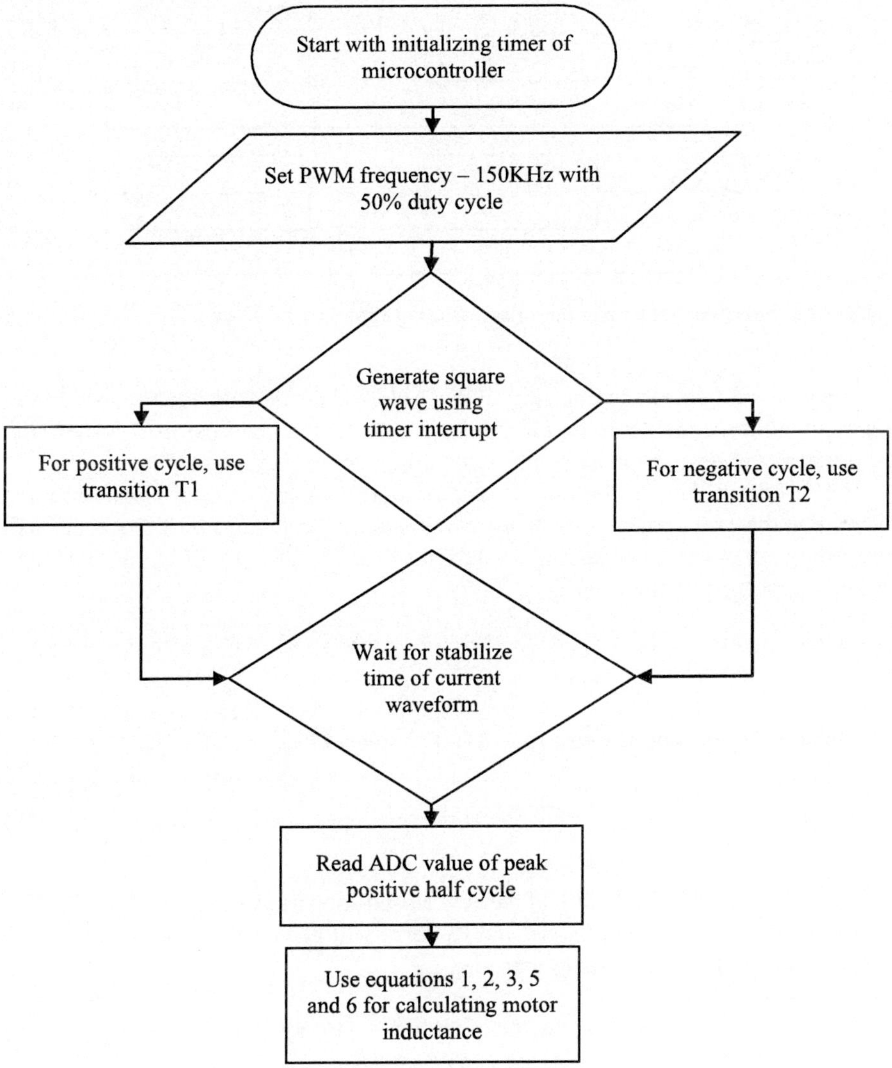

Fig. 6 Flowchart of motor resistance measurement

The open-loop transfer function of Fig. 7 is given in Eq. 7. Since tuning of these parameters is a real challenge, the K_P term sets the high-frequency gain and K_I sets the low-frequency gain. Integral gain (K_I) plays an important role in eliminating steady-state errors.

$$G_{loop}^{open}(s) = \frac{K_P + K_T/s}{R_S + sL_S} = \frac{K_P s + K_I}{s(R_S + sL_S)} \quad (7)$$

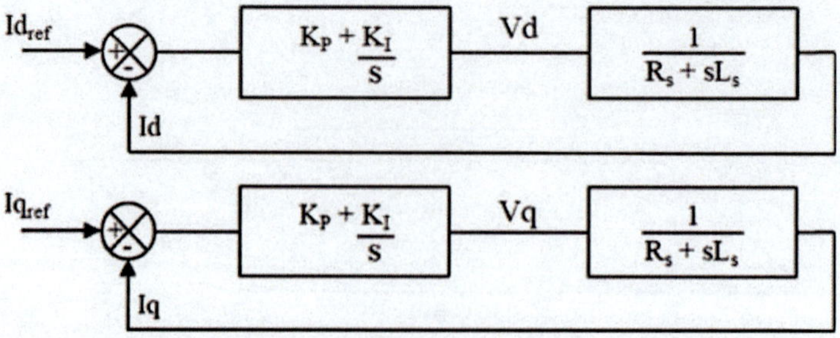

Fig. 7 The current control loop of sensorless vector control

There are two poles and one zero in the system. This depends on the location of the poles. If they are near to jω axis, this will lead to the underdamped system. So, the system may have resonant peaks, which, in practice, is undesirable.

The solution to this problem is to tune proportional and integral gains such that they are at the exact position of motor parameters, as given in Eq. 8. So, there will be a pole-zero cancellation and ultimately the system will have only one pole, i.e., setting the phase margin to 90°.

$$\frac{K_P}{K_I} = \frac{L_S}{R_S} \tag{8}$$

The new closed-loop transfer function is as given in Eq. 9.

$$G_{\text{loop}}^{\text{closed}}(s) = \frac{1}{s\frac{L_S}{K_P} + 1} \tag{9}$$

Now, calculate the bandwidth of the new closed-loop transfer function. According to Eq. 10, it is understandable that the K_P term will decide the desired bandwidth (3 dB frequency) of the current controller.

$$K_P = (\text{Bandwidth of current controller}) * L_S \tag{10}$$

It is to be noted that the desired bandwidth should be decided by keeping in mind the time constant ($\tau = L_S/R_S$) of the system. Here, the marginal stable system would be defined by keeping bandwidth = 1 time constant. So, it is preferable to keep bandwidth for 3–5 time constants. Also, the current gets stabilized to 95–98% till 3–5 time constants.

$$\text{Bandwidth of current controller} = f_{3\text{dB}} = \frac{1}{2\pi\tau_1} \tag{11}$$

where $\tau_1 = (3 \text{ to } 5) * \tau$.

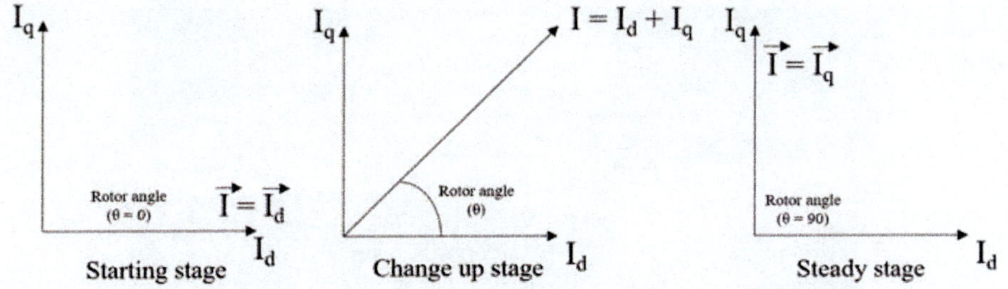

Fig. 8 Three stages of current control operation

There are three stages in motor operation, as shown in Fig. 8. The phasor is rotated from I_d to I_q in starting to steady stage for the smooth operation of motor rotation, where I_d can be kept at 40% and I_q can be kept at 20% of the maximum current rating of the motor. Under the starting stage, I_q is zero and under the steady stage, I_d is zero.

4 Experimental Results

A six-pole PMSM motor is evaluated experimentally. This is to verify the sensorless vector control using parameter estimation. The hardware model is shown in Fig. 8. The sensorless vector control algorithm is set in the microcontroller, TMPM370FYFG itself. But the parameters estimated is based on the control of the inverter board and feedback circuit. The current waveforms for resistance and inductance measurement are as shown in Figs. 10 and 11, respectively.

Table 3 shows the results of line to line stator resistance and line to line stator inductance. The tuning parameters, which are proportional gain and integral gains, are calculated at the bandwidth of 5 τ. Thus, the bandwidth of the current controller taken is 25.86 rad/s (Fig. 9).

For resistance measurement, I_{BUS} current generated in the motor windings is a DC current waveform as shown in Fig. 10. As discussed, the DC current waveform will be having ripples if the applied PWM frequency is not high enough. Here, the short pulse is for positioning of the rotor (5% duty) and a longer duration pulse is when the resistance of the motor is calculated.

Table 3 Experiment result calculations

Parameters	Calculated values
The line to line stator resistance, R_S [Ω]	0.26
The line to line stator inductance, L_S [mH]	0.32
Current controller proportional gain [K_P]	0.0087
Current controller integral gain [K_I]	6.724

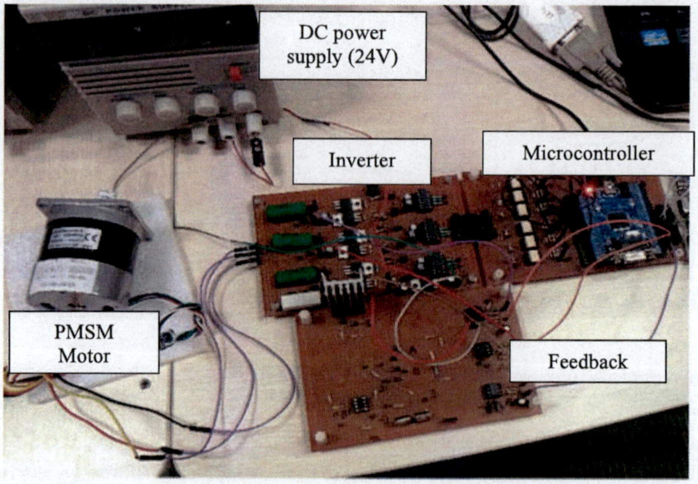

Fig. 9 Hardware setup for sensorless vector control of PMSM motor

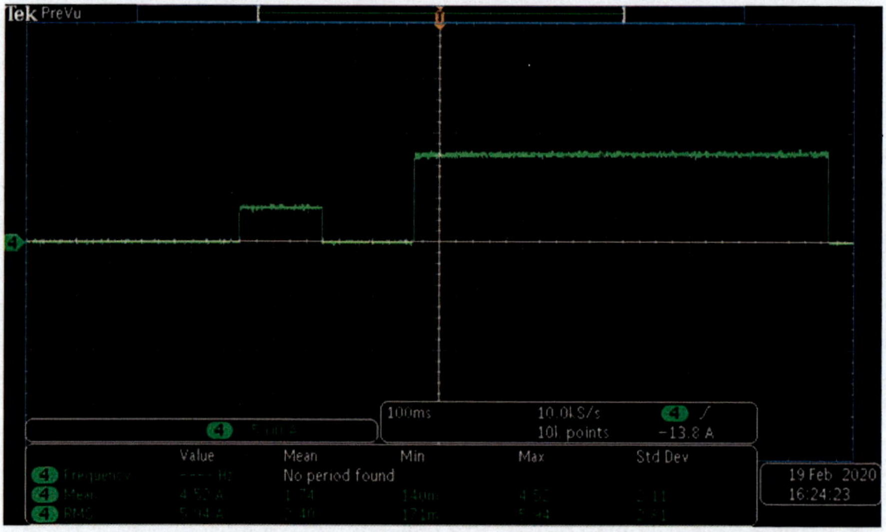

Fig. 10 Current waveform under motor resistance calculation

For inductance measurement, I_{BUS} current generated in the motor windings is a triangular current waveform with positive and negative peaks as shown in Fig. 11. This waveform is obtained under a timer frequency of 2000 Hz.

The scope graphs of the graphic user interface (GUI) are shown in Figs. 12, 13, and 14 for the starting stage, change up stage, and steady stage, respectively. These graphs are based on the current controller PI gains obtained. From Fig. 12, i.e., under the forced stage, Iq is set to zero whereas Id is set to 40% of the maximum current

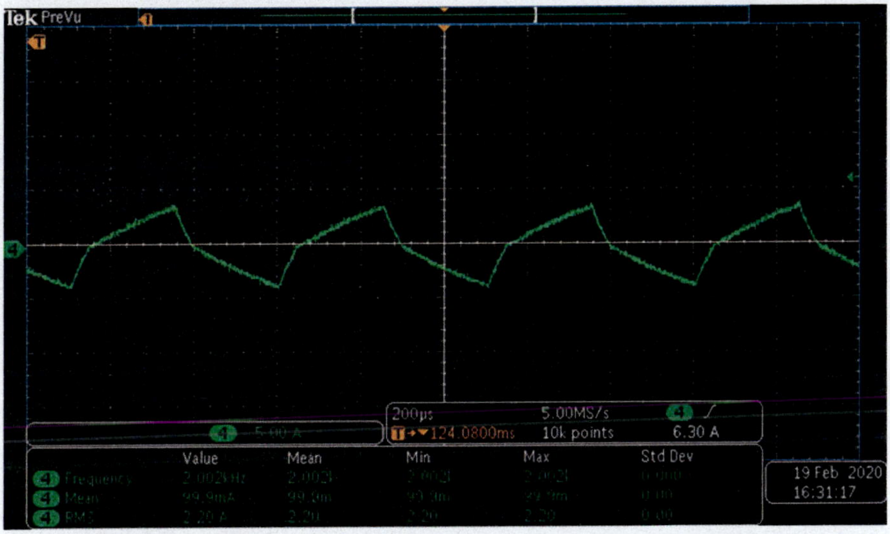

Fig. 11 Current waveform under motor inductance measurement

Fig. 12 Scope result from GUI comparing i_{d_ref} and i_{d_comm} at starting stage

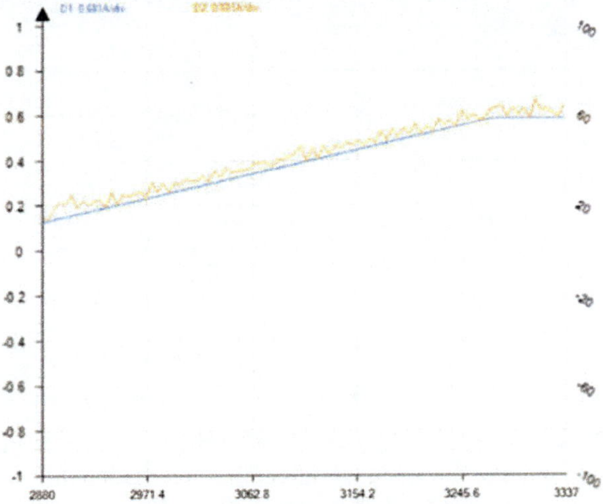

rating of the motor. As the motor is in no-load condition, so under steady stage the motor draws no-load current, that's why i_q is less in steady stage as shown in Fig. 14.

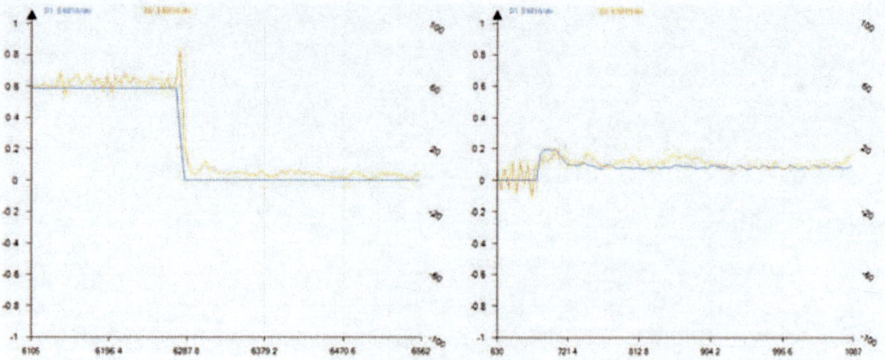

comparing $i_{d\ ref}$ and $i_{d\ comm}$ during change up stage comparing $i_{q\ ref}$ and $i_{q\ comm}$ during change up

Fig. 13 Scope result from GUI for the change up stage

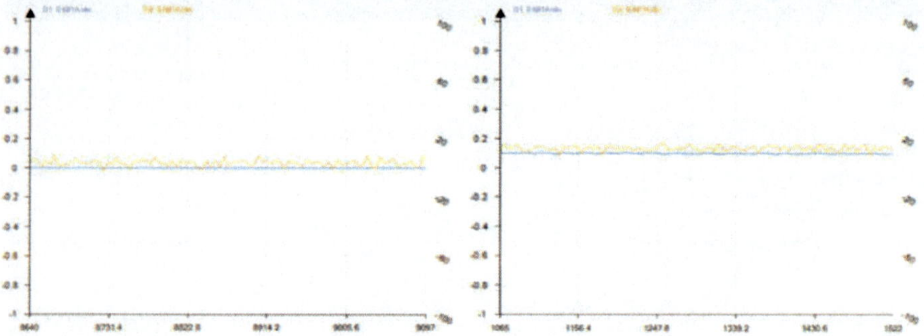

comparing $i_{d\ ref}$ and $i_{d\ comm}$ during steady stage comparing $i_{q\ ref}$ and $i_{q\ comm}$ during steady stage

Fig. 14 Scope result from GUI for steady stage

5 Conclusion

This paper presented a method to estimate the motor parameters. Based on the motor parameters, the design of a control system for the current controller for sensorless operation of a PMSM motor is being tuned. The experimental results are shown for 135 W motor. This strategy is proposed so that the gain values of the current controller for sensorless vector control operation can be tuned automatically without the use of the motor datasheet. This is done to avoid manual tuning of gain values which consumes a lot of time.

References

1. K. Renu, N.K. Kumari, D.S.G. Krishna, Sensorless control of permanent magnet synchronous motor with flux weakening operation for washing machine application, in *2018 IEEE International Conference on Power Electronics, Drives and Energy Systems (PEDES)*, pp. 1–6 (Chennai, India, 2018). https://doi.org/10.1109/PEDES.2018.8707558
2. D. Xu, B. Wang, G. Zhang, G. Wang, Y. Yu, A review of sensorless control methods for AC motor drives. CES Trans. Electr Mach Syst **2**(1), 104–115 (2018). https://doi.org/10.23919/TEMS.2018.8326456
3. M. Moradian, J. Soltani, A. Najjar-Khodabakhsh, G.R.A. Markadeh, Adaptive torque and flux control of sensorless IPMSM drive in the stator flux field oriented reference frame. IEEE Trans. Indus. Inf. **15**(1), 205–212 (2019). https://doi.org/10.1109/TII.2018.2808521
4. H. Wu, M. Wen, C. Wong, Speed control of BLDC motors using hall effect sensors based on DSP, in *2016 International Conference on System Science and Engineering (ICSSE)*, pp. 1–4 (Puli, 2016)
5. S.Y. Yun, H.J. Lee, J.H. Han, J. Lee, Position control of low cost brushless dc motor using hall sensor, in *2012 Sixth International Conference on Electromagnetic Field Problems and Applications*, pp. 1–4 (Dalian, Liaoning, 2012)
6. R.G. Krishnan, T.B. Isha, P. Balakrishnan, A back-EMF based sensorless speed control of permanent magnet synchronous machine, in *2017 International Conference on Circuit ,Power and Computing Technologies (ICCPCT)*, pp. 1–5 (Kollam, 2017). https://doi.org/10.1109/ICCPCT.2017.8074313.F; F. Genduso, R. Miceli, C. Rando, G.R. Galluzzo, Back EMF sensorless-control algorithm for high-dynamic performance PMSM. IEEE Trans. Indus. Electron. **57**(6), 2092–2100 (2010)
7. H. Oh, K.Y. Song, K.Y. Cho, H.W. Kim, B.M. Han, Initial rotor position detecting algorithm of PM synchronous motor using incremental encoder, in *2013 IEEE ECCE Asia Downunder*, pp. 681–686 (Melbourne, VIC, 2013)
8. N. Ertugrul, P. Acarnley, A new algorithm for sensorless operation of permanent magnet motors. IEEE Trans. Ind. Appl. **30**, 126–133 (1994)
9. J. Wang, S. Cheng, B. Gong, Sensorless vector control for Permanent Magnet Synchronous Motor fed by three-level inverter. ICCAS. Gyeonggi-Do **2010**, 1209–1213 (2010)

Single-Object Detection Hardware Accelerator Using XfOpenCV Library

Aman Saxena, M. P. R. Prasad, and Prashant Sivaji Sutar

1 Introduction

Real-time image and video processing plays an important role in developing algorithms for object detection, tracking [1], space technology, video surveillance and security. In electrical system where image processing is important, consumption of large electric power makes the system less power-efficient. In general, image processing contains many filters and functional operations which make the system's CPU busy and consume a lot of computational time and power [2]. However, more complex algorithm takes lots of computational power to handle the large number of pixels in video processing. On the other hand, FPGA has the capability to work with CPU [3] and make it suitable for image processing. FPGA uses hardware accelerator [4] for high-speed operations, such as filtering and thresholding, which reduce the amount of work which is to be done by CPU. Some pre-processing functions are done by FPGA before sending pixels to the CPU, which helps in reducing the processing time. The control logic can also be built and implemented on the FPGA which can control the process without interfacing the CPU [2]. By using FPGA over CPU makes the system fast and reduce power consumption.

This paper describes the design methodology and generation of an IP for detecting a single object in the image which is going to be implemented on the FPGA Zynq processor-based Zybo board. Object detection IP is generated in Vivado HLS by using xfOpenCV library APIs. xfOpenCV library provided by Xilinx is used for synthesis.

A. Saxena (✉) · M. P. R. Prasad
National Institute of Technology, Kurukshetra, India
e-mail: saxenaman1196@gmail.com

M. P. R. Prasad
e-mail: mprp823@gmail.com

P. S. Sutar
Toshiba India Software Private Limited, Bangalore, India
e-mail: prashant.sutar@toshiba-tsip.com

S. Vadhera et al. (eds.), *Latest Trends in Renewable Energy Technologies*, Lecture Notes in Electrical Engineering 760, https://doi.org/10.1007/978-981-16-1186-5_3

Generated IP contains the input and output ports that take data in the AXI stream format. AXI4-Stream is a communication protocol that works on the master–slave concept [3].

The rest of the paper is organized as follows. In Sect. 2, we discuss the functions that are used to generate the IP. In Sect. 3, we discuss the steps for implementation. The results of C-simulation and synthesis are shown in Sect. 4.1. Generated IP is shown in Sect. 4.2 and resources utilized by FPGA are explained in Sect. 4.3. Finally, Sect. 5 concludes the work (Fig. 1).

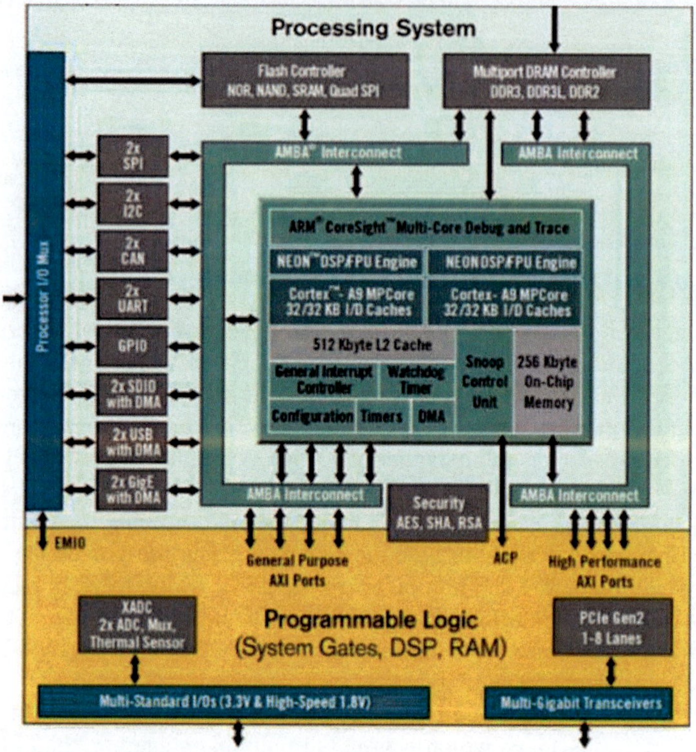

Fig. 1 System architecture of Zynq7000 SoC [2]

2 Functions Used in Object Detection Algorithm

2.1 Threshold

The threshold function changes the pixel intensity to either maximum or minimum according to the given value [5]. Many types of operation can be performed using threshold operation.

Case 1: If the pixel intensity is more than or equal to the given threshold then the pixel intensity is set to maximum, otherwise to minimum value.

Case 2: If the pixel intensity is more than or equal to the given threshold then the pixel intensity is set to threshold value, otherwise remains the same as the source image pixel.

Case 3: If the pixel intensity is more than or equal to the given threshold then the pixel intensity is set to the same source image pixel intensity, otherwise set to zero (minimum).

Case 4: If the pixel intensity is more than or equal to the given threshold then the pixel intensity set to zero(minimum), otherwise set to the same source image pixel intensity.

The mapping of this design and implementation is based on case 1 where the threshold value and maximum value set the pixel intensity to a maximum or minimum. The assumption is that the output of threshold image contains two classes of pixels, one is foreground pixels (maximum value) and the other background pixels (minimum value).

$$\text{Output(x,y)} = \begin{cases} \text{Maximum}, & \text{if input(x,y)} > \text{threshold} \\ \text{Minimum}, & \text{otherwise} \end{cases}$$

2.2 InRange

The outputs of thresholding are two pixel classes: one is background pixels and another one is foreground pixels. InRange function is used to remove the background pixels completely from the image [5]. This function uses boundary value (upper and lower value) and checks if the input image pixel intensity lies between these upper value and lower value or not. If it lies then pixel at output image is set to 255 (bright/white) and if not then output pixel is set to 0 (dark/black).

$$\text{Output(Img)} = \text{Lower Value} < \text{Input(Img)} < \text{Upper Value}$$

2.3 Corner Detection

Once the foreground pixels are identified in the image, the next process is the Fast (*Features from accelerated segment test*) corner detection. Fast function detects the corners of an object present in our image. It compares the intensity of the pixel to its neighboring pixel on a circle called Bresenham's circle [5]. If nine contiguous pixel intensities are found to be more than the given threshold, it is considered as corner; otherwise it is not the corner of an object [5]. Found corner is marked as 255 (bright/white), otherwise 0 (dark/black).

2.4 Finding Pixels Location

After getting the corners of an object, the need is to find the location of every detected corner which can be done by using XF_BITSHIFT function. XF_BITSHIFT function counts the number of times the image size shifts to right until data transfer size [5]. We use this function in a loop such that it checks every rows and columns and if it finds value (pixel) at any location, the x and y coordinates of that location are recorded into the array. The recorded coordinates are selected according to the region of interest (ROI).

Figure 2 shows the flow of the algorithm in single-object detection. The color image is first converted into an HSV image (hue–saturation–value). Extract the feature (saturation) from the HSV image. Dilation use to add some pixels in the object, which are lost in feature extraction [6].

Threshold separates the pixels into two classes: maximum and minimum pixel intensities. InRange suppresses the minimum value pixel entirely from the image.

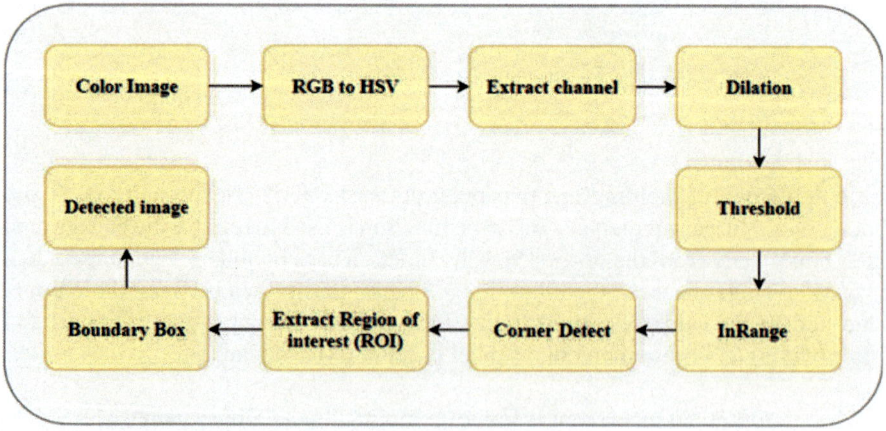

Fig. 2 Flow of algorithm in the proposed system

Fast corner detection is used to detect the corner of an object. Locate the corner location and select the region of interest (ROI). Use the region of interest to make the boundary over object present in our image using boundary function.

3 Implementation of Single-Object Detector in Vivado HLS

FPGA hardware logic is mainly written in Verilog/VHDL. For some applications, it is difficult to write complex codes in Verilog. Vivado HLS tool converts the high-level language (C/C++) into Verilog/VHDL which can be synthesized easily in Vivado for targeted FPGA [7]. So, Vivado HLS becomes very useful tool for modeling the image processing unit [4]. Figure 3 shows the design flow of Vivado HLS synthesis. In this, designers start coding in high-level language and then verify and test it using HLS tool. Once the verification and testing get over, designers synthesize the code and generate the IP.

This work implements the xfOpenCV library to generate IP in accordance with the AXI protocol [8]. In xfOpenCV, xf::Mat class is used to define the image object in Matrix format. hls::stream class is used to define the image object in AXI format. xf::AXIvideo2xfMat function is used to convert the AXI format into Matrix format

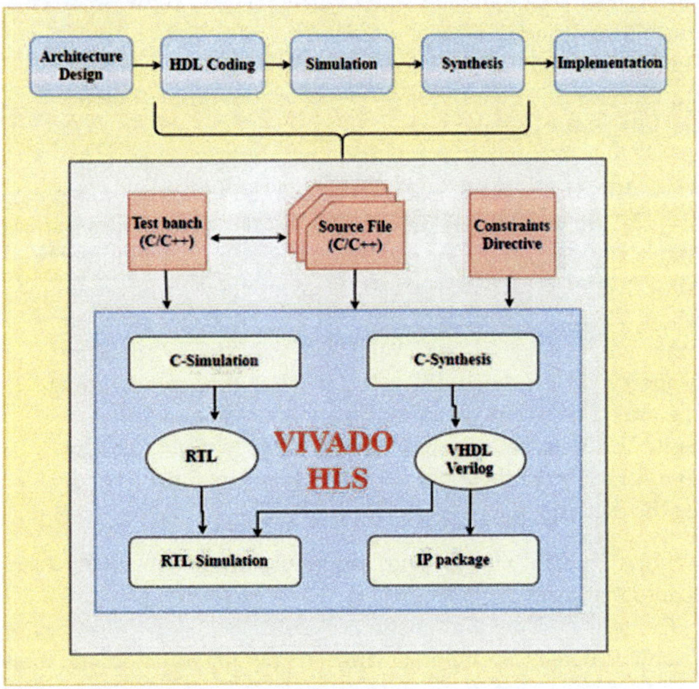

Fig. 3 Workflow for Vivado HLS synthesis [3]

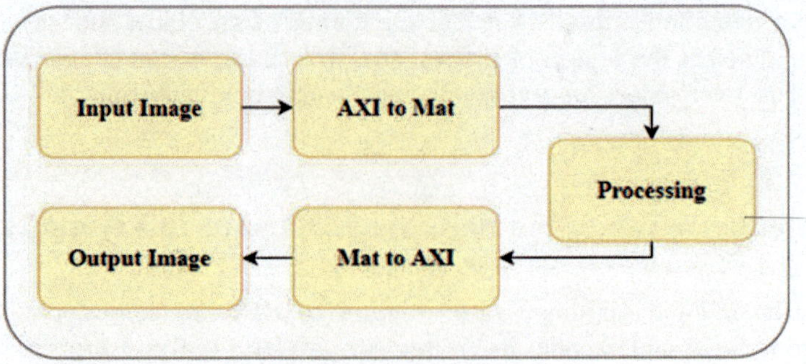

Fig. 4 Data flow in AXI and Matrix format

and xf::Mat2AXIvideo is used to convert Matrix format into AXI format [2] as shown below (Fig. 4).

Steps for algorithm:

(1) Designer writes the top-level function in C++ for synthesis:
object_detect(input image, hsv image, gray image, dilation image, threshold image, inrange image, corner image, output image, kernel matrix).
(2) Wrapper function for synthesis:
ip_object_detect(input image, output image, H, W, kernel matrix).
(3) Interface using pragma directive of HLS
#pragma HLS INTERFACE axis **register** both port = input
#pragma HLS INTERFACE axis **register** both port = output
(4) Define image object class:
xf::Mat<XF_8UC3, HEIGHT, WIDTH,XF_NPPC1> imgInput
xf::Mat< XF_8UC3, HEIGHT,WIDTH,XF_NPPC1> hsvimage
xf::Mat<XF_8UC1, HEIGHT, WIDTH,XF_NPPC1> grayimg
xf::Mat< XF_8UC1, HEIGHT,WIDTH,XF_NPPC1> dilationimg
xf::Mat< XF_8UC1, HEIGHT,WIDTH,XF_NPPC1> thresholdimg
xf::Mat< XF_8UC1, HEIGHT,WIDTH,XF_NPPC1> inrangeimg
xf::Mat< XF_8UC1, HEIGHT,WIDTH,XF_NPPC1> cornerimg
xf::Mat< XF_8UC1, HEIGHT,WIDTH,XF_NPPC1> outputimg
(5) Define the HLS stream variable using pragma directive
#pragma HLS stream variable = imgInput.data dim=1 depth=1
#pragma HLS stream variable = outputimg.data dim=1 depth=1
(6) Inside the data pragma

    (a) xf::AXIvideo2xfMat(input, imgInput) which convert AXI stream input into mat format.
    (b) Call object detection function:
    object_detect(imgInput, hsvimage, grayimg, dilationimg, thresholdimg, inrangeimg, cornerimg, outputimg, kernel)

(c) xf::xfMat2AXIvideo (outputimg, output) which convert mat format input into AXI stream.

4 Results

4.1 Simulation Results

Figure 5a, b shows the source images in Vivado HLS for C-simulation.

C-simulation in Vivado processes the source image according to the function used from xfOpenCV library.

Figure 6a, b shows the output images generated after C-simulation. The rectangle in the output image contains the detected object.

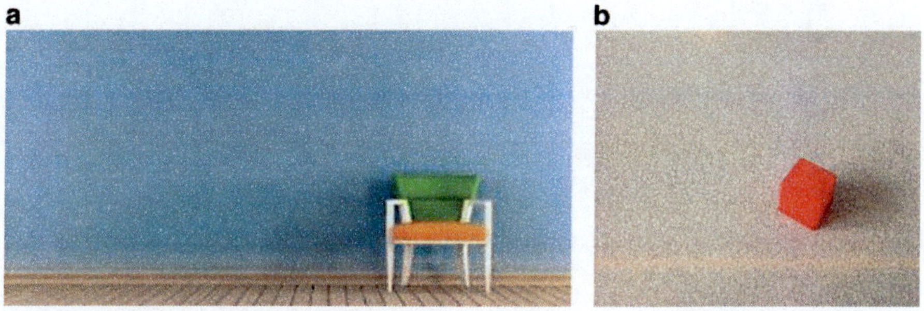

Fig. 5 a, b Source image in Vivado HLS

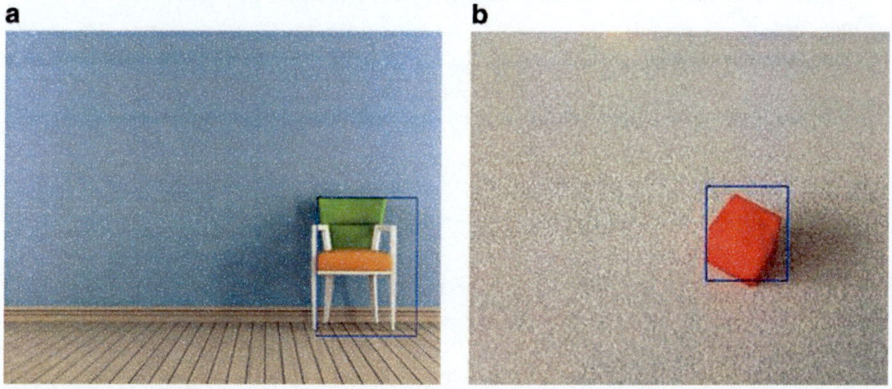

Fig. 6 a, b Output image after C-simulation

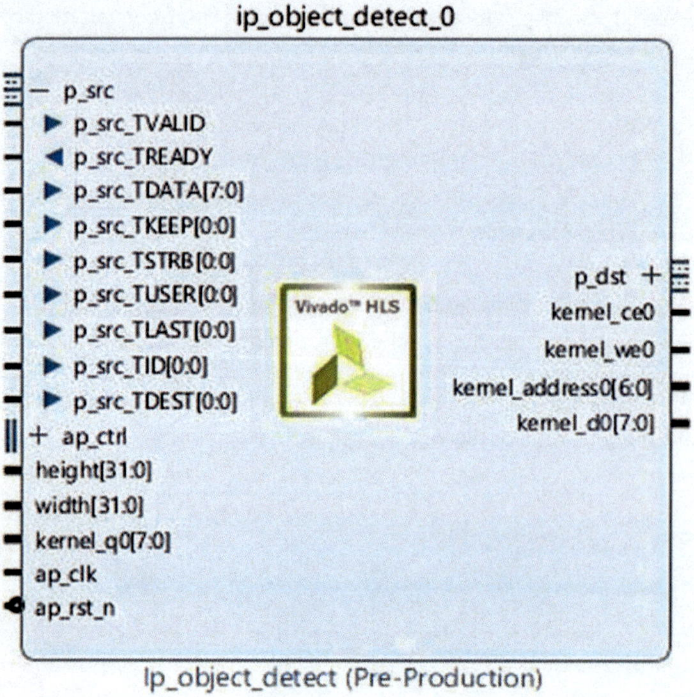

Fig. 7 Generated IP

4.2 IP Generated

The generated IP after C-synthesis and export RTL function in Vivado HLS is shown below.

The generated IP block contains AXI stream input p_src and output p_dst ports, work according to AXI standard protocol [8] (Fig. 7).

4.3 Resource Utilization

Table 1 presents a screenshot of Vivado HLS which shows the resources utilized by FPGA to generate RTL in system design. Utilization of flip flop is only 21% and the lookup table is only 49% in this implementation. The resource utilization is compared with the resource present in Zync7000 SoC-based Zybo board.

Table 1 Resource utilization of FPGA

Utilization estimates					
Summary					
Name	BRAM_18K	DSP48E	FF	LUT	URAM
DSP	–	–	–	–	–
Expression	–	–	0	72	–
FIFO	0	–	75	604	–
Instance	22	6	7370	7849	0
Memory	–	–	–	–	–
Multiplexer	–	–	–	108	–
Register	–	–	14	–	–
Total	22	6	7459	8633	0
Available	120	80	35,200	17,600	0
Utilization (%)	18	7	21	49	0

5 Conclusion

In this paper, single-object detection has been successfully done by using the xfOpenCV library in Vivado HLS. Complex algorithm for image processing is easily written on high level language (C/C++) instead of writing in Verilog. The algorithm from xfOpenCV library gives fast and better results than OpenCV library. The resource utilization table shows that less resources of FPGA are utilized. We can use different filters in between the process to get the better results such as dilation, median filter, and bilateral filter. They change the smoothness of image which helps in better finding of the pixel locations of an object. We can change the value of upper threshold and lower threshold in code to avoid the pixel loss of the targeted object. Changing threshold gives the better result in detection. Using FPGA and xfOpenCV libraries API, we can increase our system performance and reduce its power consumption and makes the system more power efficient.

References

1. M. Ning, A SoC-based acceleration method for UAV runway detection image pre-processing algorithm, in *25th International Conference on Automation and Computing (ICAC)*(2019)
2. L. Vashist, M. Kumar, *Design of Canny Edge Detection Hardware Accelerator Using xfOpenCV Library* (Springer, Berlin, 2019)
3. A. Ben Amara, E. Pissaloux, M. Atri, Sobel edge detection system design and inte- gration on an FPGA based HD video streaming architecture, in *11th International Design and Test Symposium (IDT)*(2016)
4. W.L. Wenchao Liu, H.C. He Chen, L.M. Long Ma, Moving object detection and tracking based on ZYNQ FPGA and ARM SOC, in *IET International Radar Conference*(2015)

5. Xilinx. Opencv Guide. Available: https://www.xilinx.com/support/documentation/sw_manuals/xilinx2017_1/ug1233-xilinx-opencv-user-guide.pdf
6. S. Chhabra, H. Jain, S. Saini, FPGA based hardware implementation of automatic vehicle license plate detection system, in *International Conference on Advances in Computing, Communications and Informatics (ICACCI)*(2016)
7. A. Cortes, I. Velez, A. Irizar, High level synthesis using vivado HLS for Zynq SoC: image processing case studies, in *Conference on Design of Circuits and Integrated Systems (DCIS)*(2016)
8. M. Kowalczyk, D. Przewlocka, T. Krvjak, Real-time implementation of contextual image processing operations for 4K video stream in Zynq UltraScale+ MPSo, in *Conference on Design and Architectures for Signal and Image Processing (DASIP)* (2018)

Impact Assessment of Cross-Subsidy Surcharge on Electricity Demand in Short-Term Power Market in India

Naveen Agarwal, Naqui Anwer, and Gopal K. Sarangi

1 Introduction

Competition is one of the major key to revive ailing power sector of the country. To increase competition in the sector, open access policy was enacted in the Electricity Act 2003 (EA 03). As per EA 03, open access means *"the non-discriminatory provision for the use of transmission lines or distribution system or associated facilities with such lines or system by any licensee or consumer or a person engaged in generation in accordance with the regulations specified by the Appropriate Commission".*[1] Before the open access policy, consumers did not have any choice for power supply other than buying it from the state utility. Even if he was not happy with the quality of power supply or the price of power, he was enforced to be connected with respective state utility only. However, after the enactment of EA 03 and in turn the open access policy, any consumer who has more than 1 MW load can choose their power supplier. Also, any captive power generator can supply power using states' transmission and distribution infrastructure. Power trading which was started in 2001 to cover the deficit of power in one region by supplying power from other surplus region was supposed to get boost after the introduction of open access policy [1]. There are two ways of power trading, either through bilateral market or through spot market at

[1]See Sect. 47, page 7 of EA 03.

N. Agarwal (✉)
TERI School of Advanced Studies, Delhi, India
e-mail: Naveen.agarwal@greatlakes.edu.in

N. Anwer
Energy & Environment, TERI School of Advanced Studies, Delhi, India
e-mail: Naqui.anwer@terisas.in

G. K. Sarangi
Department of Policy Studies, TERI School of Advanced Studies, Delhi, India
e-mail: Gopalkrishna.sarangi@terisas.in

S. Vadhera et al. (eds.), *Latest Trends in Renewable Energy Technologies*, Lecture Notes in Electrical Engineering 760, https://doi.org/10.1007/978-981-16-1186-5_4

electricity exchanges. IEX was the first power exchange in India which was set up in 2008 to create electricity market at national level (Economic Survey 2015–2016). At exchange, price and quantum bids are submitted by both electricity suppliers and buyers where first bids are submitted by seller and second by the buyers, and the intersection of two decides the market clearing price and market clearing volume [2]. A buyer who wants to purchase electricity in spot market or from bilateral market, first has to get open access permission from the concerned utility and regulatory commission. This is required as concerned utilities will need to assess the availability of transmission network capacity. An open access consumer has to pay different charges such as transmission charges, wheeling charges, and cross-subsidy charges. Transmission and wheeling charges need to be paid for the usage of transmission and distribution network and cross-subsidy charges are paid to compensate the loss of cross-subsidy to utility which that consumer was paying to subsidize the agricultural category of consumers.[2] Cross-subsidy charge increases the overall cost of power buying from the exchange to the consumer. It takes away all the economic benefit of buying power from the exchange [1].

The aim of this paper is to evaluate the impact of cross-subsidy surcharge on the demand of electricity in the short-term electricity market. This demand assessment is very crucial to evaluate the impact of cross-subsidy policy on the short-term electricity market which plays a very important role in filling the demand supply gap. This assessment will also help the marketers to assess the demand and supply projections in future, considering the cross-subsidy levels in the electricity tariff. Based on the results, bidders (buyers and sellers) would be able to evaluate/assess the criticality of cross-subsidy surcharge factor in assessing the future demand of electricity in the short-term electricity market.

This paper has been divided into following sections: (i) Sect. 2 reviews the literatures on electricity market and cross-subsidy to understand the level of the research done in this area and other insights into these concepts; (ii) Sect. 3 provides the methodology to measure the impact of cross-subsidy surcharge on electricity demand in bilateral market and spot market separately, along with a discussion of data incorporated in this analysis; (iii) Sect. 4 of the paper analyzes and discusses the results of this analysis and the last or the fifth section concludes the paper and provide recommendations.

2 Literature Review

Development of electricity market is necessary to create a competitive environment in the power sector. Electricity market gives the choice to buyers and sellers to buy and sell electricity. Open access is one of the most important policy measures of the EA 03, which allows consumers to choose their suppliers in the electricity market and which will lead to attract more investors to invest in the power sector (Kumar and

[2]See Sect. 2 of Sect. 42 of EA 03.

Chatterjee 2012). There are two ways of power trading; one is bilateral trading and the second is spot market at power exchange [2]. Under bilateral trading, both parties (buyer and seller) negotiate with each other for a quantum of electricity with some set of terms, and once it is finalized, electricity trading is commenced as per the contract. In spot market, double-side auction takes place in which buyers and sellers bid their price for certain quantum of electricity and based on that price is discovered [3]. In electricity trading, price follows the concept of economics means the clearing price is dependent on demand and supply of electricity. If there is reduction in supply or increase in demand, price will increase and if supply increases or demand decreases, price will decrease (Umesh Kumar Shukla).

For both the forms of short-term power trading, consumer has to get open access from respective utility and regulatory commission. The main objective of introducing open access was to create a competitive environment in the sector which will bring efficiency and motivate investors to invest in this high capitalized sector. However, the experience from the last many years suggests that the full success of open access is still far away [4]. Many of the state utilities have denied giving permission for open access. Reasons behind this are that first utility has fear to lose a high revenue consumer as mostly these are industrial consumer; second, switching of consumers impacts on their power procurement planning [4]. Other than rejections from the discoms, industries itself are not encouraged much to use open access to get electricity at better price than discom [2]. The main reason is that industries have to pay cross-subsidy surcharge and other charges also to discom if they leave the discom and choose to get power from the market using open access.[3] Paying this cross-subsidy surcharge over the cost of electricity from other sources does not give any economic benefit to consumer to prefer electricity market over discom [5]. To achieve the objective of open access, cross-subsidy should be reduced in a progressive manner [6]. National Tariff Policy 2006 (NTP 06) remarked that cross-subsidy surcharge should not be that high that it constrains the consumers to use open access. NTP 2006 also notified to prepare a roadmap to reduce cross-subsidy in a progressive manner.[4] But most of the states did not prepare roadmap for a long time. In China, in 2015, the government proposed to reduce cross-subsidy through electricity reforms but it hasn't impacted consumers in a positive manner [7]. It may be because, in China, electricity consumption is inelastic to price, and the price of appliances and temperature factors drive it (Lin and Wang 2020).

In cross-subsidy system, one category or group of consumer is subsidized by the other category or group of consumer by paying more than the actual price of the product. Highlighting the problems of cross-subsidy, a paper by Willems [8] mentioned that due to cross-subsidization, competition between any two groups of people becomes insignificant as it distorts the level-playing field. A study by Ansarin et al. [9] mentions that cross-subsidy also exists in a constant (does not change with time) electricity price system where distribution utility purchases electricity from the wholesale market at different price whereas it provides to consumers at

[3]See Sect. 2 of Sect. 42 of EA 03.

[4]See Sect. 8.5.1 of National Tariff Policy, 2006.

a constant price. Referring Electricity Act 2003, researcher Subhes C Bhattacharya (2005) also suggested to reduction in cross-subsidy, and if the government announces any subsidy, it should be paid to utility in advance. In electricity tariff structure, industrial and commercial category of consumers pays more than the actual cost of electricity to subsidize the agricultural category of consumers which reduces their irrigation cost. In this view, Ranajoy Bhattacharya (2017) proves in his study that removal or reduction of cross-subsidy may increase this irrigation cost which may further increase the inflation. Guo et al. [10] add further that to some extent cross-subsidy has a positive impact as it encourages industries to use clean energy sources.

Basically cross-subsidy level depends on the electricity demand from both the subsidizing and the subsidized group. Any change in the demand portfolio from one group will change the cross-subsidy level. With increase in rural electrification, more number of agricultural consumers are being added into the consumer list of discoms, which will increase the requirement for cross-subsidy revenue to maintain the even current cross-subsidy level for agricultural consumers. This will enforce regulators to increase the cross-subsidy level for industrial and commercial consumers to increase the overall cross-subsidy revenue or maintain the overall revenue. With the study of sub-optimality regime in cross-subsidy, researcher Chattopadhyay [11] found that there is possibility that more increase in cross-subsidy reduces the cross-subsidy revenue. This is possible because of shifting of some industrial consumers from utility to setting up their own captive power plant or purchasing power from renewable sources using open access as in both the cases consumer either has to pay no or very less cross-subsidy surcharge [12, 13].

3 Data and Methodology

To estimate the impact of cross-subsidy surcharge on short-term power trading volume, we need to analyze the relationship between cross-subsidy surcharge and electricity demand in spot market and bilateral market. To find this relationship, we estimate elasticity to see the responsiveness of electricity demand in both forms of short-term power trading to the change in cross-subsidy surcharge. The cross-subsidy surcharge elasticity η for electricity demand in short-term electricity market is

$$\eta = \frac{\text{Percentage change in electricity demand}}{\text{Percentage change in cross subsidy surcharge}} \tag{1}$$

3.1 Data

The present research is based on the secondary data of the selected states. The top 15 states in the list of highest electricity demand were selected for the research. A final sample of eight states was taken based on availability of the data. The period of study is from 2012–13 to 2017–18. Data of state-wise electricity purchase in spot market at exchanges from FY 2012–2013 to 2017–2018 was obtained from the NLDC website. For the same period, bilateral trading data was taken from the CERC monthly short-term transactions reports (CERC 2012–2013 to 2017–2018). These spot market and bilateral trading data was in the form of monthly, but as the change in cross-subsidy surcharge happens on an annual basis, we had converted monthly data into yearly data. The cross-subsidy surcharge data was sourced from tariff orders from 2011–2012 to 2016–2017 of each state, and the state domestic product at constant price data was sourced from the RBI statistics handbook (RBI 2017–2018).

As the objective of this research is to determine the impact of cross-subsidy surcharge on the electricity demand in the short-term electricity market, referring the above literature review, we will estimate the cross-subsidy surcharge elasticity of the electricity demand.

Determinants of electricity demand at short-term electricity market.

Based on the review done in the literature review section, the following determinants were identified:

1. Price at short-term electricity market: As per law of demand, if all other factors remain equal, quantity demanded is negatively related to the price. Study by Bose and Shukla [14] found a negative relationship between price and electricity consumption.
2. Cross-subsidy surcharge: The main objective of this paper is only to assess the impact of this variable on the electricity demand in short-term electricity market. As per report by Energetica India [5], cross-subsidy surcharge should discourage industrial consumers to opt short-term electricity market, which means that cross-subsidy surcharge should have negative impact on electricity demand.
3. States industrial gross domestic product (GDP): Increase in GDP will increase the demand of production which will lead to increase in the demand of electricity. In their study [14] found that sectoral electricity consumption is positively dependent on sectoral GDP.

3.2 Methodology

In this study, sample contains data across states over a period. As per paper by Guha-khashobis and Bhaduri [15], estimates from panel data are expected to be more robust. In another paper by Baltagi (2011), it is mentioned that using panel data allows the estimation of parameter in a more efficient way. Also in panel data, the impact of multicollinearity reduces both time and cross-sectional dimensions present in the data. To understand and estimate the impact of cross-subsidy surcharge on electricity demand, two panel data estimation have been used, namely, pooled ordinary least square and fixed effects panel model.

The equation for the panel data will estimate in the form given below:

$$ED_{itk} = \beta_0 + \beta_1\, price_{itk} + \beta_2 CSS_{itk} + \beta_3 SGDP_{itk} + e_{itk}$$

where.

- i number of cross-sections.
- t period of study.
- k type of market (spot market and bilateral market).
- e_{it} error-term of the equation.

4 Empirical Findings and Discussion

4.1 Descriptive Analysis

This analysis provides the idea of distribution of the data of variables using statistical analysis. It also helps in finding any outliers in the data set.

Table 1 compares the electricity prices and demand in both spot and bilateral market. As per data given in the table, prices are higher in bilateral market compared to spot market. As per demand data, it is higher in case of bilateral market but variation is higher in spot market. Here it is very interesting that though price is lower in case of spot market, but here demand is still lesser compared to bilateral. The most possible reason is the confirmation of electricity supply in case of bilateral market which is not possible in case of spot market because of pooling-based trading structure in spot market. CSS is very less in some states, whereas in some it is very much equal to price of electricity in short-term electricity market.

Table 1 Descriptive analysis

	Price (Spot)	Price (Bilateral)	Demand (Spot)	Demand (Bilateral)	CSS
Min	2.5	3.53	41.78	58	0.1
Mean	3.12	4.02	1794	2567	1.23
Max	3.67	4.33	7803	7259	3.32

Table 2 Correlation matrix—independent variables (spot market)

	CSS	GDP	Price
CSS	1	0.64	−0.06
GDP		1	−0.06
Price			1

Table 3 Correlation matrix—independent variables (bilateral market)

	CSS	GDP	Price
CSS	1	0.66	−0.33
GDP		1	−0.16
Price			1

4.2 Correlation Analysis

The correlation between independent variable is measured in this study. Tables 2 and 3 show the measured value of correlation for both bilateral and spot market and it is found that there is no high degree of correlation between independent variables. It proves that the equation formulated earlier will not be affected by problem of multicollinearity.

4.3 Regression Analysis

The regression results Tables 4 and 5 from the pooled OLS method show that in case of spot market, out of three independent variables—price, cross-subsidy surcharge (CSS), and state GDP—only GDP is a significant determinant of electricity demand at 5% level for the selected Indian states, whereas in case of bilateral market, there is no determinant that is significant. In case of spot market, price is not a significant determinant though its estimate value is -1.71, which shows that price is quite negative elastic to electricity demand. For both spot and bilateral market, cross-subsidy surcharge is not a significant determinant as estimate values are −0.04 and

Table 4 Effect of explanatory variables on electricity demand in spot market using pooled OLS estimation model

	Estimate	Std. Error	t-Value	Pr(>ItI)
Price	−1.71	1.31	−1.3	0.19
CSS	−0.04	0.25	−0.15	0.87
GDP	0.79	0.29	2.68	0.01

R^2– 0.19.

Table 5 Effect of explanatory variables on electricity demand in bilateral market using pooled OLS estimation model

	Estimate	Std. Error	t-Value	Pr(>ItI)
Price	0.04	2.11	0.02	0.98
CSS	0.37	0.24	1.55	0.12
GDP	0.44	0.25	−1.7	0.09

R^2– 0.08.

0.37, respectively, which is very low. These results are further corroborated with R-square value of 0.19 and 0.08 in spot and bilateral market, respectively, which are considerably lower.

Our data sample contains data from across states over a period of time, which may lead to cross-sectional effects. To correct it, fixed effect panel model is used. The result of fixed effect panel regression Tables 6 and 7 shows that in case of spot market, all determinants—GDP, cross-subsidy surcharge, and price—are very significant for electricity demand. In this result, the estimate value of cross-subsidy surcharge has increased to −1.56. The R-squared value also has increased to 0.82.

In case of bilateral market, fixed effect panel regression shows that GDP and price are very significant determinant of electricity demand, whereas cross-subsidy surcharge is not. Here the estimate value of cross-subsidy surcharge is −0.67. The R-squared value also has increased to 0.56.

In the step-by-step regression to understand the contribution of independent variables which were found to be insignificant in case of pooled OLS regression, it is seen that in fixed effect model most of the variables are significant. Also, the R-squared

Table 6 Effect of explanatory variables on electricity demand in spot market using fixed effect estimation model

	Estimate	Std. Error	t-Value	Pr(>ItI)
Price	−1.92	0.81	−2.37	0.02
CSS	−1.56	0.71	−2.19	0.03
GDP	0.86	0.24	2.68	0.00

R^2– 0.82.

Table 7 Effect of explanatory variables on electricity demand in bilateral market using fixed effect estimation model

	Estimate	Std. Error	t-Value	Pr(>ItI)
Price	−1.43	0.67	−2.13	0.05
CSS	−0.82	0.47	−1.74	0.09
GDP	0.96	0.38	2.68	0.02

R^2– 0.56.

value has increased significantly compared to the R-squared value in pooled OLS regression.

4.4 Discussion

Findings from the regression suggest that GDP, price, and cross-subsidy surcharge play a significant role in determining the electricity demand in short-term electricity market in the country. As the main objective of this paper is to measure the impact of cross-subsidy surcharge on the electricity demand in the short-term electricity market, findings suggests that in spot market, electricity demand is negatively elastic to cross-subsidy surcharge as the value of coefficient of CSS is −1.35 which is higher than 1. In bilateral market, results show that cross-subsidy surcharge is not a significant determinant of electricity demand and the coefficient value is −0.7 which is lower than 1. It suggests that in bilateral market electricity demand is not elastic to cross-subsidy surcharge.

The main reason for the inelastic in the bilateral markets appears to be due to the fact that bi-lateral buyers are exempted from cross-subsidy surcharge if they purchase electricity from renewable sources. Most of the states' exempt cross-subsidy surcharge in case of buying power from renewable energy sources. (Refer Annexure 1 for states' policies of cross-subsidy surcharge related to renewable energy.) In bilateral market, buyers can distinguish between thermal and renewable power seller, so cross-subsidy surcharge would not be able to impact electricity demand. However, in case of electricity exchange, this differentiation is not possible as all power is considered as thermal power.

In starting of the paper, it was discussed that cross-subsidy surcharge is an obstacle for industries as they have to pay it to the utility in case of opting open access for buying electricity from the electricity market. It means that elasticity should be negative as increase in cross-subsidy surcharge would tend to prevent industries to move away from utility and will lead to reduce in the electricity demand in short-term market. A negative coefficient of cross-subsidy surcharge confirms our hypothesis that cross-subsidy surcharge is a bottleneck for growth of short-term electricity market.

All the industrial consumers who want to leave the utility and look for short-term electricity market for power purchases have two options: one is bilateral market and second is electricity exchange. As price is a very important determinant for electricity demand, based on price, buyers can opt their market place. In the last 6 years, average price at exchange and bilateral market is INR 3.12/unit and INR 4.02/unit, respectively.[5] Because of lesser price at exchange, growth in exchange has been more than bilateral market.[6] But cross-subsidy surcharge is playing a negative role in the growth

[5] CERC, "Report on Short-term Power Market in India: 2017–2018.

[6] Statement is based on data of trading volume in CERC, Report on Short-term Power Market in India: 2017–2018.

of electricity exchange. If cross-subsidy surcharge increases, consumers either would not leave the utility or would prefer bilateral market where they can be exempted from cross-subsidy surcharge by buying electricity from renewable energy sources which is not possible in case of exchange as at exchange all power is considered as thermal power only.

Consumers will opt to buy power from market only if they get financial benefit, meaning if the price of power at market is cheaper than utility tariff. In the last few years the cost of power generation from renewable sources, especially from solar PV, has drastically come down because of technology advancement. Recently, on a record low level, in a reverse auction, tariff for solar project was discovered at Rs. 2.44 per unit.[7] This major reduction in cost of solar PV power and exemption in cross-subsidy surcharge is boosting the overall renewable energy sector in the country and luring investors to invest in the sector especially solar energy. Any reduction of price at bilateral market will surely attract buyers who are looking to buy power from market without paying cross-subsidy surcharge to the utility.

From Fig. 1, it is very clear that tariff has been reduced from above Rs. 10 per unit in 2011 to less than Rs. 3 in 2017. There is an estimate that solar tariff can be further reduced because of more technical advancement. This would increase the elasticity at exchange in future as more no of industries would like to move to bilateral market to buy cheaper solar power without paying cross-subsidy surcharge if policy makers will keep giving this benefit of exemption of cross-subsidy surcharge.

5 Conclusion and Recommendation

An efficient electricity market is required to fill the gap of demand and supply in an area by supplying electricity from one surplus area to other deficit area of the country. For an efficient market it is necessary that it attracts more and more suppliers and buyers of electricity. This paper analyzes critically a hypothesis according to which cross-subsidy surcharge is a bottleneck for industries for buying electricity from electricity market. To analyze, this paper estimates the elasticity of electricity demand in short-term electricity market in respect to cross-subsidy surcharge. Analysis was done on both forms of short-term electricity market: spot and bilateral. In the analysis we considered eight states which cover more than 50% of electricity demand of the country.

In regression analysis results, cross-subsidy surcharge was found as a significant determinant in case of spot market but in case of bilateral market it does not have significant impact on electricity demand.

The overall analysis in this paper shows that cross-subsidy surcharge has a mix impact on short-term electricity market, where in one form of market, cross-subsidy surcharge is not able to impact because of states' policies to support growth of

[7]Financial Express, 4 July, 2018 (https://www.financialexpress.com/industry/bids-for-2000-mw-solar-tariffs-revert-to-record-low-level/1230553/).

renewable energy. In spot market, it has a negative impact. Overall, cross-subsidy is the hindrance for the growth of the market as it obstruct industries to choose their supplier. Though they can choose bilateral market to get away from cross-subsidy surcharge, but availability of renewable energy is still a matter of concern. So many of the buyers either have to pay cross-subsidy surcharge to get power or have to drop plan to leave the utility which is against the objective of enacting open access policy in Electricity Act 2003.

Based on our results and analysis, it is recommended that policy makers need to reform the policies related to cross-subsidy considering all its ill-effects on electricity market which is very much required to make electricity sector more attractable for investors. If the government wants to give subsidy to farmers or any other category of consumers, it should be transferred directly into the subsidized consumer's account from the other sources of fund-like tax.

Appendix

Annexure 1 State-wise cross-subsidy surcharge regulations for renewable energy.

State	Regulation	Year	Abstract
Gujrat	Compendium of Regulations & Tariff Orders Issued by Regulatory Commissions for Renewable Energy Sources in India		Third party sale from renewable energy sources is exempted from the cross-subsidy surcharge
	GERC, Open Access Regulation, Notification No 13 of 2005	2005	The open access users, except those availing open access facility to transfer power from their captive generating plants to the destination of own use, shall pay the (cross-subsidy) surcharge to the Distribution Licensee of their area, as determined by the Commission from time to time under Sect. 42 (2) of the Act
Karnataka	Before The Karnataka Electricity Regulatory Commission, Bengaluru	5-Jan-17	The fallout of high CSS is that it opens opportunities to solar and group captive transactions
			That Group Captive is now permitted and such generators do not pay CSS and tax

(continued)

(continued)

State	Regulation	Year	Abstract
			That the impugned order suffers on account of placing the wind power plants at a disadvantageous position vis-à-vis solar power plants, to which wheeling charges, banking charges, and CSS are waived, on contrary to the Act and the National Electricity Policy
			That disallowing the CSS will defeat competition
	Wheeling charges, Banking charges & Cross Subsidy Surcharge for Solar Power Generators	2014	In the said order, the Commission has decided not to levy any wheeling charges and banking charges, or cross-subsidy surcharge for the solar generators who sell electricity on open access within the state
Rajasthan	RERC Notification [2], Sect. 38.5, Page 18	2014	The cross-subsidy surcharge as determined by the Commission from time to time shall not be applicable in case of open access transactions based on wind energy, solar pv and solar thermal power stations
Haryana	Haryana Solar Power Policy, 2016	2016	All electricity taxes and cess, electricity duty, wheeling charges, cross-subsidy charges, transmission and distribution charges and surcharges will be totally waived off for ground-mounted and roof top solar power projects
Maharashtra	MERC open access regulation 2016, Sect. 14.7.d, page 36		No relaxation on renewable energy
UP	UPERC Solar Policy 2017, 8.1.4, Page 12	2017	As Uttar Pradesh is a power importing state, cross-subsidy surcharge and wheeling charges/transmission charges will be exempted 100% for intrastate transmission system on interstate sale of solar power

(continued)

(continued)

State	Regulation	Year	Abstract
Bihar	BERC Regulation, 2017,Sect. 33.V, Page 25		In case of third party sale of solar power, the transmission licensee, distribution licensee and/or SLDC, as the case may be, will provide open access within 30 days of application on payment of required open access charges as determined by the Commission and the generator in addition to 2% banking charge, will pay to the BSPHCL/distribution licensee wheeling and transmission charges and T&D losses as determined by the Commission for the voltage level at which power is fed into the grid. However, no cross-subsidy surcharge will be paid
Punjab	Policy on Net-Metering for grid interactive roof-top solar PV, GoP	2014	To encourage the green energy, such roof top solar system installations set up under this policy would be exempted from open access regulations
		2014	The roof top solar system under net metering arrangement, whether self-owned or third party owned installed on eligible consumer premises, shall be exempted from banking and wheeling charges and losses, cross-subsidy and additional surcharge, etc. and MMC shall be applicable as per sub-clause(5) of clause 13
	PSERC, Petition No. 75 of 2016	2017	That with the setting up of a power plant as intended by the petitioner will result in significant loss of revenue for the home distribution utility as the existing cross-subsidy surcharge and the additional surcharge shall not be leviable on group captive power plant and payment of transmission and wheeling charges for non-conventional sources of energy (solar/wind power generation etc.) has already been exempted by this Hon'ble commission vide order

References

1. Energy Advisory Board, IDFC, Compendium of Proceedings Agenda Notes, Volume II (2011)
2. Y.K. Bichpuriya, S.A. Soman, Electric power exchanges: a review, in *16th National Power Systems Conference* (2010), pp. 115–120
3. R.K. Mediratta, V. Pandya, S.A. Kharpade, Power markets across the globe and indian power market. in *Fifteenth National Power Systems Conference (NPSC)* (2008), pp. 271–275
4. D. Singh, Newer challenges for open access in electricity: need for refinements in the regulations. *Brookings India* (2017)
5. Transmission and Distribution, *energetica india*, 74–76 (2009)
6. A. Agrawal, A. Kumar, T.J. Rao, Future of Indian Power sector Reforms: electricity amendment bill 2014. Energy Policy **107**, 491–497 (2017)
7. H. Yu, X. Xin, Demand elasticity, ramsey index and cross-subsidy estimation for electricity price in China. Sustain. Prod. Consum. **24**(2020), 39–47 (2020)
8. B. Willems, E. Ehlers, V.M. Fraga, Cross-subsidies in the e\Electricity Sector, in *TILEC Report* (2008)
9. M. Ansarin, Y. Ghiassi-Farrokhfal, W. Ketter, J. Collins, The economic consequences of electricity tariff design in a renewable energy era. Appl. Energy **275**, 115317 (2020)
10. S. Guo, L. Meng, W. Li, W. Yan, L. Wang, The cross subsidy of electricity price in China. Int. J. Eng. Bus. Manag. **3**(6), 97–103 (2019)
11. P. Chattopadhyay, Cross-subsidy in electricity tariffs: evidence from India. Energy Policy **32**, 673–684 (2004)
12. R.K. Bose, M. Shukla, Elasticities of electricity demand in India. Energy Policy **27**, 137–146 (1999)
13. B. Guha-khasnobis, S. Bhaduri, Determinants of capital structure in India (1990–1998). J. Econ. Integr. **17**, 761–777 (2002)
14. S. Pargal, S.G. Banerjee, More Power to India: the challenge of electricity distribution, in *The World Bank* (2014)
15. D. Heald, Contrasting approaches to the problem of cross subsidy. Manag. Account. Res. **7**, 53–72 (1996)
16. Economic Survey 2015–2016. *Powering One India*, 152–161 (2016)
17. U.K. Shukla, A. Thampy, Analysis of Competition and market power in the wholesale electricity market in India. Energy Policy **39**, 2699–2710 (2011)
18. S.C. Bhattacharyya, The Electricity Act 2003: Will it transform the Indian power sector? Utilities Policy **13**, 260–272 (2005)
19. R. Bhattacharyya, A. Ganguly, Cross subsidy removal in electricity pricing in India. Energy Policy **100**, 181–190 (2017)
20. FoR-pwc, Report on Road Map for Reduction in Cross Subsidy (2015)
21. CII-pwc, Changing rules of Indian power sector: Empowering the economy (2015)
22. D. Saha, R.N. Bhattacharya, An analysis of elasticity of electricity demand in West Bengal, India: Some policy lessons learnt. Energy Policy **114**, 591–597 (2018)
23. E. Labidi, T. Abdessalem, An econometric analysis of the household direct rebound effects for electricity consumption in Tunisia. Energ. Strat. Rev. **19**, 7–18 (2018)

An Intelligent Control Technique-Based DTC of BLDC Motor Using New Multi-level Inverter

S. Arun Naik, G. L. Pahuja, and Prakash Kulkarni

1 Introduction

BLDC is known for the density of power and rapid efficiency [1]. It has a non-sinusoidal (near trapezoidal) posterior EMF and requires special control to establish the magnetic field strength. The electric drives with high-speed activity are required to shut flux and torque regulation. This regulation is by field-based control (FOC). To control the module load in an inverter system, the FOC system requires an observer check for the change in coordinates and a current controller. However, the FOC requires accurate and complex reporting, thereby increasing the lightness and culture of the load due to the loading disturbance and uncertainty of the parameters [2]. DTC is a structure to give quick and dynamic responses of the drive even for low recurrence activity of intensity electronics. It is a consolidation of the FOC and DSC procedures. Utilizing exchanging vector look-into table, the basic thought of DTC of enlistment engine drives is to control both stator motion linkage and electromagnetic torque of the machine.

The basic concept of the DTC of motor drives is to regulate both the coupling of the stator flux as well as the electromagnetic winding of the motor and at the same time using the switching vector view table [3]. It is characterized by the lack of PI controllers, coordination changes, current controllers, and PWM signal generators. DTC works with a variable switching frequency or standard scanning techniques. Compared to the vector control scheme, DTC offers dynamic performance similar

S. A. Naik (✉) · G. L. Pahuja
Electrical Department, National Institute of Technology, Kurukshetra, India
e-mail: arn2595@gmail.com

G. L. Pahuja
e-mail: Pahuja.gl@gmail.com

P. Kulkarni
KPIT Technologies Ltd, Pune, India
e-mail: Prakash.A.Kulkarni@kpit.com

S. Vadhera et al. (eds.), *Latest Trends in Renewable Energy Technologies*, Lecture Notes in Electrical Engineering 760, https://doi.org/10.1007/978-981-16-1186-5_5

to that of the simple control architecture. The conventional strategy of DTC is the basis of the loop of hysteresis accompanied by the switching table of a single vector. Its frequency of switching deviates with speed and torque, which can result in higher torque pulses, especially at lower speeds due to lower torque frequency, which limits its use [4]. The general challenges of DTCs are: the largest torque ripples and transition of the intermediate repetition to the phase changes of the torque phase at the beginning. Therefore, intelligent methods like the fuzzy logic (FL), the sliding mode control (SMC), and artificial neural network (ANN) are used. Most of them relate to flux development and moment estimator, as well as integration in DTC with space vector modulation (SVM) technology. The main issue was the straight winding and circulation stream regulation of the BLDCM as the regulation component and the d-axis current in the rotor reference outline was regulated to diminish the air hole current, resulting in an iron loss in the motor. This issue can be resolved by an assortment of methods, for example, sliding mode control (SMC), variable structure control (VSC), self-optimization strategies, adaptive control, etc. The specific control framework above likewise relies upon the numerical model and it is hard to build up a scientific model of the framework because the framework model cannot be adjusted to obscure load and parameter factors. In this manner, the exactness of the model and the speed of the framework are influenced. This paper uses a new multi-stage inverter that is based on a straight torque regulation system and controls the velocity of the BLDC. Here, the DTC technique is proposed to reduce the torque ripple by keeping the switching frequency constant. A comprehensive review of the proposed DTC method is analyzed in the following part. The proposed approach for velocity regulation of the BLDC is described in Sects. 2 and 3. The details of the results and discussions of the proposed approach are presented in Sect. 4.

2 Utilizing the EHO Algorithm for DTC of BLDC Motor

The DTC-based variable speed drive is to create the torque command. Therefore, if multiple vectors are used with a single sampling interval, a super function is expected. This is in accordance with altered DTC. According to an altered scheme, these vectors are utilized to regulate the flux at each sampling interval, such that the torque and flux amplitudes can be controlled up to their reference values.

The proposed method is utilized to regulate the BLDC motor's speed as per the torque variation. The proposed strategy details are given as follows. Kindly assure the method proposed in the paper is employed to the torque analysis, since the referral speed and the torque values vary with the load disparity, and note the need to improve torque. The reference point for the torque displacement system shall then be determined by reference to the torque deviation from the torque. The proposed system determines the actual torque and the modified torque values are given by input of EHO. The associated output is controlled using the proposed method and these values are formed to minimize error values. Since then, this dataset is well

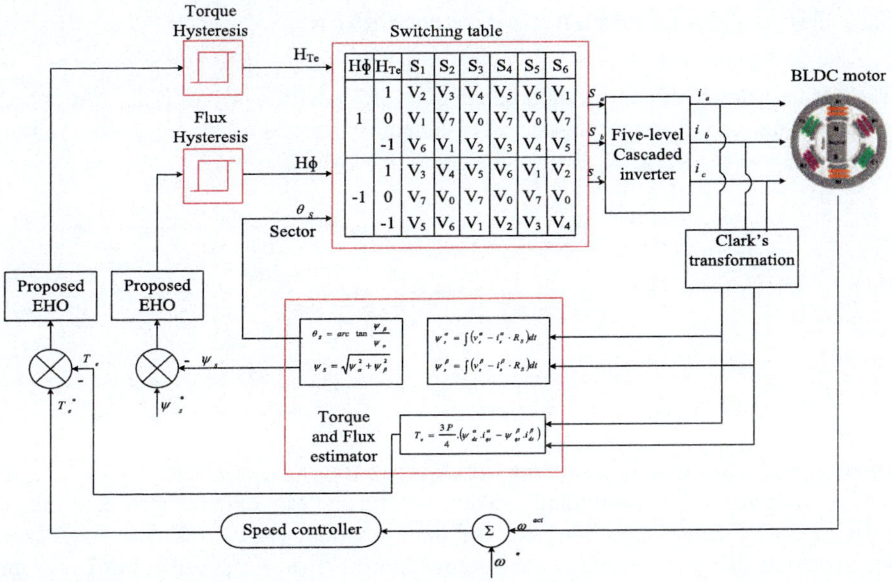

Fig. 1 Direct torque regulation system using the proposed technique

trained to generate control pulses. Afterward, the proposed method then decides to regulate the BLDC motor's DTC.

The PWM control signal is generated to control the inverter gate, as per the controlled output. The PWM with vector technique is utilized to determine the gate regulation signal. The SVM-based switching table is referred based on the output of the switch controlled by the IGBT gate. The switching pattern of the inverter can be determined according to the stator flux-linkage and torque status from the outputs of two regulators shown in Fig. 1, and the sector in which the stator flux linkage is located at that instant of time. In each sector, if the actual stator flux linkage is the same as the commanded stator flux linkage, only one non-zero-voltage space vector and a zero-voltage vector are used to control the increase or decrease of the torque, since during any 60 electrical periods only two phases are excited and controlled in a BLDC drive, as indicated in Table 11. Also, when the actual flux linkage is smaller than the commanded value, the non-zero-voltage space vector is used to increase the flux linkage, while when the actual flux linkage is greater than the commanded value, the non-zero-voltage space vector is used to decrease the stator flux linkage.

2.1 Multi-Objective Function Formulation

The optimal choice of control parameters and rotor position is to diminish the speed error, torque, and flux employing multi-objective function. The necessary multi-purpose function is detailed in the subsequent Eqs. (1), (2), and (3).

$$\text{Speed}_{\text{of}} = \omega^{e} = \omega^{*} - \omega^{\text{act}} \tag{1}$$

$$\text{Torque}_{\text{of}} = \frac{\text{Max}_t - \text{Min}_t}{\text{Mean}_t} \tag{2}$$

$$\text{Flux}_{\text{of}} = \psi^{e} = \psi^{*} - \psi^{\text{act}} \tag{3}$$

where Speedof, Torqueof, Fluxof are the objective functions of error of speed, torque ripple, and error of flux individually. Mint, Meant, and Maxt are the net electromagnetic torque of minima, the median, and the maximum values, respectively. These are to be optimized to determine the optimal controller for the speed control system. Then, a description of the DTC-based EHO algorithm is discussed in the next section.

3 DTC-Based EHO Algorithm for Gain Optimization

BLDC motor's speed control is utilized with a PI-based controller. The optimized PI gain parameters are based on the proposed techniques, which consist of the EHO algorithm. The proposed controller restricts the torque and flux, which are on the basis of the measured factors of the BLDC motor. The EHO controller delivers the control signal to the DTC, as per the error and error changing rate. The purpose of the torque and flux error function is regulated here with the PI controller with one operating condition. The switching tables provide the PWM signal for the cascaded five-stage step-up inverter for back EMF of the motor, and the regulation signal's output is on the EHO controller [12]. The behavior of the herding is bifurcated into two kinds of operators, namely, operator of renewal, and the other operator of isolating, in numerically, as depicted in (4) and (5). In the mathematical way, the steps of the algorithm are described below.

The status of elephant is renewed utilizing the next condition (4) in every clan [15],

$$E^{i}_{\text{new},cj} = E^{i}_{g}\left(+A * E_{\text{best},cj} - E^{i}_{cj}\right) * r \tag{4}$$

Afterwards, the suitable positions of elephant are renewed in every clan, shown in next condition (5).

$$E^{i}_{new,cj} = B * E_{center,cj} \tag{5}$$

$$E_{center,g} = \sum_{i=1}^{n} E^{i,d}_{cj} / \eta e \tag{6}$$

Afterward, the elephants of worst provided in the clans are divided using the next condition:

$$E^{i}_{worst,cj} = E_{min} + (E_{max} - E_{min} + 1) * rand() \tag{7}$$

The elephants and their positions are renewed based on the above equations. When using this method, convergence properties are affected by complex problems. Moreover, the poor elephant refresh functionality fails to accurately predict results and fails to predict optimal results. Therefore, there is a need for an efficient technique to update the elephant status and the associated bad elephants. The elaborate process of DTC that depends on EHO is detailed in the next.

Step 1: It is the initialization process where the algorithm parameters, such as voltage, current, speed, torque, and flux, are actuated. The population of elephants and clans are generated randomly.

$$P_{E(i)} = \left(P^E_1, P^E_2, P^E_3 \ldots\ldots\ldots P^E_n\right) \text{ Where i} = 1 \ldots \text{n} \tag{8}$$

Step 2: The fitness process is computed from the multi-objective function. Here, super fitness is appraised from the minimized value of an objective function. It is mentioned in Eq. (9),

$$FF\left(P_{E(i)}\right) = \min_{e}(\omega, \psi) \tag{9}$$

The lessened esteems are saved as the best solutions, as per the above equation. After that, the position of elephants and clans are renewed.

Step 3: The position updating of elephants in clans is renewed. For that process, the EHO parameters are randomly generated [11]. Thereafter, the renewal function of the elephants in the clans is optimized by predicting precision outcomes.

$$\begin{aligned} \frac{V\left(E^{k+1}\right)}{P(i)} = \frac{w * U\left(E^{k}\right)}{P(i)1} + \mu \times rand \frac{\left(Pbest\left[E^{k}\right) - E^{k}\right)}{P(i)P(i)} * \frac{Pbest\left(E^{k}\right)}{P(i)2} \\ + \mu \times rand() \frac{gbest(E^{k} - k)}{P(i)P(i)} \end{aligned} \tag{10}$$

$$\left(U^{k+1}_{P(i)}\right) = \left(UE^{k}_{P(i)}\right) + V\left(E^{k+1}_{P(i)}\right) \tag{11}$$

According to the equation above, the elephants' position is renewed. From the EHO [13], update the size of other elephants in the herd (11), by updating with either an evil or a male elephant (12).

$$E^{i}_{worst,cj} = E^{i}_{cj} + B * E_{center,cj} \tag{12}$$

As per that, appraise the fitness function for the updated solutions. In the fitness computation, the best and worst solutions are updated.

Step 4: Here, the evaluation of worse elephants is done [14], which are isolated by the condition (12).

Step 5: For the termination criteria until the function of fitness decreases to the least esteem, repeat the process. Then save the optimal results and store in the corresponding voltage, current, speed error, torque, and flux error.

4 Results and Discussions

In this section, the results and performance of the measured parameters are discussed. The proposed system with EHO-based PI controller technology is replicated using the MATLAB (R2018a) platform and a five-stage cascaded inverter to regulate the velocity of BLDC.

Figures 2, 3, 4 and 5 show that the analysis of the performances of the electro-

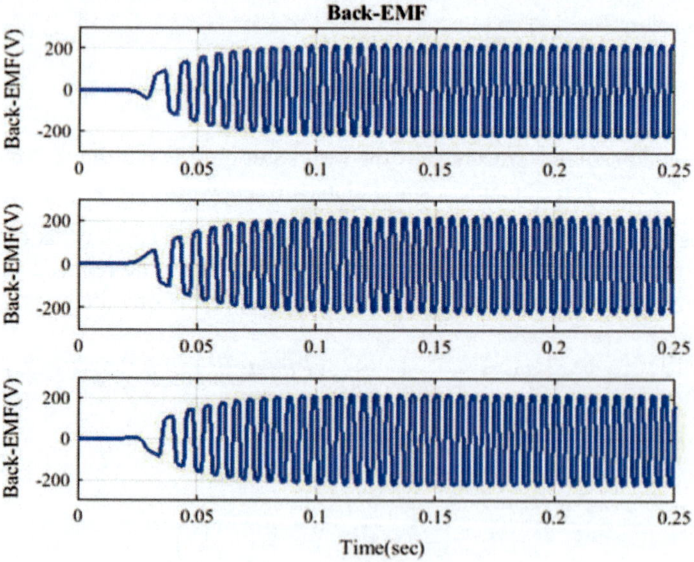

Fig. 2 Back-EMF of BLDC motor output by the proposed method

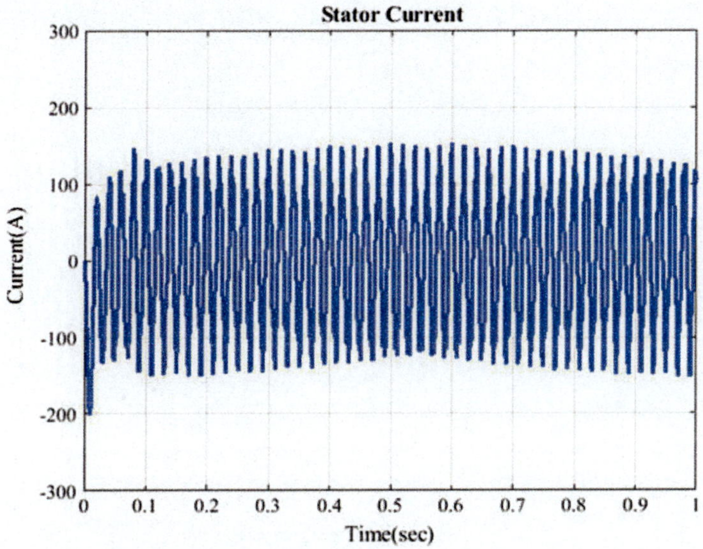

Fig. 3 Stator current of BLDC motor output by the proposed method

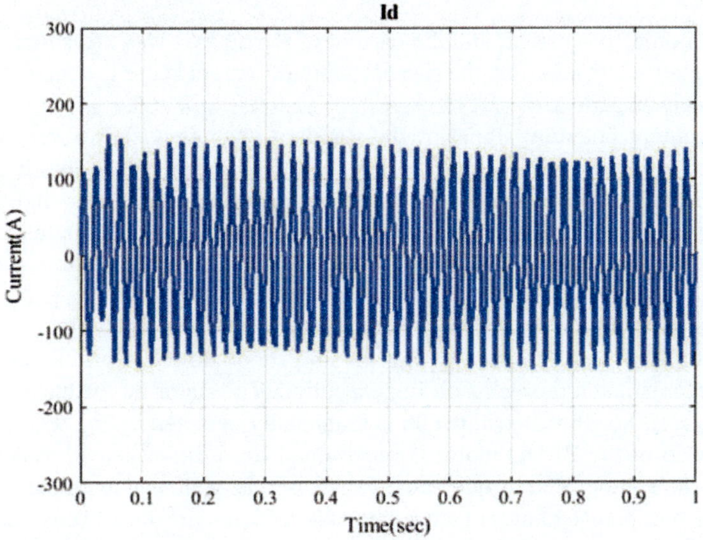

Fig. 4 Id axis current of BLDC motor output by the proposed method

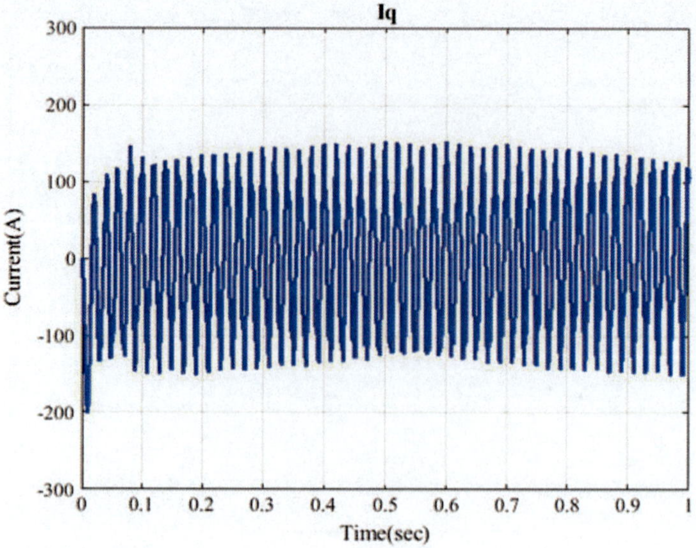

Fig. 5 Iq axis current of BLDC motor output by the proposed method

magnetic motor, the current, and the current of the dq axis was measured using the proposed method. In addition, the size of the stator current keeps the current constant for the specified period.

The resultant potential signal of the standard inverter is shown very nicely in Fig. 6. As a result, the voltage inverter is evaluated according to the duration of 0.1 s. The speed of the BLDC motor is evaluated without using the sensor, and the speed of the rotor is given in Fig. 7. Change in engine speed represents the instantaneous change in one second.

We note that the stroke time is 1.1 seconds and the stabilization time is 2.1 seconds, which changes without exception depending on torque and flow. The torque minimization of 0.14 Nm is shown in Fig. 8. The stator flow has pitfalls and certain component mismatch problems during execution. The stator flux value is estimated in the range of 0.6 wb to 1 wb, which is graphically depicted in Fig. 9.

The speed of the BLDC motor is ascertained from the evaluated transition and torque of the engine. The torque value is assessed as well as contrasted with those of the conventional techniques and the relative cases are beautifully pictured in Fig. 11. The deviation at the time of commencement is nearly 0.175 Nm, though the conventional technique exhibits elevated deviation, thus proving the superlative efficacy of the novel technique when compared to the conventional approaches. However, the proposed controller regulates the torque in the range of 0.75–0.15 Nm regarding a timeframe of 0.2–2 s in uninterrupted deviation. The novel control system evaluated flux is smaller than 1 wb. The controlled flux is evaluated and exhibited in Fig. 12, which determines the transition as well as winding, employing the transition as well as winding estimator. The control signal is capable of producing from the

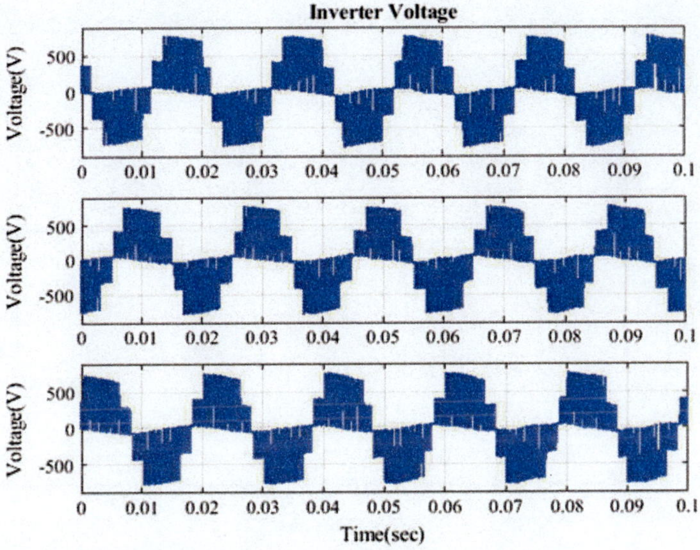

Fig. 6 Analysis of inverter voltage in the proposed method

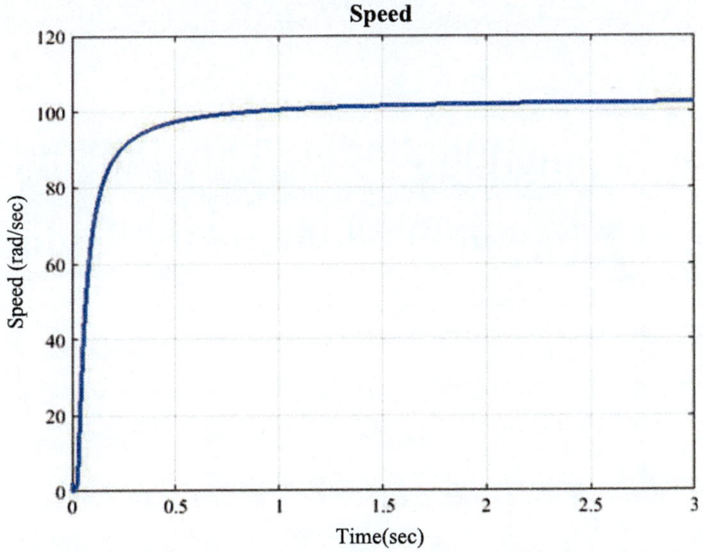

Fig. 7 Analysis of speed in the proposed method

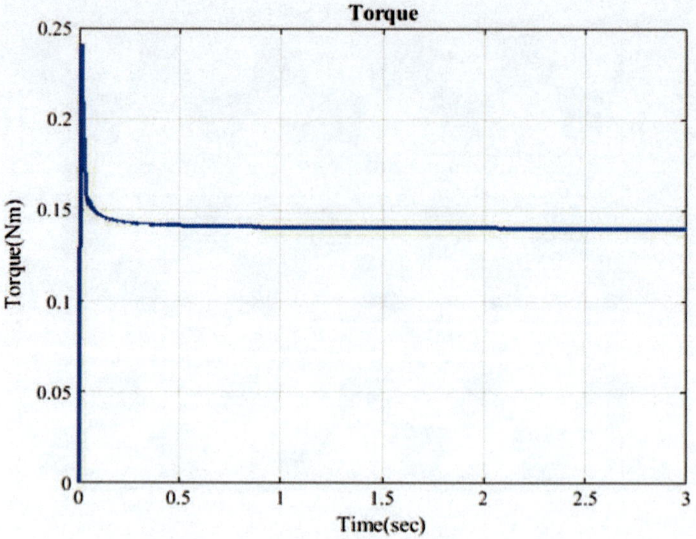

Fig. 8 Analysis of torque in the proposed method

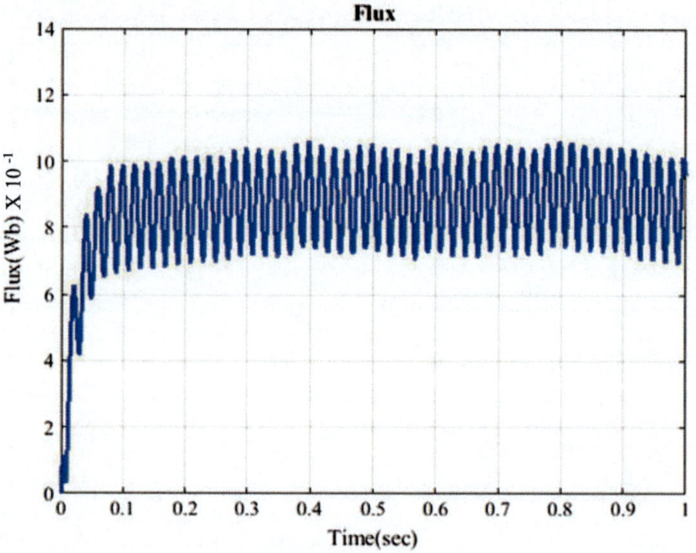

Fig. 9 Analysis of flux in the proposed method

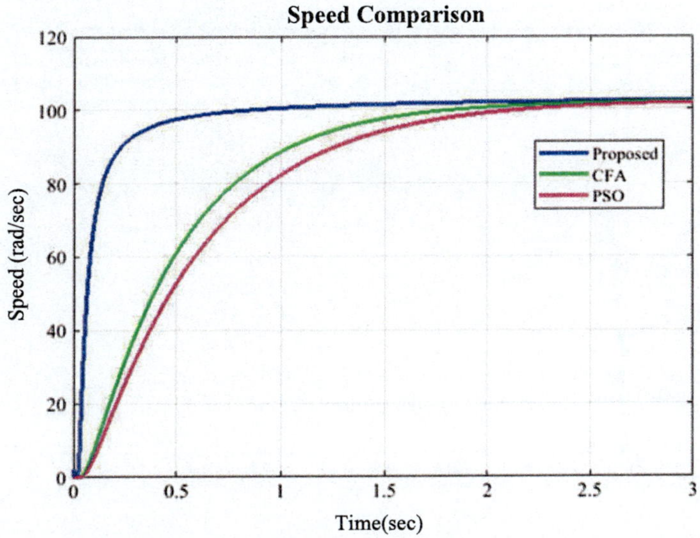

Fig. 10 Comparison analysis of speed in the proposed method

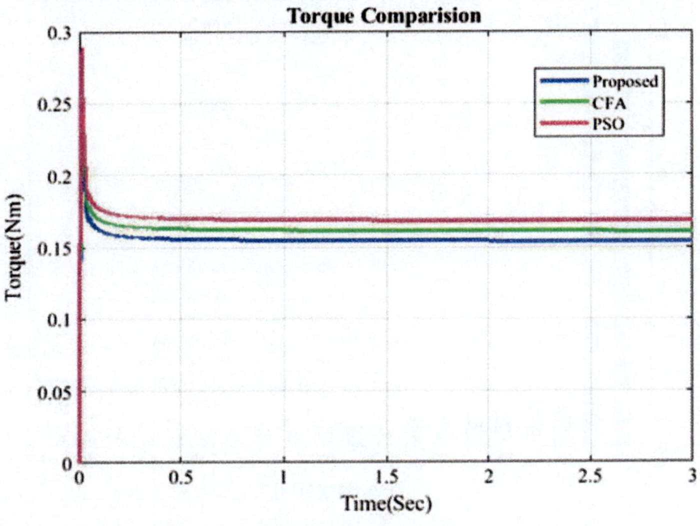

Fig. 11 Comparison analysis of torque in the proposed method

EHO-based DTC controller and furnishing the same to the five-level inverter. The examination of THD and the chosen signals are effectively exhibited in Fig. 13, from which the value of the THD level is found to be 2.41%. The decrease in the THD level of the proposed technique leads to a corresponding increase in the efficiency of the system.

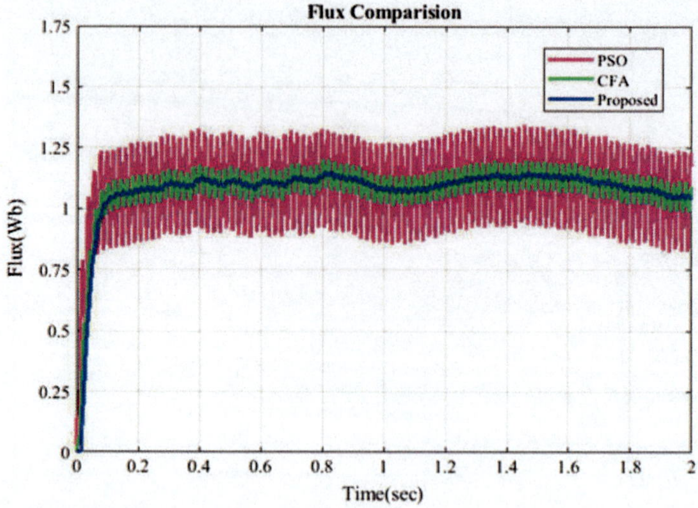

Fig. 12 Comparison analysis of flux in the proposed method

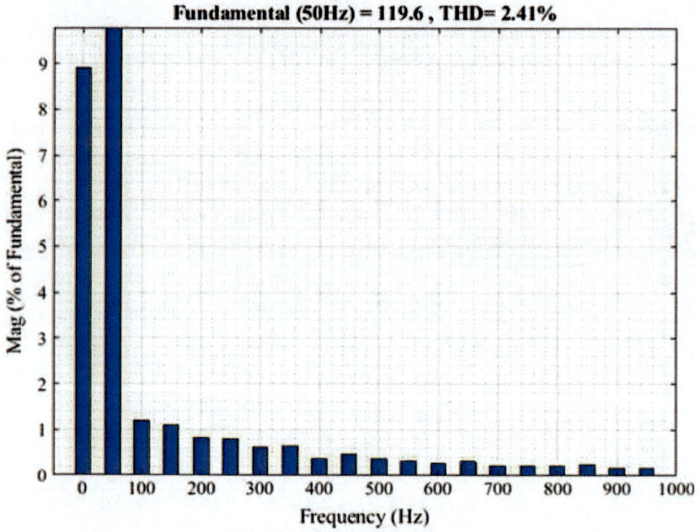

Fig. 13 Comparison analysis of THD in the proposed method

5 Conclusion

In this paper, the EHO-based DTC control technology is implemented on the MATLAB/Simulink platform. The proposed technique is utilized to regulate the velocity of the BLDC motor. First, the actual torque and touch offset are applied to the EHO. The control signal is then generated based on its inputs. These control signals are sent to the SVM, which generates the BLDC motor control pulses. The exhibition of the proposed technique is shown and contrasted and existing techniques such as the CFA and PSO methods. The dynamic behavior of the BLDC motor is analyzed in terms of speed, torque, flow, etc. The speed and torque performances of the BLDC motor are analyzed with those of the CFA and PSO method to show that the proposed method gives better results.

Appendix

Stator phase resistance Rs(ohm) = 0.2, stator phase inductance Ls(H) = 0.085, flux linkage = 0.175, back-EMF flat area (degrees) = 120, pole pairs = 4, and static friction Tf (Nm) = 0.005.

References

1. H. Liu, H. Zhang, A novel direct torque control method for brushless DC motors based on duty ratio control. J. Franklin Inst. **354**(10), 4055–4072 (2017)
2. C.K. Lad, R. Chudamani, A simple overlap angle control strategy for reducing commutation torque ripple in a brushless DC motor drive. Eng. Sci. Technol. Int. J. **20**(4), 1406–1419 (2017)
3. M. Manohar, S. Das, Current sensor fault-tolerant control for direct torque control of induction motor drive using flux-linkage observer. IEEE Trans. Ind. Inf. **13**(6), 2824–2833 (2017)
4. A. Zemmit, S. Messalti, V. Harrag, A new improved DTC of doubly fed induction machine using GA-based PI controller. Ain Shams Eng. J. **9**(4), 1877–1885 (2018)
5. V. Sudheer, S.F. Kodad, B. Sarvesh, Improvements in direct torque control of induction motor for a wide range of speed operation using fuzzy logic. J. Electr. Syst. Inf. Technol **5**(3), 813–828 (2018)
6. B.R. Vinod, M.R. Baiju, G. Shiny, Five-level inverter-fed space vector-based direct torque control of open-end winding induction motor drive. IEEE Trans. Energy Conversion **33**(3), 1392–1401 (2018)
7. N. Pimkumwong, M.S. Wang, "Full-order observer for direct torque control of induction motor based on constant V/F control technique. ISA Trans. **73**, 189–200 (2018)
8. A.G. De Castro, W.C.A. Pereira, T.E.P. De Almeida, C.M.R. De Oliveira, J.R.B. De Almeida Monteiro, A.A. De Oliveira, Improved finite control-set model-based direct power control of BLDC motor with reduced torque ripple. IEEE Trans. Ind. Appl. **54**(5), 4476–4484 (2018)
9. A. Khazaee, H.A. Zarchi, G.A. Markadeh, Loss model-based efficiency optimized control of brushless DC motor drive. ISA Trans. **86**, 238–248 (2019)
10. D. Raja, G. Ravi, Dynamic modeling and control of five-phase SVPWM inverter fed induction motor drive with intelligent speed controller. J. Ambient Intell. Humanized Comput. 1–11 (2020)
11. M.A. Elhosseini, R.A. El Sehiemy, Y.I. Rashwan, X.Z. Gao. X.Z, On the performance improvement of elephant herding optimization algorithm. Knowl.-Based Syst. **166**, 58–70 (2019)
12. W. Li, G.-G Wang, H. Amir Alavi, Learning-based elephant herding optimization algorithm for solving numerical optimization problems. Knowl.-Based Syst.105675 (2020)

13. K. Sasikumar, B. Vijayakumar, An efficient replica management based disaster recover using elephant herding optimization algorithm, in *International Conference on Computer Networks and Inventive Communication Technologies* (Springer, Cham, 2019)
14. A.E. Hassanien, M. Kilany, E.H. Houssein, H. AlQaheri, Intelligent human emotion recognition based on elephant herding optimization tuned support vector regression. Biomed. Signal Process. Control **45**, 182–191 (2018)
15. S.D. Correia, S. Beko Tomic, L.A.D.S Cruz, Energy-based acoustic localization by improved elephant herding optimization. IEEE Access **8**, 28548–28559 (2020)

Adaptive Volterra Filtered-X Logarithmic Cost Least Mean *lp*-Norm Control for Grid-Tied PV Ultracapacitor Battery Fuel Cell System

Mukul Chankaya, Aijaz Ahmad, and Ikhlaq Hussain

1 Introduction

The renewable energy systems (RES) are gaining popularities with their increasing share in the modern grid to satisfy the day-by-day increasing demand for power [1]. The high penetration of RES in the existing grid system paved a way for more stringent grid codes for maintaining the sustainability, reliability, and stability of the grid [2].

The control of voltage source converter (VSC) plays an important role in fulfilling the multi-faceted objectives. The VSC control provides the capability of load balancing, power balancing, reactive power compensation, and harmonics elimination [3]. The VSC controls can be classified as conventional controls (time domain and frequency domain), adaptive control, predictive control, adaptive-predictive control, and optimization-based control. The adaptive controls are based on a single-layer perceptron model based on neural network. The adaptive controls like least mean square (LMS) algorithm [4] and least mean fourth (LMF) algorithm [5] are the stochastic decent methods and produce the least mean square of the error signal. The LMS has the advantages of less computational burden and a faster convergence rate over conventional controls. Several advanced versions of LMS such as recursive LMS (RLMS) [6], normalized LMS (NLMS) [7], sign-error LMS (S-LMS) [8], sign-sign LMS(SS-LMS) [9], resized zero attracting LMS (RZA-LMS) [10], LMS-LMF [11], robust least mean logarithmic square (RLMLS) [12], and many more have

M. Chankaya (✉) · A. Ahmad
Electrical Engineering Department, NIT Srinagar, Srinagar, India
e-mail: mukulchankaya@gmail.com

A. Ahmad
e-mail: aijaz54@nitsri.net

I. Hussain
Electrical Engineering Department, Kashmir University, Srinagar, India
e-mail: ikhlaqiitd2015@gmail.com

S. Vadhera et al. (eds.), *Latest Trends in Renewable Energy Technologies*, Lecture Notes in Electrical Engineering 760, https://doi.org/10.1007/978-981-16-1186-5_6

been mentioned by the researchers. The adaptive LMS algorithm can be further modified by incorporating another filtering technique. There have been many studies of combining the kernel tricks with the LMS algorithm and generating new algorithms such as kernel LMS (KLMS) [13], extended kernel recursive least square (KRLS) [14], kernel normalized LMS (KN-LMS) [15], Gaussian kernel LMS (GK-LMS) [16], zero-attracting Gaussian kernel LMS (ZA-GK-LMS) [17], maximum correntropy criterion (MCC) [18], variable kernel width MCC (VKW-LMS) [19], and many more. Likewise, by incorporating filtered-x least mean *lp*-norm (FxLMP) [20] and filtered-x logarithmic LMS (FxlogLMS) [21] into the Volterra filtering results in Volterra FxLMP (VFxLMP) and Volterra FxlogLMS (VFxlogLMS) [22]. The proposed adaptive Volterra filtered-x logarithmic cost least mean *lp*-norm (AVFxlogCLMP) algorithm for the VSC is utilized in the presented work. The proposed system minimizes the logarithmic cost function w.r.t. weight vector, and then the *lp*-norm function generates AVFxlogCLMP algorithm.

The multi-carrier energy storage system (MESS) is implemented with a lead-acid battery, proton exchange membrane fuel cell (PEMFC and ultracapacitor (UC)) [23]. The MESS helps in maintaining the stability, reliability, and sustainability of the grid-tied PV system during bad weather conditions and unavailability of the sunlight. The diverse energy storage element with different load handling characteristics complement each other in handling the disturbance occurring on the grid side, the PV side, and load side [24]. The PEMFC has high power density but not suitable for satisfying sudden change in demand. The lead-acid battery has high energy density and UC has an ultrahigh energy density that makes them handling sustained and sudden nature of irregularities.

The proposed system is a three-phase three-wire grid-tied dual-stage PV-MESS system with the nonlinear load. The proposed system is utilizing the incremental conductance (InC) algorithm for maximum power point tracking (MPPT). The main contribution of the presented work is as follows:

- MESS: The MESS with UC, lead-acid battery-bank, and PEMFC can handle sudden sustainable and prolonged duration of system variations.
- VSC control: The proposed is derived from the steepest descent method and can be considered as the variable-step version of the VFxlogLMS. The proposed system minimizes the logarithmic cost function w.r.t. weight vector.
- Multi-faceted VSC: The VSC performs multiple operations including harmonics suppression, reactive power management, load and power balancing.
- Fixed and variable power mode: During fixed power mode a predefined amount of power need to be supplied to the grid by utilizing the PV and MESS. Whereas in variable power mode, the variable amount of power during irradiation variation and load unbalancing is being supplied to the grid.

2 Schematic Diagram

The schematic diagram of the presented system is shown in Fig. 1. The dual-stage PV system of 32 kW is simulated with the MESS in which PEMFC is directly connected with the DC bus and battery bank and UC are connected to DC bus via the buck-boost bi-directional converter. The VSC as a compensator is provided with gating pulses and its current is supplied to the nonlinear load of 12.5 kW. The interfacing inductor and the RC filter is implemented to reduce the current ripples.

2.1 *Designing and Selection of Battery, UC, and PEMFC Parameters*

The equivalent circuit diagram of the battery bank, UC, and the PEMFC is shown in Fig. 2a–c. E_b is the open-circuit voltage of the battery, I_{batt} is the battery charging/discharging current, as shown in Fig. 2a.

R_p and C_p are the polarization resistance and capacitance, respectively, which govern the dynamic characteristics, and voltage inside this loop is polarization voltage E_0. R_s is the series resistance and governs the static characteristics of the battery bank. The C_p is chosen as per (1) and R_p and R_s are chosen as 0.1 Ω and 10 kΩ, respectively [25].

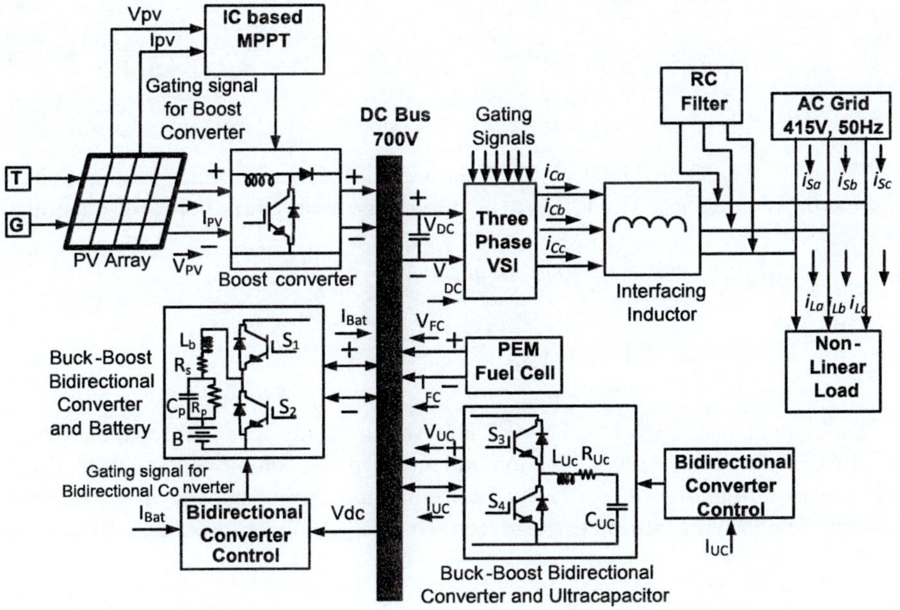

Fig. 1 Block diagram of the presented system

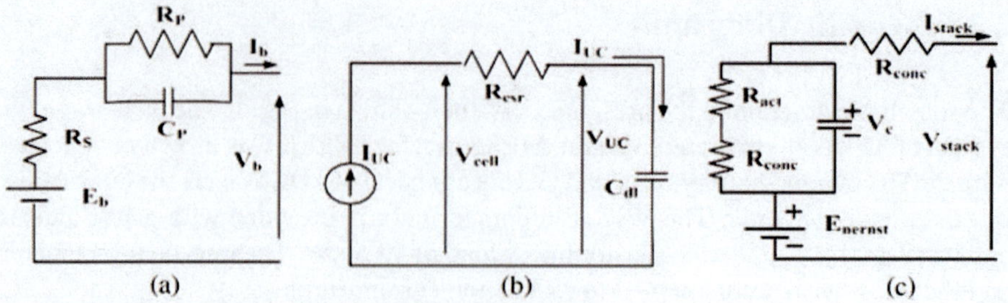

Fig. 2 Equivalent circuit **a** battery bank, **b** UC, **c** PEMFC

$$C_p = \frac{kWh * 3600 * 1000}{0.5 * \left(V_{bmax}^2 - V_{bmin}^2\right)} \tag{1}$$

The equivalent circuit of the UC in Fig. 2b shows $I_{\rm UC}$ as charging/discharging current, and $V_{\rm UC}$ is the terminal voltage across double-layer capacitor $C_{dl} = 3.375$ F. Each call of UC produces 2.7 V and collectively 148 cells in series produce 400 V. The equivalent series resistance $R_{\rm esr} = 10\ {\rm m}\Omega$. The UC does not require the overcurrent protection because as the charging of UC increases the charging current starts decreasing, respectively. The mathematical modeling of the UC is as per (2)–(5) [26].

$$at\,the\,commencement\,of\,the\,charging\,process\;V_{UC} = \frac{1}{C_{dl}}\int_{t=0}^{t=1} i_{UC}dt = \frac{I_{\rm cell}}{C_{dl}}t \tag{2}$$

$$during\,the\,charging\,process\;V_{\rm cell} = I_{\rm cell} * {\rm R}_{\rm esr} + V_{UC} = I_{\rm cell} * R_{\rm esr} + \frac{I_{\rm cell}}{C_{dl}}t \tag{3}$$

When UC if fully charged to $V_{\rm cell}^{\rm max}$ the $_{01}$ becomes zero and reverse its direction as discharging commence. The V_{UC}^0 will be the voltage across the $_{dl}$ before discharging.

$$during\,the\,discharge\;\,process\;V_{UC}^0 = V_{\rm cell}^{\rm max} - I_{\rm cell} * R_{\rm esr} \tag{4}$$

$$V_{\rm cell} = V_{UC}^0 - I_{\rm cell} * R_{\rm esr} - \frac{I_{\rm cell}}{C_{dl}}t \tag{5}$$

The equivalent circuit of the PEMFC is shown in Fig. 2c. The static characteristics of PEMFC are governed by activation voltage drop $V_{\rm act}$, ohmic voltage drop $V_{\rm ohm}$, and concentration losses $V_{\rm conc}$. I_{FC} is the discharging current and V_{FC} is the terminal voltage. The PEMFC attributes are described as $V_{\rm stack}$ and $I_{\rm stack}$ as per (6) and (7) [27].

$$V_{\rm stack} = n * V_{FC} = E_{\rm Nernst} - V_{\rm act} - V_{\rm ohm} - V_{\rm conc} \tag{6}$$

$$V_{stack}(I) = E_0 - I_{stack} * R - A * \ln(I_{stack}) + m * \exp(n * I_{stack}) \tag{7}$$

where n is number of cells; E_0, R, A, and m are empirical coefficients.

3 Control Algorithm

Several controlling strategies have been implemented in the presented work. The InC algorithm is utilized for achieving the MPPT during the insolation variation. The bi-directional converter control for the battery bank and UC is shown in Fig. 3a. The VSC control is shown in Fig. 3b.

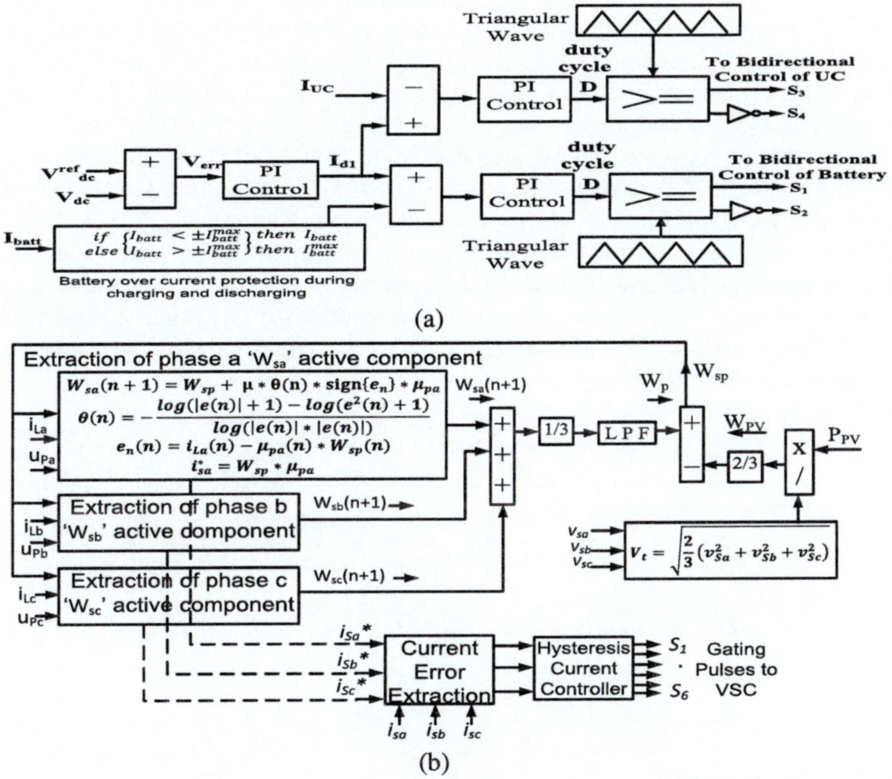

Fig. 3 Control algorithm **a** battery and UC current control **b** VSC control

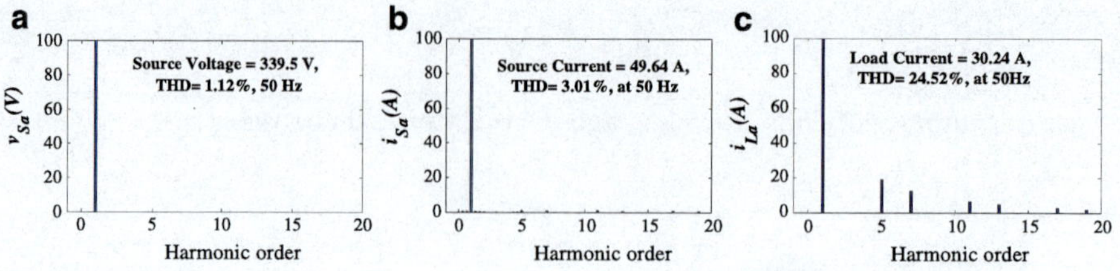

Fig. 4 Harmonics spectrum analysis of **a** source voltage **b** source current **c** load current

3.1 Bi-directional Buck-Boost Converter Control for Battery and UC

The boost converter maintains the V_{dc} at the desired level throughout the operation. The bi-directional buck-boost converter control is shown in Fig. 3a. The $V_{DC}^{\text{ref}} = 700$ V and the V_{dc} are sensed form of the coupling capacitor that connects DC bus and VSC as per (8). The difference in V_{DC}^{ref} and V_{dc} which is V_{err} is fed to the PI controller with gains $K_p = 1$ and $K_i = 0$ to generate the loss component I_{D1}. The battery current protection is provided by not allowing the maximum charging/discharging current to exceed $I_{\text{batt}}^{\text{max}} = 30$ A. The comparison of I_{D1} and I_{batt} is fed to another PI controller with gains $K_p = 1$ and $K_i = 0.8$ to generate the duty cycle D as per (9) and further switching pulses for the bi-directional converter of battery [28]. Likewise, the comparison of I_{D1} and I_{UC} is fed to PI controller with gains $K_p = 1$ and $K_i = 0$ to generate the D value as per (10) and further generate the switching sequence for the bi-directional converter of UC. D in both cases is compared with the triangular signal of 10 kHz of frequency.

$$I_{d1}(n+1) = I_d(n) + K_p\left(V_{DC}^{ref} - V_{DC}\right) + \frac{K_i}{s}\left(V_{DC}^{ref} - V_{DC}\right) \tag{8}$$

$$D(n+1) = d(n) + K_p(I_{d1} - I_{UC}) + \frac{K_i}{s}(I_{d1} - I_{UC}) \tag{9}$$

$$D(n+1) = d(n) + K_p(I_{d1} - I_{\text{batt}}) + \frac{K_i}{s}(I_{d1} - I_{\text{batt}}) \tag{10}$$

3.2 VSC Control

The VSC control is provided by AVFxlogCLMP algorithm as shown in Fig. 3b. The Volterra filtered-x algorithm is utilized to minimize the logarithmic cost function generated by the least mean *lp*-norms [22]. The $\mu_{px}(x = a, b, c)$ are the unit template or in-phase component generated as $v_{sx}/V_t(x = a, b, c)$. The voltage magnitude is

V_t, error signal is $e_n(_n)$ as per (11), θ_n is the logarithmic cost function as per (12), and $\mu = 0.01$ is the step signal. The W_{PV} is the feed-forward term of the PV system as per (13), and $W_{sx}(x = a, b, c)$ is the weight signals generated of each phase as per (14)–(16), and W_p is the average weight signal as per (17). The W_{sp} is the overall weight calculated as per (18). The reference current signals $i^*_{sx}(x = a, b, c)$ is generated as per (19) and compared with the source current of each phase, and further supplied to the hysteresis current controller for producing the VSC switching pulses. During the fixed power mode, each phase weight component will be taken over by $W_{sp_{\text{fix}}}$ and $i^*_{Sa_{\text{fix}}}, i^*_{Sb_{\text{fix}}}, i^*_{Sc_{\text{fix}}}$ are calculated as per (20).

$$e_n(n) = i_{La}(n) - \mu_{pa} * W_{sp}(n) \tag{11}$$

$$\theta(n) = -\frac{\log(|e(n)| + 1) - \log\left(e^2(n) + 1\right)}{\log(|e(n)|) * |e(n)|} \tag{12}$$

$$W_{PV} = \frac{2}{3}\left(\frac{P_{PV}}{\sqrt{\frac{2}{3}\left(v_{sa}^2 + v_{sb}^2 + v_{sc}^2\right)}}\right) = \frac{2}{3}\frac{P_{PV}}{V_T} \tag{13}$$

$$W_{sa}(n+1) = W_{sp} + \theta(n) * \mu * \mu_{pa} * \text{sign}\{e_n\} \tag{14}$$

$$W_{sb}(n+1) = W_{sp} + \theta(n) * \mu * \mu_{pb} * \text{sign}\{e_n\} \tag{15}$$

$$W_{sc}(n+1) = W_{sp} + \theta(n) * \mu * \mu_{pc} * \text{sign}\{e_n\} \tag{16}$$

$$W_p = \frac{1}{3}\left(W_{pa} + W_{pb} + W_{pc}\right) \tag{17}$$

$$W_{sp} = W_p - W_{PV} \tag{18}$$

$$i^*_{sa} = W_{sp} * \mu_{pa}, i^*_{sb} = W_{sp} * \mu_{pb}, i^*_{sc} = W_{sp} * \mu_{pc} \tag{19}$$

$$i^*_{sa_{\text{fix}}} = W_{sp_{\text{fix}}} * \mu_{pa}, i^*_{sb_{\text{fix}}} = W_{sp_{\text{fix}}} * \mu_{pb}, i^*_{sc_{\text{fix}}} = W_{sp_{\text{fix}}} * \mu_{pc} \tag{20}$$

4 Results and Discussion

The performance of the proposed system is analyzed in the steady-state and dynamic state, i.e., fixed power mode, variable irradiation, and load unbalancing in MATLAB/Simulink environment.

4.1 Steady-State Performance

During steady-state, the solar irradiation level is kept at 1000 W/m^2. The source voltage v_{Sabc}, source current i_{Sabc}, load current of phase 'a' i_{La}, compensator current i_{Ca}, active and reactive power delivered to the grid which is P_g and Q_g are analyzed as the AC quantities of the system. The v_{Sabc} and i_{Sabc} are maintained the unity power factor (UPF) in-phase opposition as the power is being transferred to the grid as P_g. The reactive power requirement of nonlinear load and VSC switches is compensated by VSC itself; hence the Q_g will be around zero as no reactive power is supplied from the grid as shown in Fig. 5a. The DC quantities of the system are voltage, current, and power of the PV system represented as V_{PV}, I_{PV}, and P_{PV}. The voltage and current of the battery, UC, and PEMFC are V_{Bat}, I_{Bat}, V_{UC}, I_{UC}, V_{FC}, and I_{FC}.

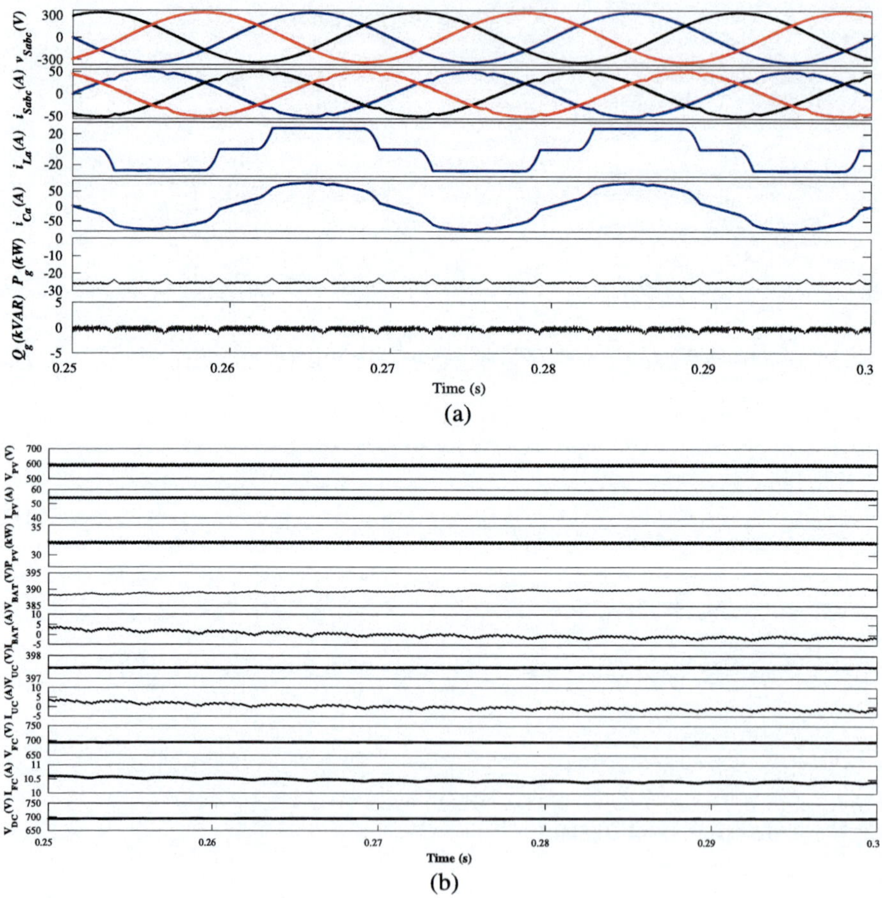

Fig. 5 Steady-state performance of **a** v_{Sabc}, i_{Sabc}, i_{La}, i_{Ca}, P_g & Q_g. **b** V_{PV}, I_{PV}, P_{PV}, V_{bat}, I_{bat}, V_{FC}, I_{FC} V_{UC}, I_{UC} & V_{DC}

V_{PV} is maintained at 600 V and around 32 kW of PPV is being generated. The V_{Bat} and V_{UC} are around 400 V. The PEMFC is directly connected to DC bus, so V_{dc} and V_{FC} are maintaining the 700 V, as shown in Fig. 5b. The harmonics analysis of phase 'a' of the source voltage v_{Sa}, source current i_{Sa}, and load current i_{La} is shown in Fig. 4a–c.

4.2 Dynamic-State Performance During Fixed Power Mode

During the fixed power mode a fixed amount of power is needed to be supplied from the PV and MESS collectively to the grid. The fixed power mode is simulated from 0.25 to 0.35 s of simulation time. P_g after 0.25 s decreases to around 20 kW which was earlier 27 kW, as shown in Fig. 6a. The excess power is being supplied to the MESS so that the battery and UC start charging and their current goes negative, as shown in Fig. 6b. The V_{PV}, I_{PV}, and P_{PV} remain as it was in steady-state. The DC bus voltage is maintained at 700 V during fixed power mode.

4.3 Dynamic-State Performance During Insolation Variation

The insolation variation is at 0.3–0.4 s of simulation time during which the insolation level reduced from 1000 W/m^2 to 600 W/m^2. With the irradiation change, the P_g also reduces to around 12 kW and the Q_g delivered by the grid remains zero as VSC provides the necessary reactive power for the nonlinear load. The v_{Sabc} and i_{Sabc} maintain the UPF in-phase opposition as shown in Fig. 7a. The V_{PV} is maintained at 600 V with the help of boost converter controlled by MPPT. The subsequent reduction in the I_{PV} and P_{PV} with irradiation change is shown in Fig. 7b. During irradiation change, the battery and UC start discharging to satisfy the load requirement, while the system maintains the V_{dc} at 700 V.

4.4 Dynamic-State Performance During Load Unbalancing

The load unbalancing is simulated at 0.51–0.61 s of simulation time. The *vSabc* and *iSabc* are in-phase opposition and maintain UPF. After 0.51 s, phase 'a' of the load is disconnected from the system. As the power delivered to the load decreases, the compensation current requirement also reduces, as shown in Fig. 8a. The decrease in the p_g is also expected but the excess amount of power is used for charging battery bank and UC. The V_{PV}, I_{PV} and P_{PV} remain unchanged, but the I_{Batt} and I_{UC} become more negative. The V_{dc} is maintained at the desired level, as shown in Fig. 8b.

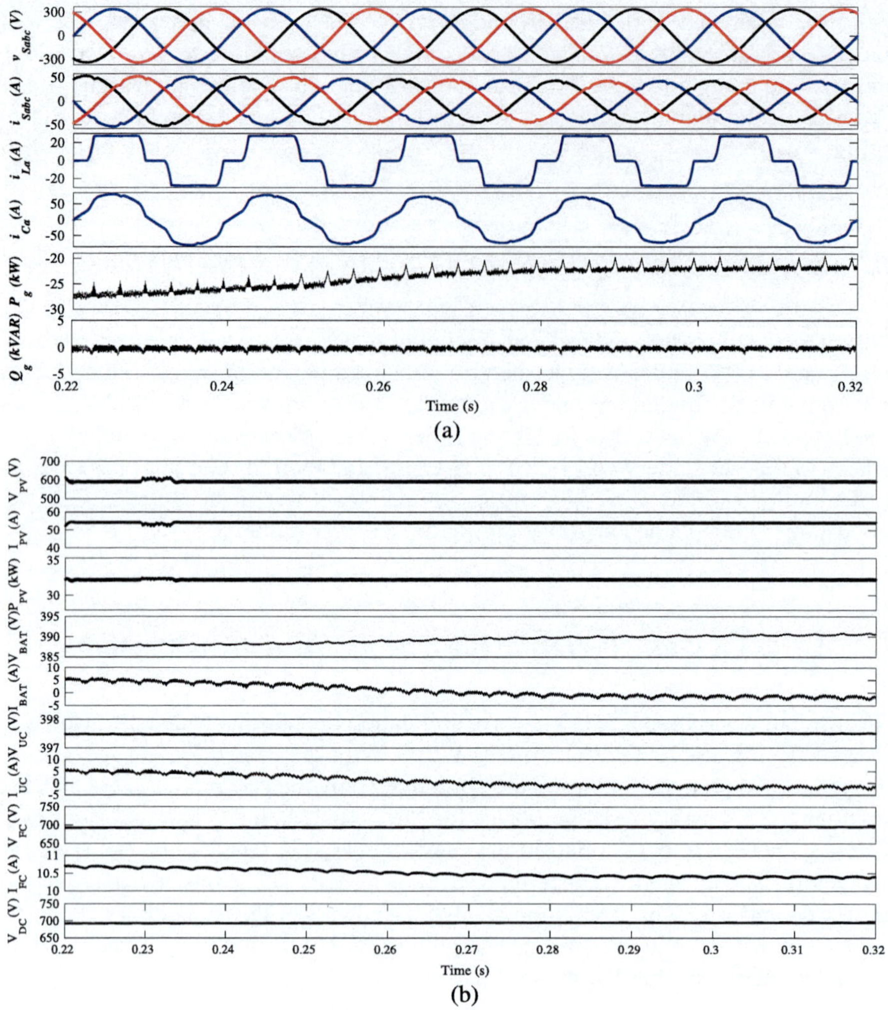

Fig. 6 Dynamic-state performance during fixed power mode of **a** $v_{Sabc}, i_{Sabc}, i_{La}, i_{Ca}, P_g$ & Q_g. **b** $V_{PV}, I_{PV}, P_{PV}, V_{\text{bat}}, I_{bat}, V_{FC}, I_{FC}$ V_{UC}, I_{UC} & V_{DC}

5 Conclusion

The three-phase three-wire grid-tied dual-stage PV-MESS system has been controlled by adaptive VFxlogCLMP algorithm. The VSC control algorithm has reduced the logarithmic mean error in the proposed system in terms of cost function by Volterra filtered-x algorithm. The system has performed satisfactorily under steady-state and various dynamic-state such as fixed power mode, irradiation variation and load unbalancing. The system has successfully handled the PV side and load side anomalies.

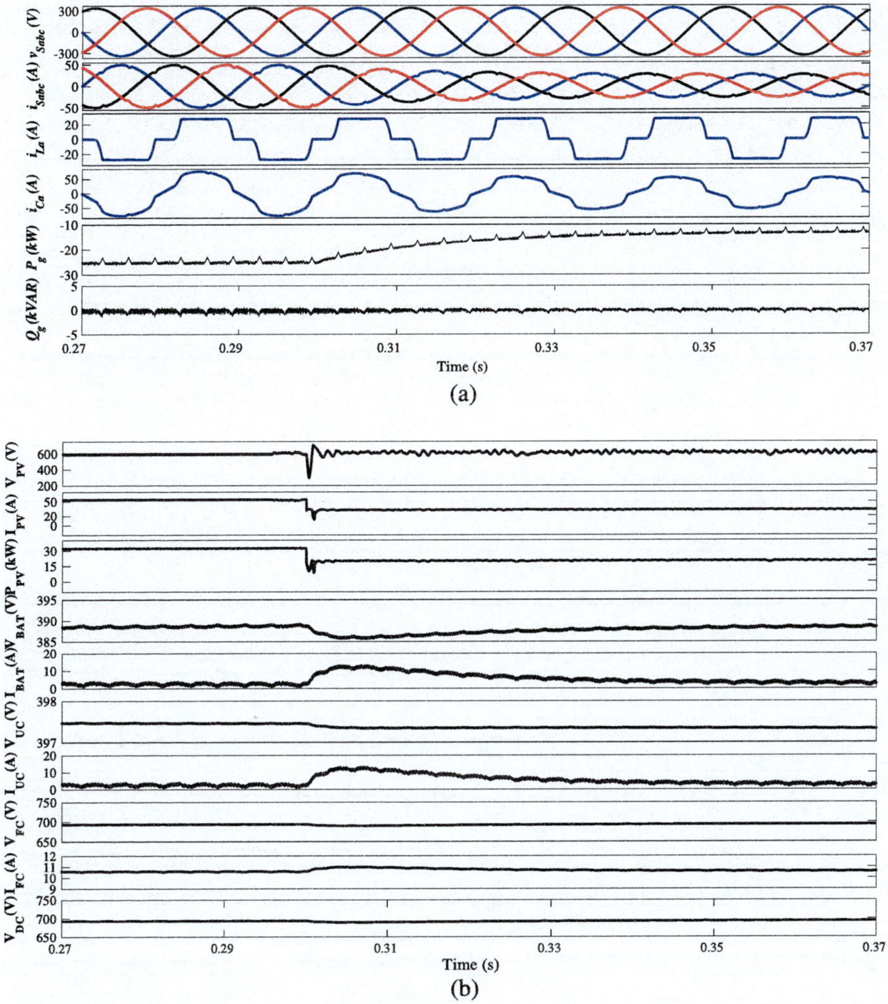

Fig. 7 Dynamic-state performance during irradiation variation of **a** v_{Sabc}, i_{Sabc}, i_{La}, i_{Ca}, P_g & Q_g. **b** V_{PV}, I_{PV}, P_{PV}, V_{bat}, I_{bat}, V_{FC}, I_{FC} V_{UC}, I_{UC} & V_{DC}

The contrasting nature of energy storage elements in MESS has enabled the proposed system to handle sudden, sustained and prolonged duration of irregularities.

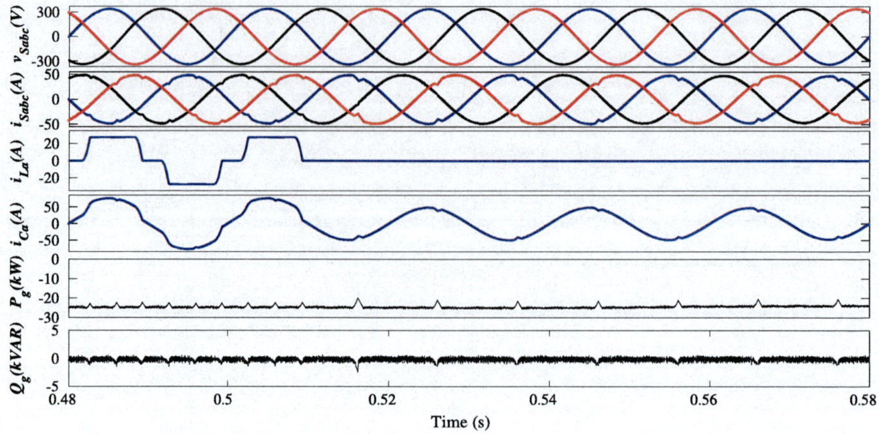

(a)

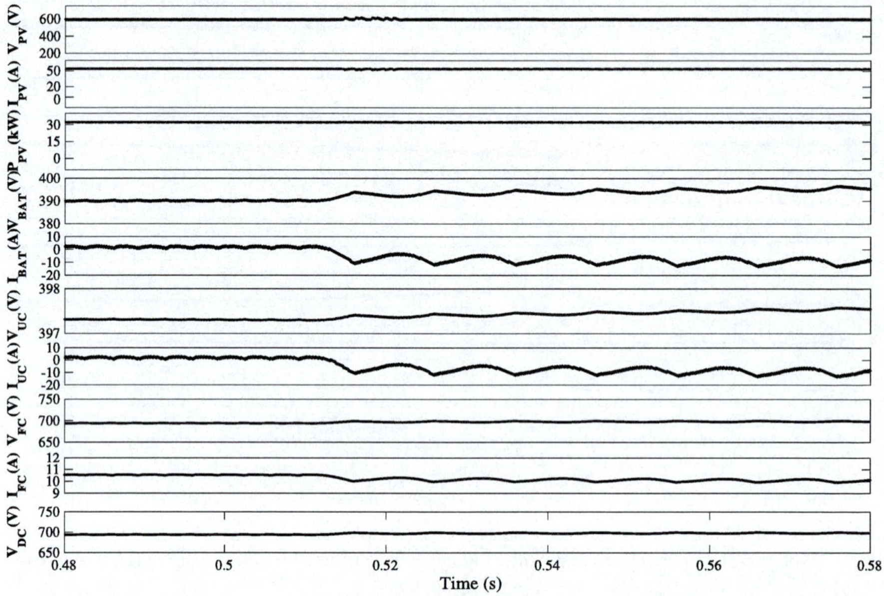

(b)

Fig. 8 Dynamic-state performance during load unbalancing of **a** *vSabc, iSabc, iLa, iCa, Pg & Qg*. **b** $V_{PV}, I_{PV}, P_{PV}, V_{\text{bat}}, I_{bat}, V_{FC}, I_{FC}\ V_{UC}, I_{UC}$ & V_{DC}

References

1. E. Kabir, P. Kumar, S. Kumar, A.A. Adelodun, K.H. Kim, Solar energy: Potential and future prospects. Renew. Sustain. Energy Rev. **82**, 894–900 (2018). https://doi.org/10.1016/j.rser.2017.09.094
2. Rathbun, M., Xu, Y., nejad, R. R., Qu, Z. & Sun, W.: Impact Studies and Cooperative Voltage Control for High PV Penetration. IFAC-PapersOnLine **51**, 684–689 (2018). https://doi.org/10.1016/j.ifacol.2018.11.784

3. B. Singh, A. Chandra, K. Al-Haddad, *Power Quality Problems and Mitigation Techniques* (Wiley Online Library, 2016)
4. R.K. Agarwal, I. Hussain, B. Singh, Application of LMS-based NN structure for power quality enhancement in a distribution network. IEEE Trans. Neural Netw. Learn. Syst. **29**, 1598–1607 (2018). https://doi.org/10.1109/TNNLS.2017.2677961
5. R.K. Agarwal, I. Hussain, B. Singh, LMF-based control algorithm for single stage three-phase grid integrated solar PV system. IEEE Trans. Sustain. Energy **7**, 1379–1387 (2016). https://doi.org/10.1109/TSTE.2016.2553181
6. M.S. Salman, O. Kukrer, A. Hocanin, Recursive inverse algorithm: mean- square-error analysis. Digit. Signal Process. A Rev. J. **66**, 10–17 (2017). https://doi.org/10.1016/j.dsp.2017.04.001
7. R.K. Agarwal, I. Hussain, B. Singh, Grid Integration of single stage SPV system using NLMS filtering control technique, in *2016 IEEE 6th International Conference on Power System* (2016), pp. 1–6. https://doi.org/10.1109/icpes.2016.7584234
8. S. Shivam, I. Hussain, A. Jain, B. Singh, Sign-error based adaptive control technique for single stage grid tied SPV system, in *2016 National Power System Conference* (2016), pp. 1–6. https://doi.org/10.1109/npsc.2016.7858872
9. Y. Singh, I. Hussain, B. Singh, S. Mishra, Real time implementation of sign- sign lms algorithm in single phase SPV grid integrated system, in *2016 IEEE 7th Power India International Conference on PIICON 2016*, vol. 2 (2017), pp. 3–7. https://doi.org/10.1109/POWERI.2016.8077281
10. M. Ibrahim, M.A. Mallick, M. Chankaya, A. Iqbal, I. Hussain, Variable parameter resized zero attracting least mean fourth control for grid-tied pv system. Int. J. Innov. Technol. Explor. Eng. **8**, 2489–2493 (2019). https://doi.org/10.35940/ijitee.K1869.0981119
11. M. Srinivas, I. Hussain, B. Singh, Combined LMS – LMF-based control algorithm of DSTATCOM for power quality enhancement in distribution system. IEEE Trans. Ind. Electron. **63**, 4160–4168 (2016). https://doi.org/10.1109/TIE.2016.2532278
12. M. Chankaya, I. Hussain, A. Ahmad, *Control of Multifunctional PV Battery Grid Tied System*. Lecture Note Electrical Engineering, vol. 669 (Springer, Singapore, 2020). https://doi.org/10.1007/978-981-15-5374-5
13. P.P. Pokharel, L. Weifeng, J.C. Principe, Kernel LMS. ICASSP, in *IEEE International Conference on Acoustics, Speech, and Signal Processing – Proceedings*, vol. 3 (2007), pp. 1421–1424. https://doi.org/10.1109/icassp.2007.367113
14. W. Liu, I. Park, Y. Wang, J.C. Principe, Extended kernel recursive least squares algorithm. IEEE Trans. Signal Process. **57**, 3801–3814 (2009). https://doi.org/10.1109/TSP.2009.2022007
15. K. Nishikawa, F. Albu, D. Coltuc, H. Coanda, Adaptive kernel bandwidth method for kernel normalized LMS adaptive algorithm, in *2017 21st International Conference on System Theory, Control and Computing ICSTCC 2017* (2017), pp. 477–481. https://doi.org/10.1109/icstcc.2017.8107080
16. W. Gao, O. Université, Gaussian kernel least-mean-square : design, analysis and To cite this version : HAL Id : tel-01285514 Docteur en Sciences (2016)
17. W. Gao, J. Chen, Transient performance analysis of zero-attracting Gaussian kernel lms algorithm with pre-tuned dictionary. IEEE Access **7**, 135770–135779 (2019). https://doi.org/10.1109/ACCESS.2019.2942088
18. B. Chen, X. Wang, J.C. Principe, Maximum correntropy criterion-based kernel adaptive filters, in *Adaptive Learning Methods for Nonlinear System Modeling* (Elsevier Inc., 2018). https://doi.org/10.1016/b978-0-12-812976-0.00007-5
19. F. Huang, J. Zhang, S. Zhang, Adaptive filtering under a variable kernel width maximum correntropy criterion. IEEE Trans. Circuits Syst. II Express Briefs **64**, 1247–1251 (2017). https://doi.org/10.1109/tcsii.2017.2671339
20. M. Tahir Akhtar, W. Mitsuhashi, Improving performance of FxLMS algorithm for active noise control of impulsive noise. J. Sound Vib. **327**, 647–656 (2009). https://doi.org/10.1016/j.jsv.2009.07.023
21. L. Wu, H. He, X. Qiu, An active impulsive noise control algorithm with logarithmic transformation. IEEE Trans. Audio Speech Lang. Process. **19**, 1041–1044 (2011). https://doi.org/10.1109/tasl.2010.2061227

22. L. Lu, H. Zhao, Adaptive Volterra filter with continuous lp-norm using a logarithmic cost for nonlinear active noise control. J. Sound Vib. **364**, 14–29 (2016). https://doi.org/10.1016/j.jsv.2015.11.029
23. Y. Yuan et al., An advanced multicarrier residential energy hub system based on mixed integer linear programming. Int. J. Photoenergy **2019**, 1–12 (2019). https://doi.org/10.1155/2019/1384985
24. S. Pazouki, M.R. Haghifam, Impact of energy storage technologies on multi carrier energy networks, in *Smart Grid Conference 2014, SGC 2014* (2014) https://doi.org/10.1109/sgc.2014.7090854
25. M.A. Casacca, Z.M. Salameh, Determination of lead-acid battery capacity via mathematical modeling techniques. IEEE Trans. Energy Convers. **7**(3), 442–446 (1992). https://doi.org/10.1109/60.148564
26. R. Suhane, M.K. Chopra, V.K. Sethi, Mathematical modeling of battery and ultra capacitor for photo voltaic system **7**, 41–46 (2018)
27. Y.F. Guo, H.C. Chen, F.C. Wang, The development of a hybrid PEMFC power system. Int. J. Hydrog. Energy **40**, 4630–4640 (2015). https://doi.org/10.1016/j.ijhydene.2015.01.169
28. N. Beniwal, I. Hussain, B. Singh, Control and operation of a solar PV- battery-grid-tied system in fixed and variable power mode. IET Gener. Transm. Distrib. **12**, 2633–2641 (2018). https://doi.org/10.1049/iet-gtd.2017.1095

Electrical Energy Storage Influencing Shift in Grid Balancing Approach

Asif Nazar and Naqui Anwer

1 Introduction

The contribution of variable renewable energy sources (VRES) in the energy mix of a country is getting prominence. This increased penetration of VRES is the result of countries' commitment to climate change mitigation under the Paris Agreement [1]. Grid balancing is a real challenge in this new energy mix dominated by VRES. Flexible options that are agile enough to ensure stability and reliability of grid are worth to explore, probe and deploy. Certainly, flexible generation is one of the ways of grid balancing. Conventionally countries are using resources at their disposal to balance the grid. Hydro power and natural gas are popular flexi options exercised now [2].

Interconnection of the grid of a country with neighbouring countries' grids, a sort of supergrid, is another way of grid balancing. With the widening of areas, the impact of VRES gets smoothened. European interconnection may be cited as an example of transmission enhancement strategy. Many supergrids have been envisaged to fructify the concept of regional supergrids. Some of the envisaged supergrids are: Asian Supergrid, Southeast Asian Supergrid, and BIMSTEC Supergrid. This strategy will help a lot in integration of VRES [3].

Demand-side management (DSM) is another method of grid balancing. The examples of DSM include industrial demand, appliance demand and electric vehicle demand.

A. Nazar (✉)
Research Scholar, Department of Energy and Environment, TERI School of Advanced Studies, New Delhi, India
e-mail: asif.nazar2@terisas.ac.in

N. Anwer
Associate Professor, Department of Energy and Environment, TERI School of Advanced Studies, New Delhi, India
e-mail: naqui.anwer@terisas.ac.in

S. Vadhera et al. (eds.), *Latest Trends in Renewable Energy Technologies*, Lecture Notes in Electrical Engineering 760, https://doi.org/10.1007/978-981-16-1186-5_7

Storage deployment is another strategy to balance the grid. Electrical energy storage (EES) has been classified into mechanical (pumped hydro storage (PHS), compressed air energy storage (CAES) and flywheel), electrochemical (batteries, flow battery and hydrogen storage) and electrical (supercapacitor) [4].

Traditionally, PHS is participating in managing grid imbalances. However, PHS has inherent locational limitations. The response time of battery energy storage (BES) is in milliseconds. Further, it can be used in all locations of the power system: generation, transmission and distribution. But BES is facing various barriers due to less responsive regulatory framework, and the electricity market, mainly due to historical and legacy factors.

Electricity markets are constantly evolving. They are at various stages of reform. Some may be rigidly regulated, and some may be deregulated. But most of the electricity markets are in between two stages. The electricity market is historically inclined toward conventional generators. Inherent inertia of conventional generators makes it a popular choice in primary frequency regulation. BES due to its sub-second response time can provide 'synthetic inertia'. So BES can participate in electricity market not only in supplying bulk energy services but also in ancillary services and renewable specific services.

Section 2 of this paper discusses existing grid balancing strategy of VRES-dominated representative countries. Section 3 discusses electrical energy storage and electricity market product for BES. Section 4 is the conclusion.

2 Grid Balancing Strategy of Countries with Substantial VRES

2.1 Methods of Identifying Representative Countries

The selection of countries is based upon the contribution of VRES in the energy mix of a nation. Five representative countries and a state of the USA, developed as well as emerging economies, which qualify these criteria have been selected. Geographically these countries represent Europe, Asia and the USA. The chosen countries/states are: Germany, Spain, Denmark, China, India and California.

2.2 Grid Balancing Strategies

Countries like Germany and Denmark are vying for 100% renewables, and countries like India and China are adding renewables in a significant way. Their future intent can be had from their nationally determined contributions (NDC) commitment for COP 21 [1]. Currently, Germany has 45% VRES (wind and solar) in its energy mix [5]. Spain has 28% share of VRES [6]. Denmark has 47% share of variable renewables, mainly wind energy (40%) in its energy mix [7]. The share of VRES in California is nearly 20% [8]. The share of wind and solar PV combined for China and India is 19% [9] and 18% [10], respectively.

Germany and Denmark have a good flexible generation option in the form of coal plants. Apart from that, these two countries have a very good ancillary market, and these have advanced weather forecasting system in place [11].

In contrast to the above nations, China is significantly curtailing its 15% of wind energy and 9% of solar PV, respectively. China has a surplus capacity of 200 GW. Further, China's energy market is rigid. China's grid balancing strategy has yet to be progressive in terms of renewable integration [12].

Recently, India has become power surplus country (ignoring peak power deficit of 0.8% in 2018–2019). Even then she is curtailing wind energy. Load shedding is one of the strategies of grid balancing. India's energy market is evolving continuously. A recent addition is regulation service with the introduction of reserves regulation ancillary service (RRAS), a tertiary frequency control service. Grid balancing strategy of six countries/state has been summarized in Table 1.

Table 1 Balancing strategy of VRES dominated representative countries

Country/state	Renewable share (%)	Balancing strategy
Germany	45% (Wind and solar PV)	1. Flexible operation of coal and nuclear plants 2. Good ancillary market 3. Good weather forecasting
Spain	28% (Wind and solar PV)	1. of installed power 2. Hydro and natural gas plants 3. Frequent scheduling 4. Advanced control centres
Denmark	47% (mainly wind)	1. Flexible operation of coal plants 2. CHP plants for balancing 3. Advanced day ahead forecasting 4. Day ahead market and ancillary market 5. Exports electricity to Sweden, Norway and Germany
China	19% (Wind and solar PV)	1. Curtailment of a significant portion of wind and solar 2. Power market extremely rigid 3. Flexibility in generation mix (hydro and pumped hydro storage) 4. Capacity surplus of 200 GW
India	18% (Wind and solar PV)	1. Hydro power and pumped hydro storage 2. Gas-based plant (25 GW, 7% of installed capacity) 3. Progressively evolving energy market 4. Load shedding
California	20% (Wind and solar PV)	1. Flexible natural gas (61.3%) and hydropower (7.1%) 2. Sharing power with ten other states 3. Additional power import from other states

Grid balancing strategy of Germany, Spain and Denmark is based upon good weather forecasting. The immediate shortfall of power due to forecast errors would need the mobilization of costly resources at disposal and rescheduling. To overcome this extreme forecasting error, more flexi options like BES need to be part of grid balancing strategy [13]. Further curtailment of VRES is not a solution. The role of BES for renewable integration will be more appropriate in such cases. In the changing and changed energy mix in the future amidst increased VRES penetration, nuclear and coal plants may be phased out. BES will be needed to provide 'synthetic inertia'.

3 Electrical Energy Storage and Electricity Market

Electricity market encompasses within it an energy market, an ancillary service market and a capacity market. Ancillary service market caters to the need of stability and reliability of the grid. Here we will have a look at ancillary service market and its mandate in some representative countries to have a look and understanding of the procurement mechanism and compensation so as to link this market to electrical energy storage applications.

3.1 Ancillary Services and Their Procurement

Primary frequency regulation is a mandatory ancillary service in India, and as such no compensation for this service. In China, ancillary services have been categorized into basic ancillary services and paid ancillary services. Primary frequency regulation has been identified as basic ancillary service, so there is no compensation for this service [14]. In European interconnection, primary frequency regulation is procured through bilateral contracts, and compensation is offered as per bid [15].

Secondary frequency regulation involving AGC is absent from the Indian power system. In China, AGC has been identified as paid ancillary service [14]. Reserves Regulation Ancillary Services (RRAS) represent tertiary frequency control in India [16]. In China, only spinning reserve falls under paid ancillary service [14]. In the European grid, tertiary regulation is procured on spot market or through bilateral contract or by tender [15].

3.2 Electrical Energy Storage

Differentiation amongst different EES technologies are done based upon size, discharge duration, cycle life and response time [4, 17]. These characteristics have been tabulated in Table 2. Pumped hydro power (PHS) is matured EES technology,

Table 2 Key characteristics of EES

EES	Power rating (MW)	Cycle life	Response time	Discharge time
PHS	100–1000	30–60 years	Seconds–minutes	4–12 h
NaS	10–100	2500–4400	10–20 ms	1 min–8 h
Li-ion	0.1–100	1000–10,000	10–20 ms	1 min–8 h

Table 3 Installed capacity of storage

Country	Installed capacity of PHS as of 2018	Installed capacity of BES as of 2018
China	30 GW	1.0 GW
US	22.9 GW	0.87 GW
Germany	6.8 GW	0.37 GW
India	4.8 GW	In nascent stage

and its response time is also commensurate with required grid balancing applications, and discharge duration is more than 8 hours. But it has geographical location limitation. Hence the need for other advanced energy storage systems like battery (NaS, Li-Ion) comes into the picture. PHS is the best for bulk storage. As of 2018, the share of PHS was 160.3 GW [18], and the share of grid scale battery storage was 5.3 GW in 2019 [19]. USA, China and South Korea have appreciably deployed battery storage [20–22]. The first utility scale battery storage of 10 MW was installed in India in 2019 [23]. The installed capacity of PHS and battery storage is given in Table 3.

3.3 *Energy Market Product for EES*

Regulatory changes have been made in some markets so as to accommodate BES as per their merits and limitations. National Grid of Great Britain introduced enhanced frequency response, PJM introduced regulation D fast ramping based upon AGC communication signal, EirGrid of Ireland introduced fast frequency response, MISO brought up-ramp capability and down-ramp capability, and CAISO introduced flexibility ramp-up and flexibility ramp-down uncertainty awards [24, 25].

Battery energy storage by virtue of its performance and ability (like response time) is indispensable in increased penetration of VRES. Hence level-playing field based upon its merit is warranted. Response time in milliseconds is one of the characteristics of BES, so its application in frequency regulation service is apt. A study has shown that the compensation for participation in regulation is better as compared to other ancillary services like reserve. Regulation is compensated both for capacity ($/MW/h) and energy ($/MWh). Further discharge duration and state of charge (SOC) is crucial for battery storage. So new ancillary service market product for battery storage is very much needed. Some of the independent system operators

(ISO), as discussed above, paved the way for energy products tailor-made for battery energy storage. The introduction of these market product will boost the penetration of innovative technology like battery storage, and their economic viability, in the long run, may be ensured. These experiences from good ancillary markets of developed economies may help emerging economies in reshaping and evolving their ancillary services market.

4 Conclusion

The existing grid-balancing strategy of representative advanced economies with substantial VRES share relies on good weather forecasting and good ancillary market. If there is an extreme error in forecasting, other resources need to be mobilized, and rescheduling is required. This problem has a solution in flexi option like battery energy storage. Further coal and nuclear plants are likely to have less presence with the increase of VRES share in the energy mix. In this case, BES can provide 'synthetic inertia'.

Ancillary service market has a pivotal role. Hence responsive ancillary market taking fast frequency response of BES and SOC of BES should introduce innovative market products as done by many developed economies. The idea is to acknowledge the merit and performance of BES vis-à-vis conventional generators.

References

1. United Nations Climate Change. The Paris Agreement. 2015. https://unfccc.int/process- and-meetings/the-paris-agreement/the-paris-agreement
2. B.A. Frew, S. Becker, M.J. Dvorak, G.B. Andresen, M.Z. Jacobson, Flexibility mechanisms and pathways to a highly renewable US electricity future. Energy **101**, 65–78 (2016)
3. A. Nazar, A. Haldar, The BIMSTEC supergrid: renewable energy mix and regional economy, in *1st International Conference on Large-Scale Grid Integration of Renewable Energy in India* (Delhi, 2017)
4. IEC (International Electrochemical Commission). White paper on Electrical Energy Storage. 2011. https://www.iec.ch/whitepaper/pdf/iecWP-energystorage-LR-en.pdf
5. Federal Ministry for Economic Affairs and Energy. Electricity Market of the Future. 2020. https://www.bmwi.de/Redaktion/EN/Dossier/electricity-market-of-the- future.html
6. RED Electrica de Espana. The Spanish Electricity System: Preliminary Report 2018. 2019. https://www.ree.es/sites/default/files/11_PUBLICACIONES/Documentos/InformesSistemaElectrico/2019/Avance_ISE_2018_en.pdf
7. Danish Energy Agency. Energy Statistics2018. 2018. https://ens.dk/sites/ens.dk/files/Statistik/energy_statistics_2018.pdf
8. California Energy Commission. Data on Renewable Energy Markets and Resources. 2019. https://www.energy.ca.gov/data-reports/energy-almanac/data-renewable-energy-markets-and-resources
9. Reuters. China's 2018 renewable power capacity up 12 percent on year 2019. January 28, 2019. https://www.reuters.com/article/us-china-renewables/chinas-2018-power- capacity-up-12-percent-on-year-idUSKCN1PM0HM

10. CEA (Central Electricity Authority). All India Installed Capacity of Power Stations as on 31.12.2018. 2018. https://www.cea.nic.in/reports/monthly/installedcapacity/2018/installed_capacity-12.pdf
11. E. Martinot, How is Denmark Integrating and Balancing Renewable Energy Today? (2015). https://www.martinot.info/Martinot_DK_Integration_Jan2015.pdf
12. A.F. Finamore, *Big Plans for Integrating Renewable Energy Into China's Electricity Grid* (The Huffington Post, 2016)
13. P. Komarnicki, P. Lomardi, Z. Styczynski, Future power systems, in *Electric Energy Storage Systems* (Springer, Heidelberg, 2017)
14. Z. Ming, L. Ximei, P. Lilin, The Ancillary Services in China: an overview and key issues. Renew. Sust. Energ. Rev. **36**, 83–90 (2014)
15. H. Safiullah, G. Hug, R. Tongia, Design of load balancing mechanism for indian electricity markets. Energy Syst. **8**, 308–350 (2017)
16. CERC (Central Electricity Regulatory Commission). Notification No. 18/1/2013/Reg.Aff.(AS Regul.)/CERC dated Aug 13, 2015. 2015. https://www.cercind.gov.in/2015/regulation/Noti13.pdf
17. World Energy Council. Energy Storage Monitor: Latest Trends in Energy Storage.2018
18. IHA (International Hydropower Association). Hydropower Status Report: Sector Trends and Insight. 2019
19. IEA (International Energy Agency). Energy Storage: Tracking Report-June 2020. 2020. https://www.iea.org/reports/energy-storage
20. CNESA (China Energy Storage Alliance). CNESA's 2018Year in Energy Storage. 2019. https://en.cnesa.org/latest-news/2019/2/23/cnesas-2018-year-in-energy-storage
21. US Energy Information Administration (eia). Battery Storage in the United States: An update on market trends. 2020. https://www.eia.gov/analysis/studies/electricity/batterystorage/pdf/battery_storage.pdf
22. GTAI (Germany Trade & Invest). The Energy Storage Market in Germany. 2019. https://www.gtai.de/resource/blob/64514/5e629c11f7d8ea20e814d95d84a866d6/fact- sheet-energy-storage-market-germany-en-data.pdf
23. Business Line. Grid-scale Energy Storage System Set up in Delhi. 2019. https://www.thehindubusinessline.com/economy/grid-scale-energy-storage-system-set-up-in-delhi/article26260871.ece Battery
24. PJM, PJM State & Member Training Dept. PJM Regulation Market. 2018. https://pjm.com/-/media/training/nerc-certifications/gen-exam-materials-feb-18- 2019/training-material/02-generation/4–1-regulation-market.ashx?la=en
25. California ISO. Flexible ramping product, Nov. 2016. https://www.caiso.com/informed/Pages/StakeholderProcesses/CompletedClosedStakehol derInitiatives/FlexibleRampingProduct.aspx

A New Modular Multilevel Converter Topology Using Flying Ultra-Capacitor and Cascaded H-bridges

Yawar Irshad Badri, Ikhlaq Hussain, Zaid Ahmad, Mustufa Usman, and Junaid Ali

1 Introduction

The conversion of power from DC to AC has widespread applications in industrial scenarios. The converters employed for such purposes must withstand electrical stresses and ensure that the output AC waveforms adhere to low total harmonic distortion (THD) standards such as IEEE-519 [1]. The lower THD of the output waveform has a lot of advantages; it reduces the chances of premature saturation of magnetic flux in the cores of electromagnetic devices such as transformers and motors due to harmonic components, and also the losses due to these harmonic components are reduced which would otherwise result in derating of the device [2]. This results in reduced heating and higher operating efficiency of such devices. For motors the vibrations caused by harmonic components are reduced. The switches employed in these converters must also face electrical stresses and their dV/dt rating should be high for higher operating frequencies. This problem commonly occurs with traditional high-frequency PWM inverters [3]. This leads to deterioration of winding insulation in electromagnetic devices and produces electromagnetic interference (EMI). The *dV/dt* stresses experienced by a single switch are large, thus the switches need to be

Y. I. Badri (✉) · I. Hussain · Z. Ahmad · M. Usman · J. Ali
Department of Electrical Engineering, University of Kashmir, Srinagar, India
e-mail: dev.yawar@gmail.com

I. Hussain
e-mail: ikhlaqb@gmail.com

Z. Ahmad
e-mail: mohammadzaid0409@gmail.com

M. Usman
e-mail: mustafaiot01@gmail.com

J. Ali
e-mail: mjmystic4@gmail.com

S. Vadhera et al. (eds.), *Latest Trends in Renewable Energy Technologies*, Lecture Notes in Electrical Engineering 760, https://doi.org/10.1007/978-981-16-1186-5_8

of higher voltage rating, reducing the life of the converter [4, 5]. The higher switching frequencies also affect the insulation strength as it must continuously face high electric stress. Modular power converters have the advantages of being scalable as well [5]. To avoid the problems discussed above, the use of multilevel converters becomes advantageous [6]. Multilevel converters have a lot of advantages in these types of applications. They can be operated at lower frequencies to reduce interference [7]. The introduction of more steps with appropriate number of switches allows for use of switches with lower dV/dt rating [8]. The THD of output waveform depends on the number of levels being produced by the converter and using filters improve the output waveform by removing higher frequencies [9]. Multilevel inverters are employed in industries for applications such as power systems and AC motor drives [15]. The multilevel design is efficient in case of electric motor drives like those in electric vehicles, where energy recovery using regenerative braking is employed [10].

The use of isolated DC sources presents an opportunity of diversifying the sources. Instead of just using batteries, multiple sources like solar photovoltaic, batteries, wind electricity generators can be integrated and independently operated [11, 12].

There are many types of multilevel inverters topologies, each having their own advantages and disadvantages. Some of the common types of multilevel inverters include *diode-clamped multilevel inverter*, *flying-capacitor multilevel inverter* and *cascaded multilevel inverter*. New designs which focus on reducing number of components, increasing number of levels and more optimizations in control are also being used in different areas [13, 14]. Diode-clamped inverters use a combination of switches, diodes and capacitors to make multiple conduction paths to the voltage taps on capacitive dividers. This approach is advantageous in cases where simplicity and redundancy are required; however, this topology is good for low-level counts. The control method for this design is also simple. The flying capacitor topology, on the other hand, uses multiple capacitive dividers and switches and diodes to produce the multilevel output. Excessive number of storage capacitors is used in this case, however, when the number of output levels is high. The control scheme is also complicated. [16–18]

In comparison to these designs, the design discussed in this paper finds the best of both of these converter topologies and combine that with the simple construction improved operation of cascaded design. The ability to integrate multiple types of energy sources and using the internal capacitor charging circuits for purposes like maximum power point tracking (MPPT) also makes this design more advantageous. The ability to swap out modules keeps the inverter up-time high as well.

This paper introduces a novel multilevel converter topology. The configuration is based on multiple cascaded units composed of a step generator and an h-bridge. The h-bridges are used to cascade multiple units and they also perform the task of sharing the voltage stress each switch is exposed to. The step generators are comprised of ultra-capacitor voltage dividers and a latching switch array (LSA). The use of the step generator is advantageous in scenarios where smaller voltage sources are used for a single unit. The proposed topology is modeled and simulated in MATLAB.

2 Proposed System

A. *Configuration.*

The general design of the proposed modular topology employs a capacitive voltage divider which is used to generate multiple voltage steps from a single DC source, as shown in Fig. 1. This is followed by a LSA, which connects the voltage taps from the capacitive divider to a single voltage rail in a controlled and unidirectional manner preventing back-flow of power into lower voltage region. This stage is followed by a standard h-bridge arrangement of power switches. These modules are then cascaded optimally as is required for a given application. This results in lower *dV/dt* rating of the switches being used. Also, the voltage of DC sources for each individual unit can be smaller with increasing number of h-bridge units [19].

The number of switches used in the setup being presented is also less as compared to the normal cascaded multilevel converter (CMLC) configuration. The capacitive voltage divider also doubles as a high energy buffer during high demand situations such as starting of a motor, and during processes like regenerative braking. This reduces the direct stress faced by the DC source, resulting in better source performance, and reducing maintenance and downtime [20].

The proposed topology is modular; multiple modules each having an LSA can be added to the setup as needed to increase the performance of the system, resulting in lower initial cost of the overall setup. LSA by itself comprises switches providing the link for full DC voltage of the DC source and for sub-voltage values of the DC source. The number of switches and the voltage steps for a given module is determined based on the *dV/dt* and voltage ratings of switches. The following analysis provides details

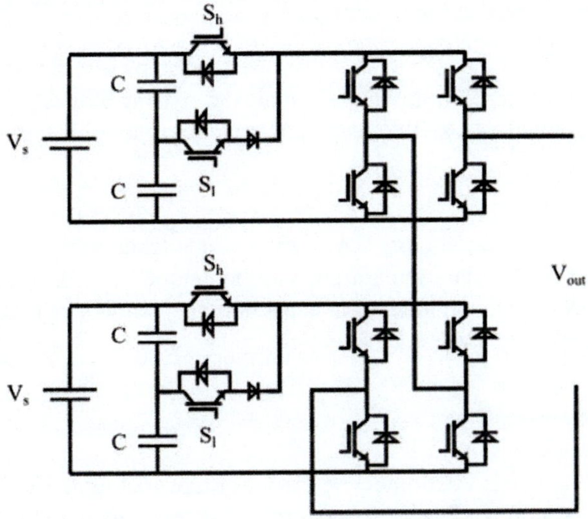

Fig. 1 Proposed system implementation where $n = 2$ and $H = 2$

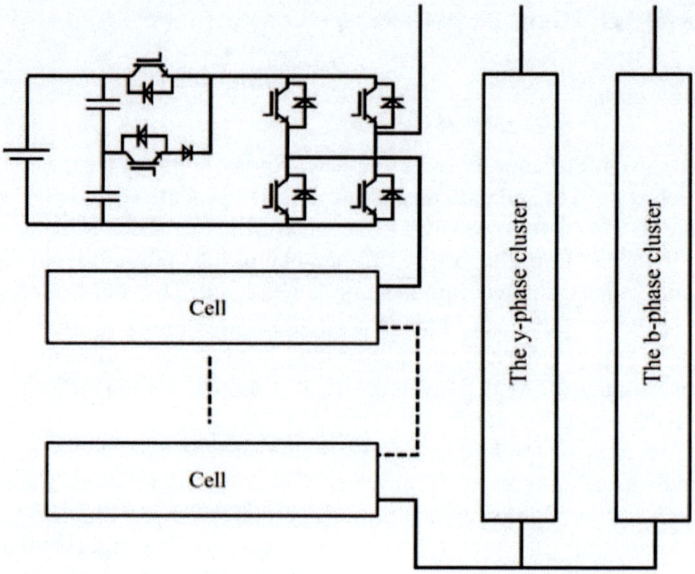

Fig. 2 Three-phase star connection ($n = 2$)

on how the number of levels relates to switch count and dependency on the number of switches in the LSA.

Figure 1 gives the general overview of the setup with an example. The capacitors C are used to obtain sub-voltage values from the DC source V_s followed by the LSA which connects the capacitor taps to the main bus, supplying power to h-bridge in voltage-steps. For three-phase configurations, the possible star and delta connections are shown in Figs. 2 and 3, respectively, where each cell or module is part of a cluster in a phase.

The control of this setup is provided using levels shifted PWM, which is supervised by a module monitor, to simulate the missing blocks. This also helps in understanding the resilience of the design toward faults or module failures or source failures [21, 22].

B. *Basic Analysis.*

In the general case, the number of switches in the LSA depends upon the number of levels that are required, and the number of h-bridge units involved. The equation relating to the number of capacitor taps, number of h-bridge units and the number of resulting levels is given by (1).

$$2nH + 1 = m \tag{1}$$

where n is the number of capacitor taps including the full DC source voltage tap, in a single module, H is the number of modules in the system, m is the number of the output levels including both positive, negative and zero voltage steps.

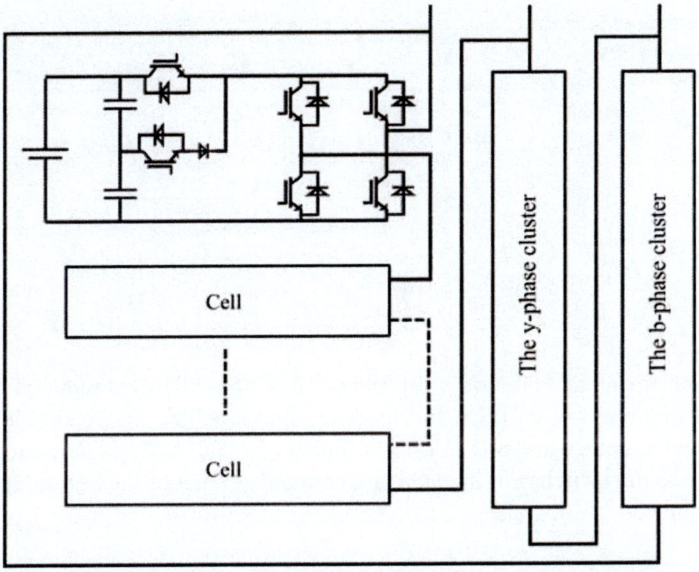

Fig. 3 Three-phase delta connection ($n = 2$)

For $n = 2$, Eq. (1) is reduced to $4H + 1 = m$. The configuration for $n = 2$ is shown in Fig. 4. For asymmetric cases where the LSA is not used in all modules the number of levels is again given by (1).

Each voltage step in LSA requires a switch and the following h-bridge unit employs four additional switches. Thus, the total number of switches used is given by (2).

$$H * (n + 4) \tag{2}$$

or in terms of number of levels m as

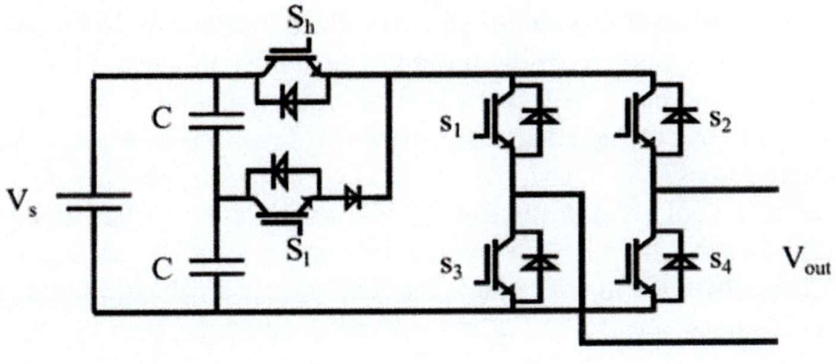

Fig. 4 A single module with $n = 2$ and $H = 1$

Table 1 Number of levels at output for given switch configurations

Number of levels m	Number of switches	
	CMLC	Proposed topology ($n = 2$)
5	8	6
9	16	12
13	24	18
21	40	30

$$\frac{m - 1 + 8}{2} \tag{3}$$

However, for an asymmetric setup the number of switches is reduced further as not all the modules use an LSA. In this case some modules can be simple h-bridge units with a DC source and no LSA. The number of switches for such a setup is given by (4), where the switches in the LSA are counted in each of the kth module.

$$\sum_{k=1}^{H} (n + 4)k \tag{4}$$

For a simple CMLC the number of switches is given by $2(m - 1)$. Table 1 shows a comparison between the number of switches used for the number of levels obtained in the simple CMLC and the proposed topology.

3 Operating Principle

The switching sequence of this converter is illustrated in Table 2, see Fig. 1. The h-bridge units are used to alternate the polarity of waveform as well as to connect multiple modules in cascade electrically. The switches in the h-bridge also share the voltage applied across them, and the LSA is used to latch voltage steps, generated by ultra-capacitor voltage divider, feeding into h-bridge units in a module. These two switching sequences work in coordination to produce a multilevel voltage waveform.

While the output voltage is going through a positive half cycle, the h-bridges all stay in the same switching state except when the requirement for steps requiring more or less modules to be connected arises. During this time, the LSA actively switches between voltage steps.

The switching configuration is also flexible as it can be shifted between more used modules and less used ones to even out the stress faced by each module. This helps in reducing overheating of a single module. In case a voltage source connected to a module is out of service and/or not operating correctly, then that module can

Table 2 Switching table for proposed topology ($n = 2$, $H = 2$)

Switches	V_{out}								
Module 1	0VDC	0.5VDC	1VDC	1.5VDC	2VDC	−0.5VDC	−1VDC	−1.5VDC	−2VDC
S_h	0	0	1	1	1	0	1	1	1
S_l	0	1	0	0	0	1	0	0	0
S1	1	1	1	1	1	0	0	0	0
S2	1	0	0	0	0	1	1	1	1
S3	0	0	0	0	0	1	1	1	1
S4	0	1	1	1	1	0	0	0	0
Module 2									
S_h	0	0	0	0	1	0	0	0	1
S_l	0	0	0	1	0	0	0	1	0
S1	1	1	1	1	1	1	1	0	0
S2	1	1	1	0	0	1	1	1	1
S3	0	0	0	0	0	0	0	1	1
S4	0	0	0	1	1	0	0	0	0

be isolated from the system by using the bypass switching configuration. For the h-bridge of that specific module albeit the performance is reduced. This can be advantageous in applications like electric vehicles where the ability of power converter to continue to operate during 'battery segment maloperation' is a very big advantage.

The bypass switching configuration is a state of switches in an h-bridge where either both of upper switches or both lower switches are in the on-state. In Fig. 4 the switches $S_1(S_3)$ and $S_2(S_4)$ are in the on-state and $S_3(S_1)$ and $S_4(S_2)$ are in the off-state. During such a switching state the DC source of the module is bypassed and the switch state of LSA is irrelevant.

The following case discusses the operation of the setup during a battery maloperation and the subsequent steps are taken to bring the system to a balanced operating condition albeit with a performance penalty.

A. *Isolating a source of a module*

As discussed earlier, the h-bridge of a module can be used to bypass the same module and its DC source by using the bypass switching configuration. Figure 5 uses a block diagram to explain the steps that are taken in the bypass process. The source failure or maloperation is detected by the voltage sensor. The sensor informs the *'module controller'* about the issue. Module controller then issues a fresh switching table and the process of monitoring is resumed.

After an h-bridge is set to bypass, in case of multiple phases the system may become unbalanced. To avoid this imbalance, the corresponding modules in the other two phases may also be bypassed to restore the balance between phases. Section 4 shows the result of removing a DC source from a module of a phase in the simulation

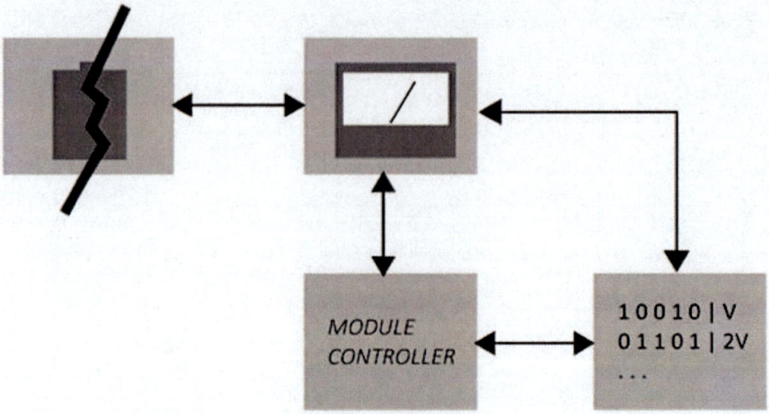

Fig. 5 Source isolation handling block-diagram. *Top left*, Battery (maloperating); *Top center*, Voltage sensor; *Bottom center*, Module controller; *Bottom right*, New switching configuration table

setup. The capacitive voltage divider acts as a temporary buffer as it is depleted of the stored energy.

4 Simulation and Results

The system is simulated in MATLAB with a simple three-phase R-L load. The individual phases are connected in delta configuration and the number of modules in each phase is three. Each module is composed of one h-bridge, thus $H = 3$, and the LSA for each module contains three taps, thus $n = 3$. The number of levels is thus obtained by using Eq. (1) to be $m = 19$ for a single phase. The number of switches is thus given by Eq. (3) to be 21 for a single phase. There are three capacitors in each module rated 500F. The voltage of each of the SDCS is 100 V, giving a peak voltage of $\pm$ 300 V for a single phase. The voltage stress on each switch in one module in this case is shared by two switches of the h-bridge, thus each switch faces 50-V steps. This reflects one of the goals of the proposed topology, to reduce switch voltage rating and *dV/dt* stress faced by them. Another advantage of the setup is the use of fundamental level carrier frequency reducing radio frequency interference (RFI). Level-shifted carrier modulation is used to generate a reference waveform for a single phase. This reference wave is used to drive the switches using pre-fed instantaneous switch states. The carrier frequency is set to 100 Hz and the modulating sine wave frequency is set to 50 Hz. The single-phase voltage waveform is shown in Fig. 6. Figure 7 shows the line current and line voltage of the setup described above. No filters are used in the setup.

The THD of output line current waveform is obtained to be 0.38% and the FFT analysis is shown in Fig. 8. Also, the THD of line voltage is obtained to be 4.67% and the FFT analysis of the same is given in Fig. 9. The voltages across the capacitive

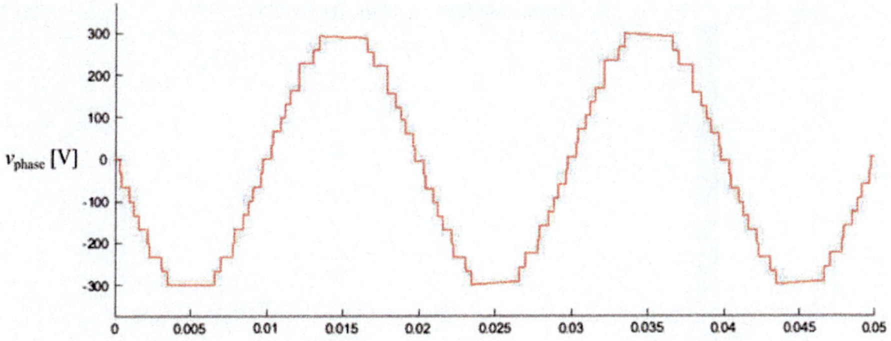

Fig. 6 Single-phase voltage output of the converter

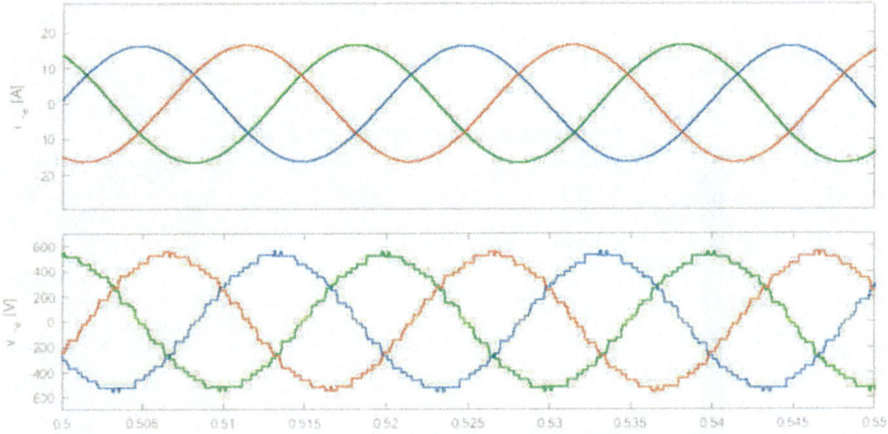

Fig. 7 Output line currents and line voltages of three phases in the simulation setup

buffer of a single module in a phase are shown in Fig. 10. The transient spikes as seen in Fig. 10 are generated during the switching process of LSA. The magnitude of these spikes is small as the voltage levels within a module are small in this case, and this directly corresponds to reduced EMI of the converter. Also, Fig. 11 shows the voltage on the main-bus (which connects LSA to h-bridge in a module) as it alternates between different "stepped voltage levels".

A. *Dynamic response and battery failure*

The dynamic response of the setup to a step change in load is shown in Fig. 12. It can be observed that the system retains the stability very fast.

The case of a battery failure is simulated by isolating the battery from a test module. The capacitors for this module are started with appropriate initial conditions. This causes a collapse of DC voltage supply to LSA of this specific module; however, the voltage across the capacitive divider does not collapse instantly, but

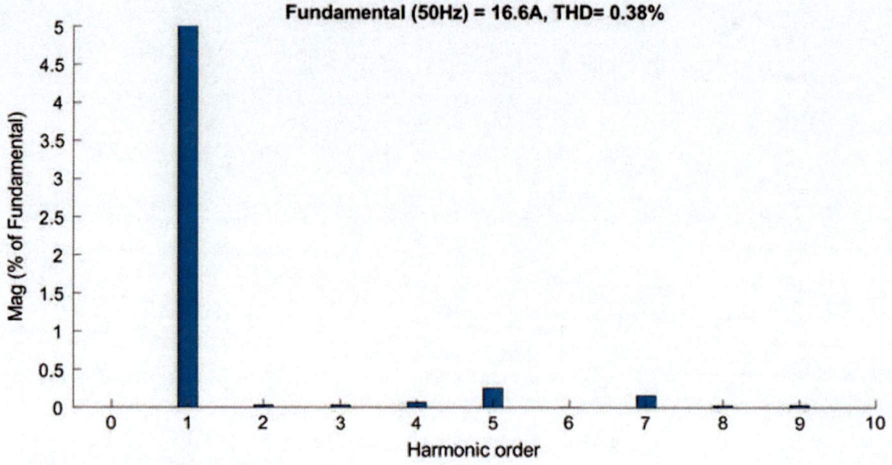

Fig. 8 THD in output line current

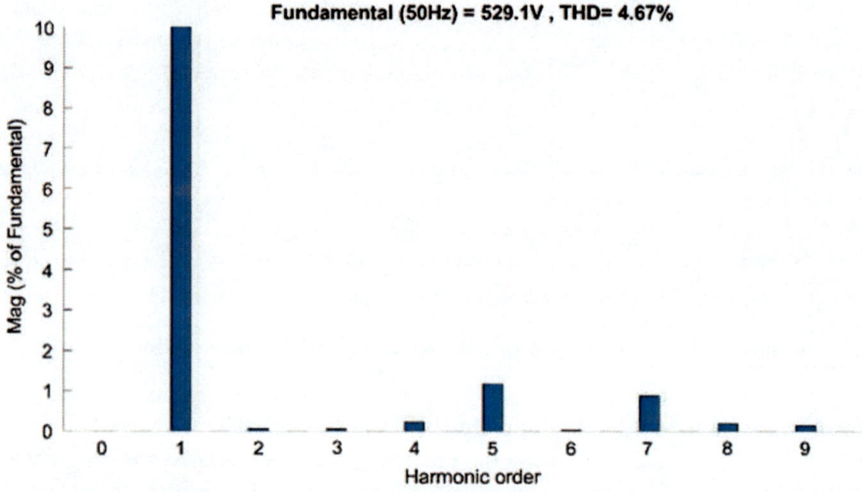

Fig. 9 THD in output line voltage

decays exponentially at a rate proportional to the instantaneous load on the module. This allows the system some time to respond to the sudden change. Figures 13 and 14 show the voltage across the capacitive buffer and, voltage and current being fed to the load, respectively.

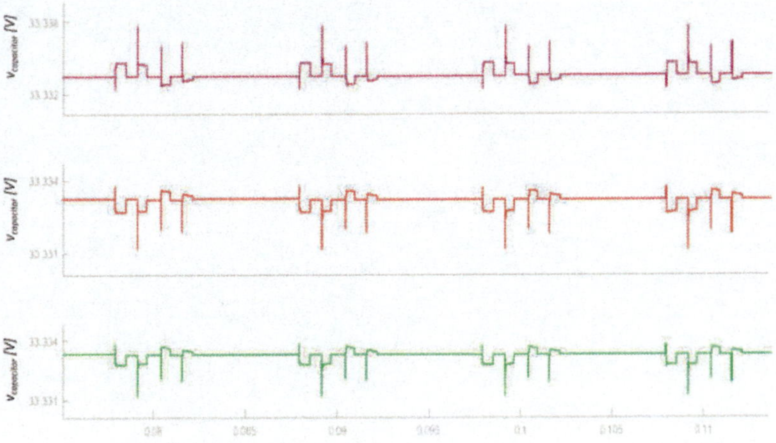

Fig. 10 Voltages across the capacitors of a single module in a phase leg, during normal operation

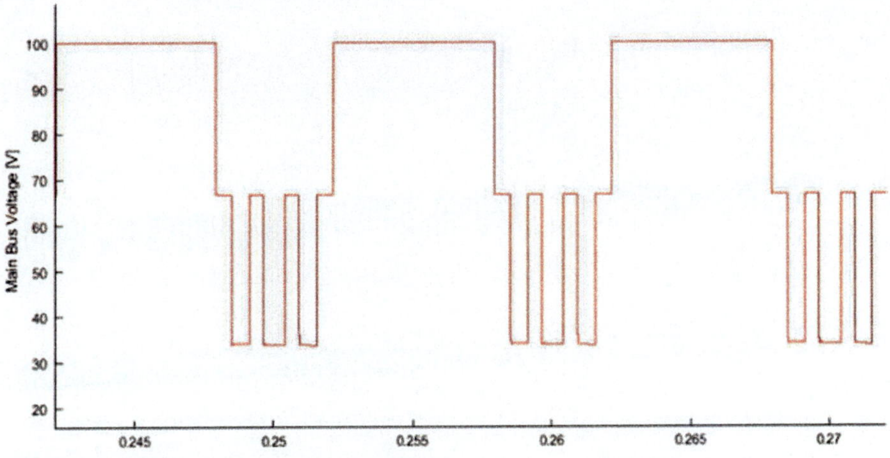

Fig. 11 Voltage of the main bus of a single module in a phase leg, during normal operation

B. *Controller synchronization failure*

At times the controller of the multilevel inverter can lose synchronization which may in turn be caused by feedback sensor failure. In such situations the AC waveform goes out of sync and the phase shift between phases no longer remains 120°. In such a scenario the output voltages and currents behave in an erratic manner. This is more pronounced in the multilevel inverter design because the line voltages are referenced between two lines, and in this case the difference is computed between two fixed voltage steps. Therefore, the synchronization failure of the controller can lead to collapse of output line voltages. Figure 15 shows the load line currents and

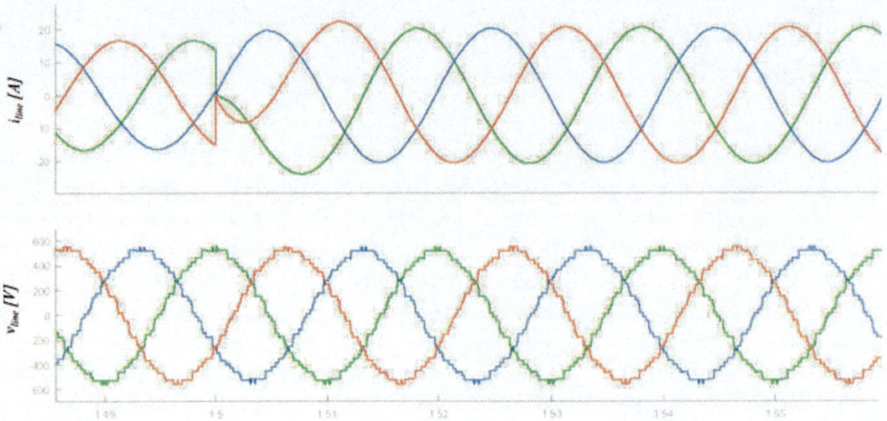

Fig. 12 Variation in three-phase output line current and line voltage subjected to a step change (at $t = 1.5$) in the load

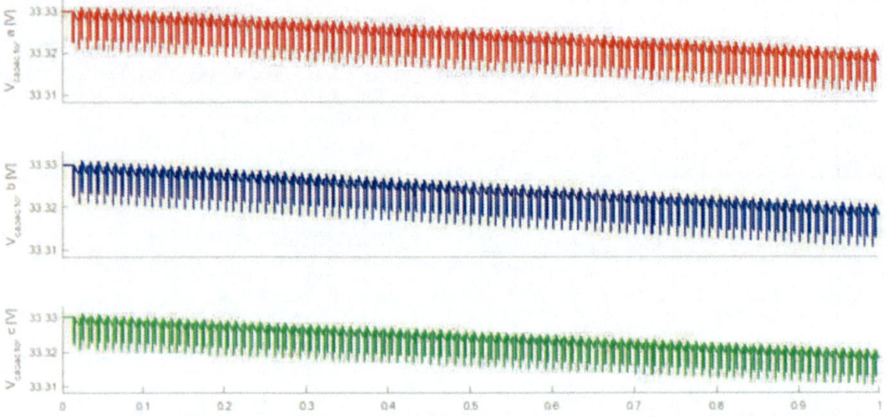

Fig. 13 Voltages across the capacitors of the capacitor buffer of a single module in a phase leg, with the DC source (battery) of the module disabled

voltages in case of such a failure.

III. Application: Integration of solar PV (photovoltaic) as a source and increase in load.

Figure 16 shows the use of PV energy source powering some of the modules. The use of multiple smaller separate DC sources in this design can be used to integrate multiple energy sources and not just batteries. The batteries can be replaced by DC sources such as MPPT (maximum power point tracking) controlled solar photovoltaic systems. This is another advantage of the design, as such sources can be integrated

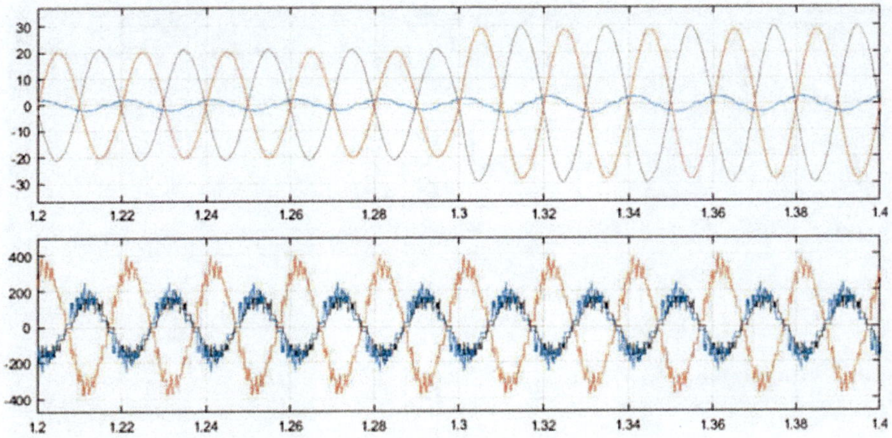

Fig. 14 Output three-phase line current (above) and line voltages (below), with the DC source (battery) of the test module disabled

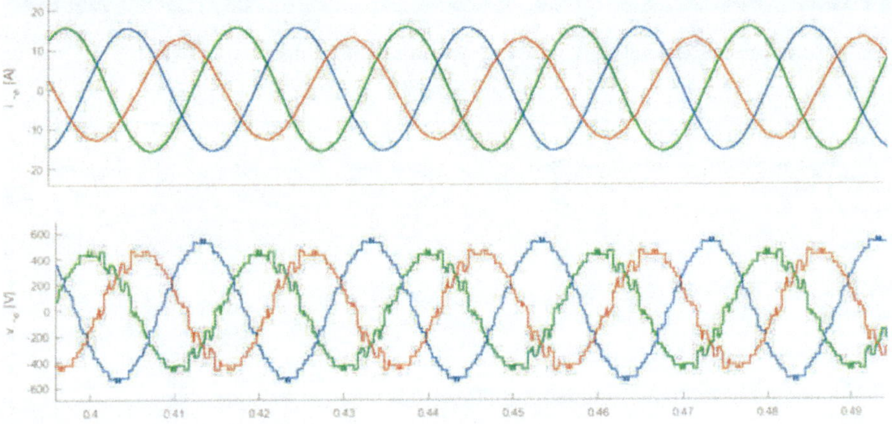

Fig. 15 Block diagram showing some of the batteries replaced with solar PV systems

as well as isolated from the modules; they provide power by dynamically adjusting the controller output.

The following case uses a 7.3 kW solar PV system to replace a battery and provides power to the inverter module. The output of solar PV module is more than the DC source. This is, however, adjusted to the required value by the splitter mechanism, where the capacitors are charged to the required fixed voltage set by the inverter controller. Figures 17, 18 and 19 show the power supplied by the solar PV system to the connected module. Note that the voltage provided by the PV is more than that set for the module, 100 V. The output of the inverter is provided to a switched load which goes from 41.5 to 51 kW as is shown in Fig. 20.

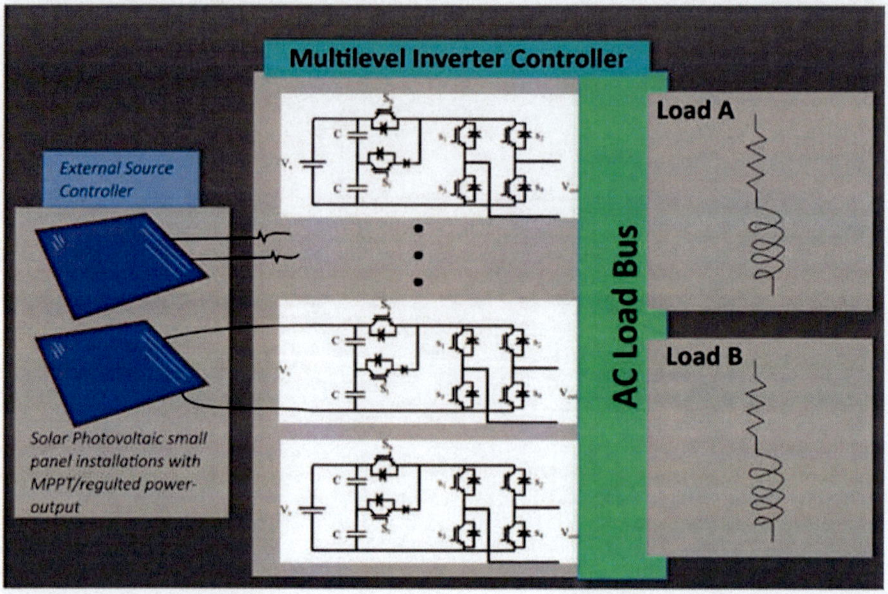

Fig. 16 Line currents and voltages in case of synchronization failure

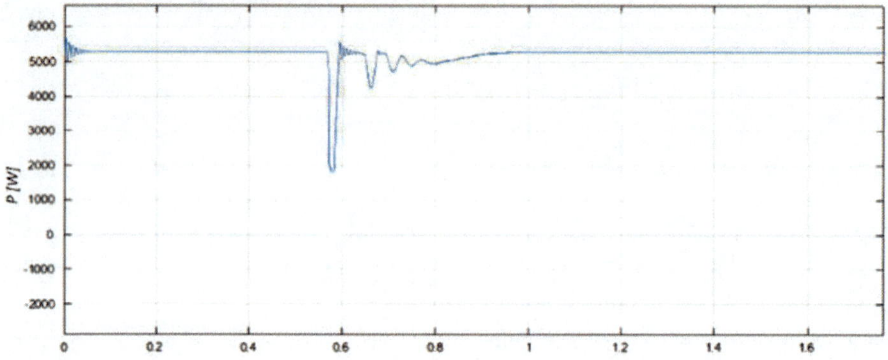

Fig. 17 Power supplied by solar panel. The dip in power is caused by reduced light input on the panel (clouding simulation)

5 Conclusion

A new multilevel converter topology using cascaded h-bridges with ultra-capacitor sub-DC voltage step generator has been presented. From the analysis and results of simulation the proposed topology has attained reduction of THD without filters while operating at low carrier frequency or fundamental frequency and using low *dV/dt* rated switches. The EMI and RFI produced by the setup are also low because the

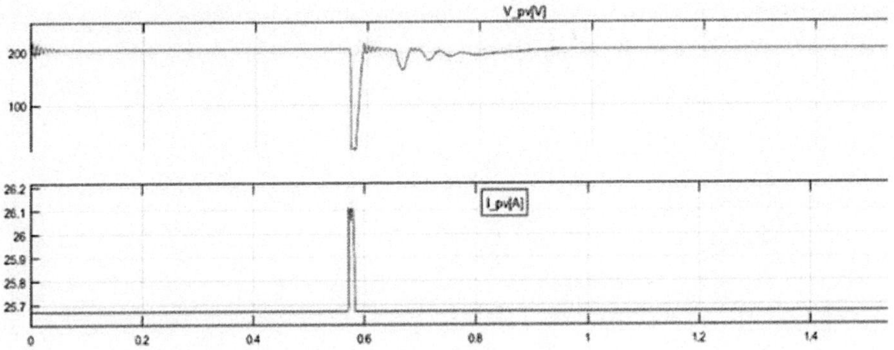

Fig. 18 V and I of the solar photovoltaic installation

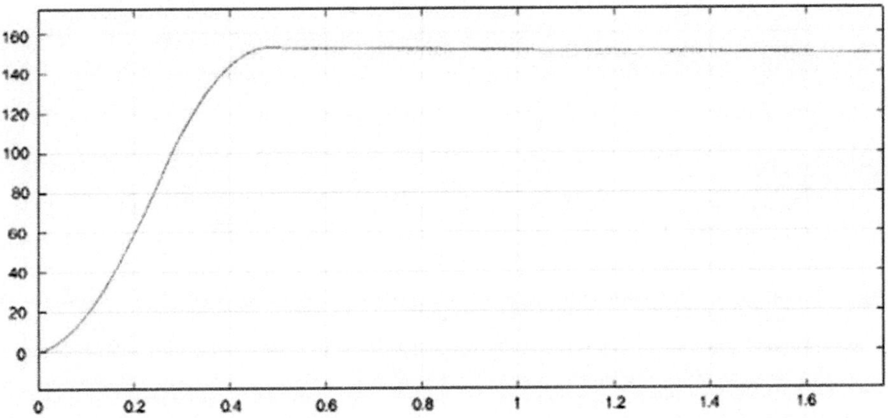

Fig. 19 MPPT voltage from solar PV system; this is reduced by capacitive divider to 100 V

switching frequency is kept low. The number of switches used is also low in comparison to a simple CMLC, while retaining the stress distribution among switches and the ability to have smaller separate DC sources. This topology can find applications in situations where lower THD is required without filtering, in renewable energy and automotive applications as well.

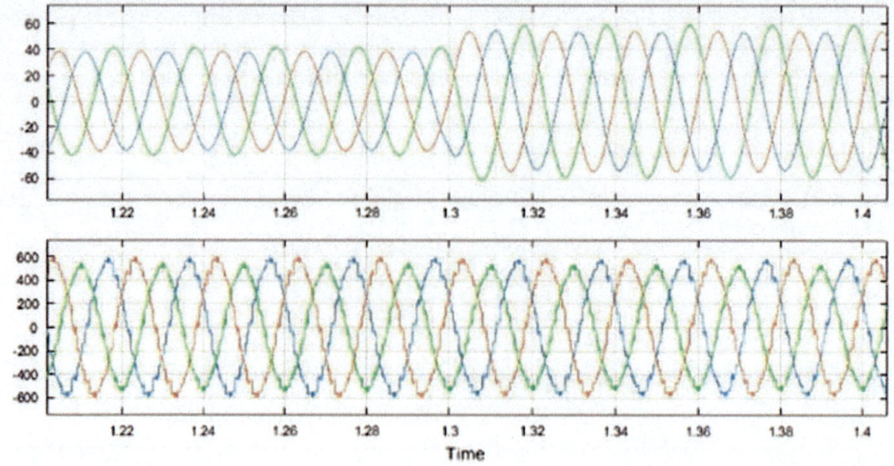

Fig. 20 Simulated load switch from 41.5 to 51 kW, with one module power source being solar PV system

References

1. IEEE Std 519–1992, in *IEEE Recommended Practices and Requirements for Harmonic Control in Electrical Power Systems.*
2. S. Bell, J. Sung, Will your motor insulation survive a new adjustable frequency drive? IEEE Trans. Ind. Appl. **33**(5), 1307–1311 (Sep. 1997)
3. J.K. Steinke, Use of an LC filter to achieve a motor-friendly performance of the PWM voltage source inverter. IEEE Trans. Energy Conversion **14**, 649–654 (1999)
4. L.M. Tolbert, F.Z. Peng, T.G. Habetler, Multilevel converters for large electric drives. IEEE Trans. Ind. Appl. **35**(1), 36–44 (1999)
5. I. Hasegawa, H. Akagi, M. Hagiwara, Speed motor driven by a Modular Multilevel Cascade Inverter (MMCI). IEEE Trans. Power Electron. **26**(11) (2011)
6. A. Nabae, I. Takahashi, H. Akagi, A new neutral-point clamped PWM inverter. IEEE Trans. Indus. Appl. **IA-17**(5), 518–523 (1981)
7. J. Erdman, R. Kerkman, D. Schlegel, G. Skibinski, Effect of PWM inverters on AC motor bearing currents and shaft voltages. IEEE Trans. Ind. Appl. **32**(2), 250–259 (1996)
8. W.A. Hill, C.D. Harbourt, Performance of medium voltage multilevel inverters. IEEE Indus Appl Soc (IAS) Conf **2**, 1186–1192 (1999)
9. B. Wu, S. Dewan, G. Slemon, PWM-CSI inverter induction motor drives. IEEE Trans. Ind. Appl. **28**(1), 64–71 (1992)
10. L. Liu, H. Li, S.-H. Hwang, J.-M. Kim, An energy efficient motor drive with autonomous power regenerative control system based on cascaded multilevel inverters and segmented energy storage. IEEE Trans. Indus. Appl. **49**(1), 178–188 (2013)
11. B. Xiao, L. Hang, J. Mei, C. Riley, L.M. Tolbert, B. Ozpineci, Modular cascaded H-Bridge multilevel PV inverter with distributed MPPT for grid-connected applications. IEEE Trans. Indus. Appl. **51**(2), 1722–1731 (2015)
12. A.A. Nazar, R. Jeyabharath, M.D. Udayakumar, Cascaded multilevel inverters for reduce harmonic distortions in solar PV applications. Asian J. Res. Soc. Sci. Human.
13. E. Babaei, S. Alilu, S. Laali, A New General Topology for Cascaded Multilevel Inverters with Reduced Number of Components Based on Developed H-Bridge. IEEE Trans. Industr. Electron. **61**(8), 3932–3939 (Aug. 2014)

14. P. Bhatnagar, R. Agrawal, K. Kumar Gupta, Reduced device count version of single-stage switched-capacitor module for cascaded multilevel inverters. IET Power Electron. **12**(5), 1079–1086 (2019)
15. L.M. Tolbert, F.Z. Peng, T.G. Habetler, Multilevel converters for large electric drives. IEEE Trans. Indus. Appl. **35**(1), 36–44 (1999)
16. E. Ozdemir, S. Ozdemir, L.M. Tolbert, Fundamental frequency-modulated six-level diode-clamped multilevel inverter for three-phase stand-alone photovoltaic system. IEEE Trans. Indus. Power Electron. **56**(11) (2009)
17. Y. Lei et al., A 2-kW Single-phase seven-level flying capacitor multilevel inverter with an active energy buffer. IEEE Trans. Power Electron. **32**(11), 8570–8581 (2017)
18. M. Malinowski, K. Gopakumar, J. Rodriguez, M.A. Perez, A survey on cascaded multilevel inverters. IEEE Trans. Industr. Electron. **57**(7), 2197–2206 (2010)
19. L.M. Tolbert, F.Z. Peng, T.G. Habetler, Multilevel inverters for electric vehicle applications, in *WPET '98*, pp. 79–84 (Dearborn, Michigan, 1998)
20. M. Cheng, Z. Wang, K.T. Chau, Y. Wang, W. Wang, A hybrid-source switched-capacitor multilevel converter for electric vehicles, in *2010 International Conference on Electrical Machines and Systems* (2010)
21. S. Kim, J. Lee, K. Lee, A modified level-shifted PWM strategy for fault-tolerant cascaded multilevel inverters with improved power distribution. IEEE Trans. Industr. Electron. **63**(11), 7264–7274 (Nov. 2016)
22. M.K. Sahu, J.M.R. Malla, M. Biswal, S. Behera, THD analysis and comparison of different cascaded multilevel inverters for improving the quality of energy. *IJAER* **14**(10), 2422–2429 (2019). ISSN 0973–4562

An Analytical Study on Electric Generators and Load Control Schemes for Small Hydro Isolated Systems

B. V. Murali Krishna and V. Sandeep

1 Introduction

To meet peak demand of energy and improve reliability in energy supply, the emerging economies are focusing on power generation using renewable energy sources (RES) in recent times, which maximizes the utilization of locally available energy resources [1–3]. The share of renewable energy generation crosses 20% in total installed capacity in India by 31 August 2018, which shows an impressive growth and sets a target of 175 GW by the year 2022. Total installed capacity of RES systems that include small hydro, solar, wind, bio-mass and urban and industrial waste to power is 87384.02 MW [4]. We can find many advantages with RES-based electric power generation systems as compared to the conventional coal, oil and gas-based systems, such as eco-friendly, non-polluted, high reliable, simple and safe operation and easy installation [5, 6]. The distributed generation is the main feature which directly involves the consumers and local utilities. Nowadays, distribution generation and isolated mode of operation systems are gaining popularity with RES for many merits and few of the technical advantages like minimizing the transmission and distribution (T&D) losses, high installation cost, maintenance and its losses. For small-scale applications, like domestic loads, offices and micro industries, the distribution generation systems are more attractive and promising technologies than grid-connected systems.

All the major RE sources like, wind, hydro and bio-mass systems depend on electric generators for generation of electric power. In such a system, the electric

B. V. M. Krishna (✉)
Department of Electrical Engineering, Central University of Karnataka, Gulbarga 585367, India
e-mail: muralikrishna.cuk@gamil.com

V. Sandeep
Department of Electrical Engineering, National Institute of Technology, Andhra Pradesh, Tadepalligudem 534101, India
e-mail: sandeep@nitandhra.ac.in

S. Vadhera et al. (eds.), *Latest Trends in Renewable Energy Technologies*, Lecture Notes in Electrical Engineering 760, https://doi.org/10.1007/978-981-16-1186-5_9

generator converts mechanical energy from the couple turbine to electric energy. As of now, various electric generators-based RES are reported in the literature for the isolated mode of operation [1, 7–16]. Much difference can be found for the same generators, while they are operating in grid-connected mode and isolated mode. When any EGs are operated in a grid-connected mode, their reactive power compensation is supplied by the grid, whereas the EGs are operated in the isolated mode of operation, with the lack of reactive power supply from gird; therefore we need to provide the same externally. Most of the EGs are not self-started machines and one more major disadvantage is the poor voltage regulation with low efficiency of the system on direct unexpected load or dynamics in operation. Certain requirements and operations make the machine as a self-started/self-excited generator with good voltage regulation, such as excitation capacitance and dedicated power electronic circuits. Sometimes, simple static converter systems (non-power electronic) may work for improvement of power quality if the nature of loading is not so critical and dynamic.

Small/micro hydro-based systems are well reported in the literature due to many advantages like low cost, simple operation, eco-friendly, suitable for almost major existing generators [2, 7, 17–20]. These are one of the best choices for the most cost-effective systems for long-term use because of low budget required at the construction time and can withstand more than 50 years. The efficiency of small/micro hydro systems is 60–80%, which is higher than the solar and wind systems. Typically, these are equipped with constant speed turbine to rotate the shaft of the generators to generate the constant power. The output power (P) of hydro turbine is given by (1)

$$P = \rho \cdot g \cdot H \cdot Q \tag{1}$$

In (1), ρ is the water density in kg/m^3, g is the acceleration due to gravity in m/s^2, H is the water head in meters (m), and Q is the discharge of water in m^3/s.

A schematic operational layout diagram of small/micro hydro system is shown in Fig. 1. In the hydro system layout, the major mechanical components are the dam, penstock, turbine and tailrace, and electrical components are generator power house. The selection on generator depends on the mode of operation and applications.

As the input of the generator is a constant speed turbine in the small/micro hydro systems, the output of the generator is also constant. In off-grid applications, the loads are always susceptible and vary from minimum to maximum. When micro hydro systems are adapted to isolated loads, one should be carefully designing the system through a load controller to balance the load power. Power electronic-enabled switches are used for the logical control of load power; such a featured controller is called electronic load controller (ELC) and well reported in the literatures [21–26].

This paper deals with the popular generators for three-phase load supply; those are self-excited induction generator (SEIG), permanent magnet synchronous generator (PMSG), brush-less synchronous generator (BLSG), switched reluctance generator (SRG) and synchronous reluctance generator (SynRG) for RES-based micro hydro systems. The basic physical phenomena, configurations and minimum excitation requirement for isolated operation are briefly discussed. Further, the importance and

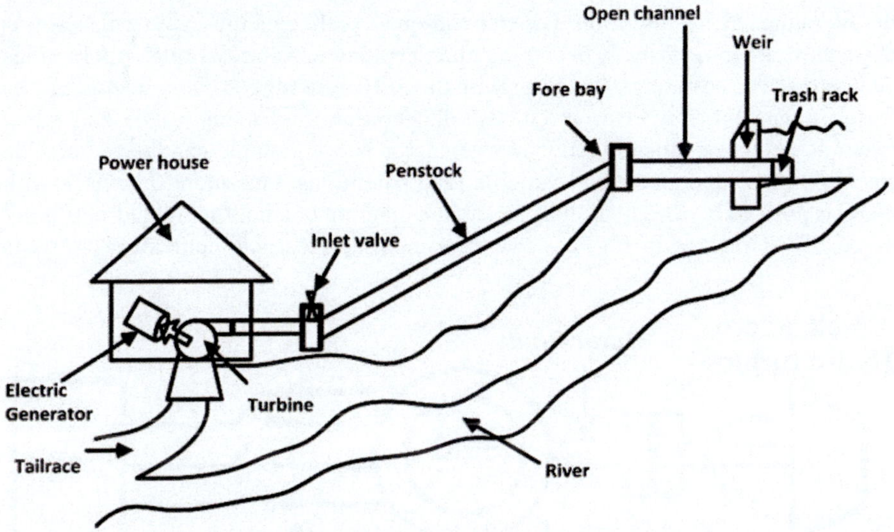

Fig. 1 Layout diagram of small/micro hydro system

working of electronic load controllers and methods of controlling the load power in isolated system are discussed. The study gives an idea on how to work with isolated generator for the small/micro hydro-based isolated systems with electronic load controllers for power balance.

2 Electric Generators for Small/Micro Hydro Isolated Systems: Phenomena, Configurations and Performance

This section deals with the working phenomena and schematic configurations of SEIG, PMSG, BLSG, SRG and SynRG in off-grid operation. The working principle is explained by the schematic diagram, and the performance analysis is given by the equivalent circuit for each electric generator separately.

Self-Excited Induction Generator (SEIG)

As mentioned earlier, the SEIG is not a self-stated generator in isolated mode operation. To run the induction generator as a self-started generator in a standalone mode of operation, a suitable rating of minimum reactive power through pre-charged capacitors must be supplied to the stator terminals of induction generator (IG) and this phenomenon in induction generator is called capacitor excitation phenomenon, and such a featured IG is called self-excited induction generator (SEIG) [27]. In SEIG, terminal voltage and frequency are not constant. At varying voltage and frequency, the air-gap flux varies and tends to operate at wider range of magnetic flux density, assuming different values are dependent on capacitor, speed and load [5, 28]. As per the literature, SEIGs are more popular for small-scale generation systems. Nowadays, these are progressively becoming more popular than conventional synchronous generators. The advantageous things with the SEIGs are rugged construction, brushless arrangement, low cost, simplicity of operation, less maintenance and quick dynamic response. But the induction generator is not a stable candidate for wide speed range of operations and variable load conditions. One more drawback with SEIG is poor voltage regulation. A schematic diagram of small/micro hydro turbine-based SEIG is given in Fig. 2a, and the simplified electrical equivalent circuit is

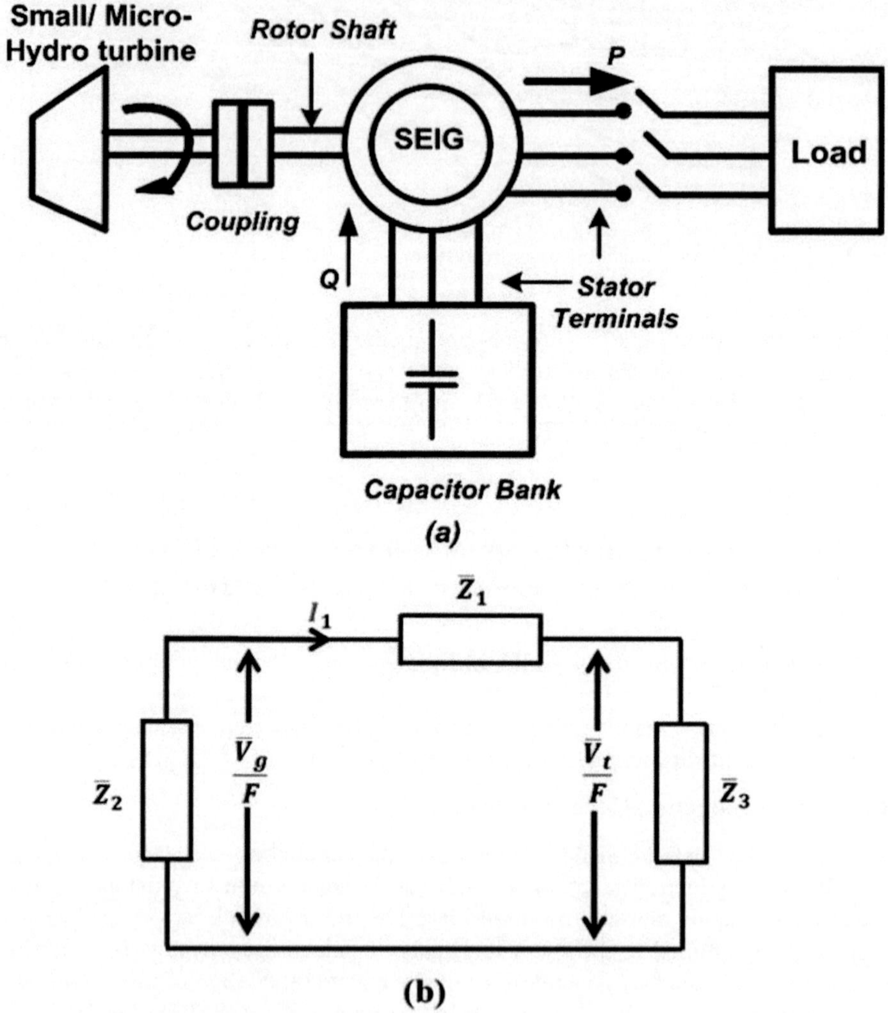

Fig. 2 **a** Schematic diagram of micro/pico- hydro-based SEIG system. **b** Simplified equivalent circuit of SEIG

presented in Fig. 2b [17]. Figure 2b will be useful for finding the key parameters of the considered machine to work in isolated mode of operation.

The minimum capacitance value is chosen in such a way to run the induction machine as a SEIG and is given by the Eq. (1) [8, 29].

$$C_{min\_\Delta ph} = \frac{Q_g}{3 \times v^2 \times 2 \times \pi \times f_{base}} \tag{2}$$

In (2), Qg is the reactive power consumed by the induction generator to build the desired phase voltage.

Permanent Magnet Synchronous Generator (PMSG)

Nowadays, the PMSG systems are becoming popular because of many advantages, like no chance for loss of core excitation, self-started generator in both grid and off-grid applications, more efficient than SEIG and no need of slip-rings and brushes, and can be operated at high power factor (pf). Apart from the advantages, the major disadvantage is unregulated output voltage rating [30, 31]. The PMSG can be able to possess a capability of inherent voltage compensation and which is desirable for maintaining a constant load voltage. According to the power electronic converter topologies and BESS control, various blocks may be added. A schematic diagram of small/micro hydro turbine-based PMSG-fed three-phase load is given in Fig. 3a and the simplified per-phase electrical equivalent circuit is given in Fig. 3b, which is used for the performance analysis.

Brush-less Synchronous Generator (BLSG)

For the off-grid/isolated applications, the small/micro hydro turbine-based power generation system with BLSGs is becoming popular. The output voltage will generate the proper amount of excitation to the field winding of the machine. Brushed and brush-less SGs are of two types, which are based on scheme of excitation. Small rating of BLSGs is much suitable for temporary/emergency power sources, like a very well-known example of diesel generator. To make the error voltage zero in closed-loop operation of BLSG, the exciter arrangement adjusts itself accordingly, so voltage across the load is regulated for all load conditions. Generally, the excitation elements are pre-charged capacitors of DC voltage source in BLSGs. The DC excitation-based BLSGs are well suited for variable speed operation also. A schematic diagram of small/micro hydro turbine-based BLSG-fed three-phase load is given in Fig. 4a and the simplified electrical equivalent circuit per phase is given in Fig. 4b, which is used for the performance analysis [2, 12, 13].

Synchronous Reluctance Generator (SynRG)

As like the SEIG, the synchronous reluctance generator (SynRG) is having all the same benefits like simple in design, robust nature, inexpensive operation, core loss is less, less noise in working condition, smooth starting torque because of no cogging torque and in addition to these advantages, we can avoid the rotor winding losses because of lack of rotor bars in its construction. The SynRG-based isolate systems

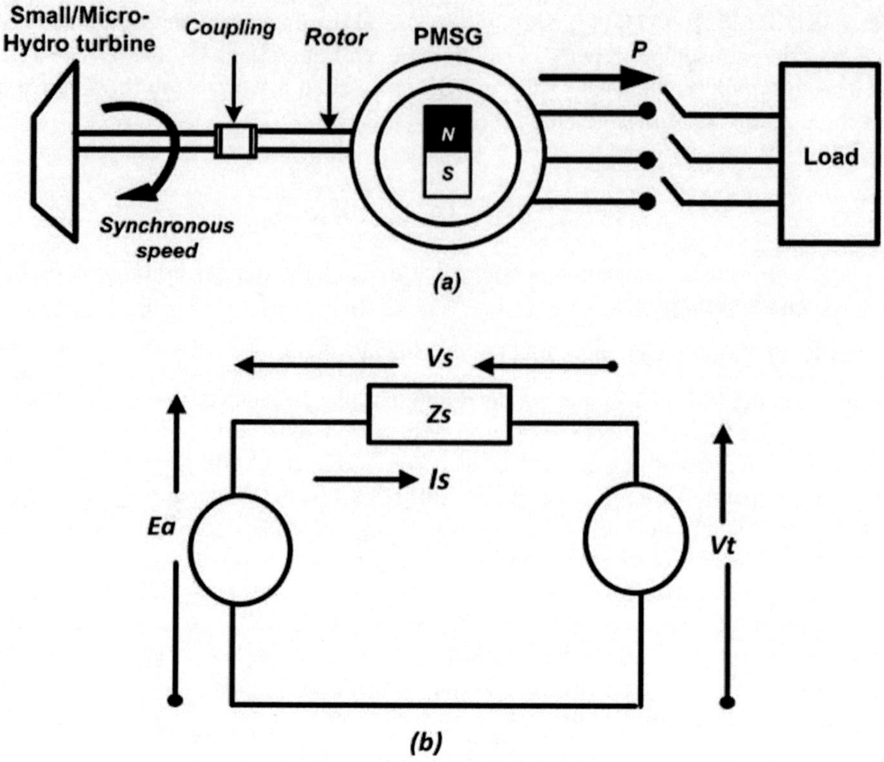

Fig. 3 **a** Schematic diagram of small/micro hydro-based PMSG system. **b** Simplified equivalent circuit of PMSG

of RE systems represent an alternative solution in isolated systems instead of being used for many advantages than SEIG; moreover, it is low cost as compared to other brushless generators. The SynRG is not a self-excited machine like SEIG in standalone operation and it also needs the necessary excitation to build-up the rated terminal voltage. However, the existence of a minimum residual flux in the core of machine is not enough to get the self-excitation. Generally, the excitation elements are pre-charged capacitors, which are needed to connect across the machine terminals. A schematic diagram is shown in Fig. 5a, and the simplified SynRG's per-phase equivalent circuit diagram is shown in Fig. 5b, which is used for the performance analysis through dq-axis circuit analysis. The minimum capacitance value for the self-excitation of SynRG is given by the Eq. (3) [32, 33].

$$C_{\min} = \frac{Q}{6\pi f v^2} \tag{3}$$

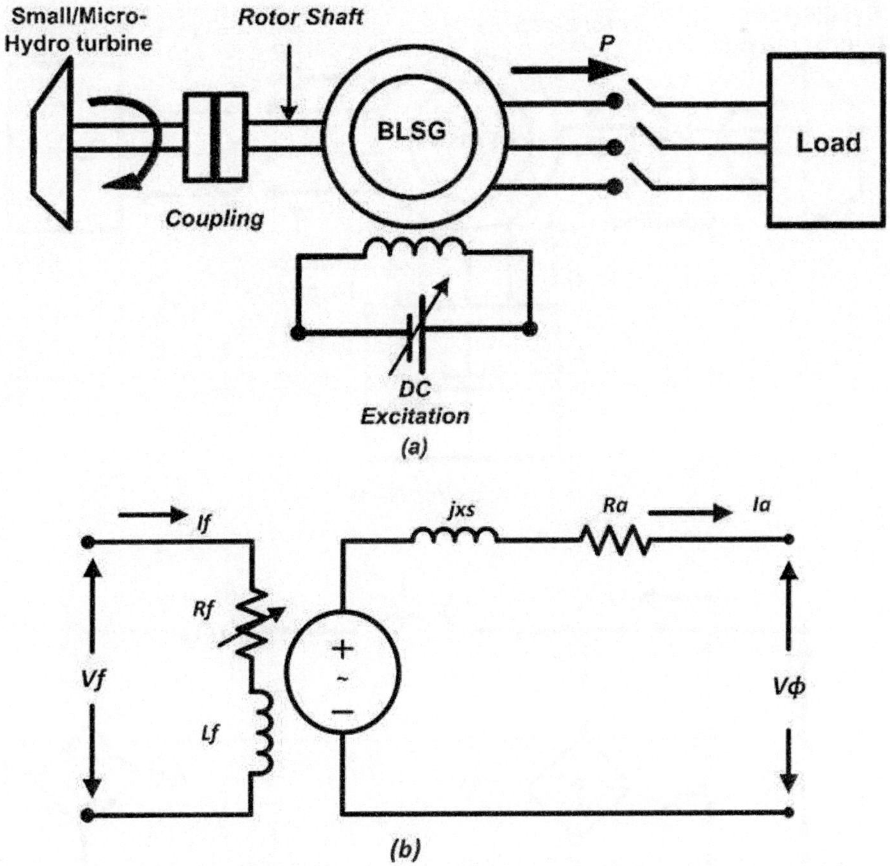

Fig. 4 **a** Basic schematic diagram of small/micro hydro-based BLSG system. **b** Simplified per-phase equivalent circuit of synchronous generator

Switched Reluctance Generator (SRG)

For the generation operation of a switched reluctance machine during the generation mode, a proper synchronization of stator phase currents is performed by means of an external DC source. At its rotor position an electro motive force (m.m.f) will induce in SRG. The SRG produces negative torque which is in opposite to the rotor rotation, hence extracts the energy from connected RES. Through an asymmetric half bridge converter, the phase winding will be energized. The turn-on/off triggering pulse angles of DC excitation unit are the variables. The common control methods association with SRGs are angle position control (APC), soft chopping current control (SCCC) and voltage chopping control (VCC) method. These methods are meant for controlling of phase voltage, currents and switching on/off. The SRGs are very simple in construction, rugged, cheap in cost, can be operated at higher speed, highly fault tolerant capability, no need of starting torque, high efficiency and can be suitable for

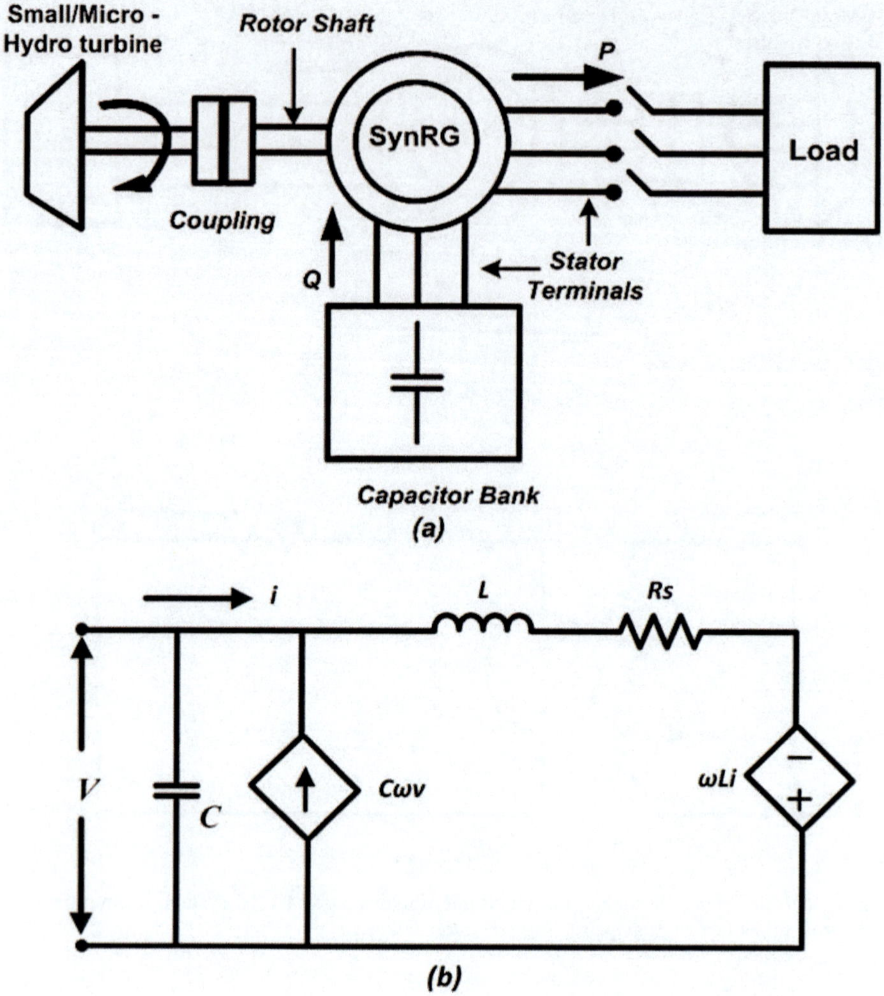

Fig. 5 **a** Schematic diagram of small/micro hydro-based SynRG system. **b** Simplified equivalent circuit of SynRG

high temperature environments also. As like PMSG, the SRG does not require high torque to start and run and can be used in hard environmental conditions too. These are not recommended for low-speed application because of its larger torque ripples. A schematic diagram small/micro hydro turbine-based SRG is shown in Fig. 6a and the simplified per-phase equivalent circuit diagram is shown in Fig. 6b, which is used for the performance analysis [14, 34].

Based on the investigation study and analysis, the comparison among the popular generators for operation, cost, performance, and all reported in Table 1.

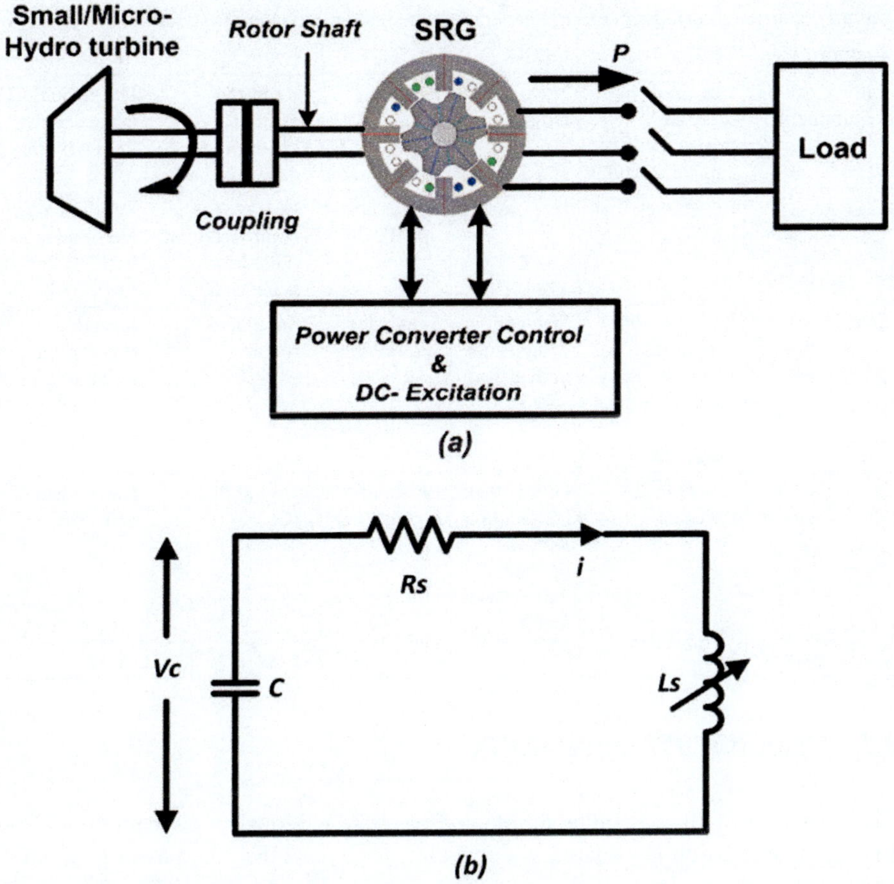

Fig. 6 **a** Schematic diagram of small/micro hydro based standalone SRG system. **b** Simplified equivalent circuit of SRG

3 Electronic Load Controller (ELC) for Load Balance: Design and Control

An electronic load controller is a power electronic-enabled circuit unit to balance the power in the isolated systems. Many researchers have reported the various control approaches of ELC in the literature [21, 6].

Table 1 Comparison study on electric generators for small/micro hydro systems

Feature	SEIG	PMSG	BLSG	SynRG	SRG
Self-excitation requirement	Required through pre-charged capacitor bank	Not required	Required through DC excitation for field winding	Required through pre-charged capacitor bank	DC excitation is needed for field winding
Cost	Lower cost than PMSG	High cost	Lower cost than PMSG	Low cost as compared to all other brushless generators	Very low cost as compared to SEIG and PMSG
Size	Smallest and low weight simplest machine among Induction machines	Size and weights are higher than SEIG	Small size and reliable	Small and robust	Simple, robust, and reliable
Power factor	Lower than PMSG	Higher than SEIG and almost operate at unity power factor	PF value is higher	Higher than IGs	Lower than SEIG
Complexity	Less	Less	Less	Less	More

3.1 Operating Principle of ELC

The operation of ELC is quite simple and easy for practical implementation. The ELC implementation in isolated systems aims to divert/block the power generated power into a dump load. If the load is less than the generation, then the excess amount of electrical power is diverted to dump load; and the DC chopper switch is in ON position. On the other hand, if it is equal and less than the generation, then the logical switch is OFF position to block the generated power into the dump load. Power balance phenomenon of ELC is given in (4). One should note that from (a), the main load and dump loads are varying from zero to maximum values, but the generation is always a constant and maintains its rated value in small/micro hydro isolated systems. To satisfy the operation of (4), the reference signal for turn ON of the power electronic switch is either the voltage or current value. Various controllers, such as analog, digital, microprocessor, micro controller, proportional (P), proportional and integral (PI), proportional and derivative (PD), proportional integral derivative (PID) and fuzzy logic controllers, are reported in the literature to enhance the efficiency. The working flowchart of ELC is given in Fig. 7. The role of the ELC in small/micro hydro systems is only to control the power electronic-based switch to consume or divert the excess load power into dump load resistor but did not show any effect on the power generation.

Fig. 7 Operating principle of ELC

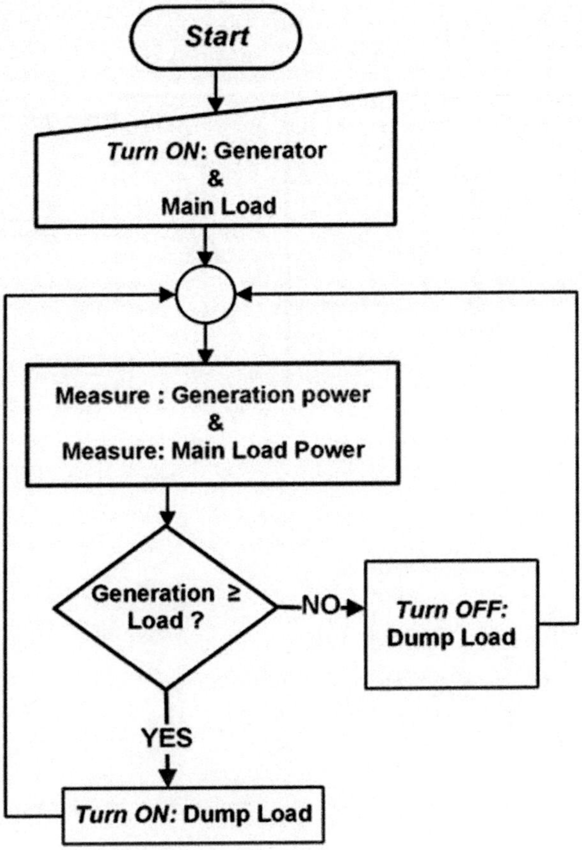

$$\text{Generated power} = \text{Main load power} + \text{Dump load power} \tag{4}$$

3.2 Design Aspects of ELC

The ELC mainly consists of (i) an uncontrolled bridge rectifier/universal bridge rectifier for rectification purpose of generated voltage, (ii) a controllable DC-chopper switch for allowing or blocking the generated power into dump load, (iii) a DC-link capacitor for voltage leveling, and iv) a resistor, which is also referred as dump load for consuming the excess load power. The schematic diagram of ELC is shown in Fig. 8. Power electronics switches, such as metal–oxide semiconductor field-effect transistor (MOSFET) or insulated-gate bipolar transistor (IGBT), are commonly used as DC chopper switch in ELC design. The dump resistor (R_D) should be designed in such a way that to withstand to the rated capacity of the system. The R_D is given

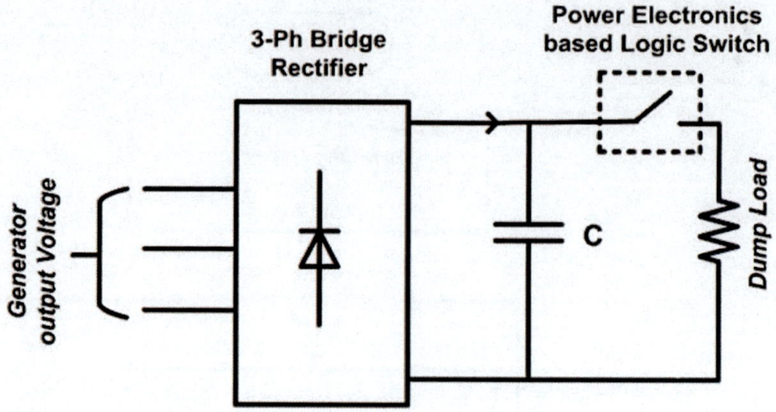

Fig. 8 Block diagram of ELC

by (5)

$$R_D = \frac{(V - cap)^2}{P_{rated}} \tag{5}$$

In (5), $V_{\_cap}$ is the output voltage of three-phase full-diode bridge rectifier and $Prated$ is rated power of generator. The $V_{\_cap}$ is given by (6)

$$V_{-cap} = \frac{3\sqrt{2}}{\pi} \times V_{LL} \tag{6}$$

In (6), V_{LL} is the line voltage of generator. The selection of the DC-link voltage (V_{DC}) is given by (7), where R.F is the ripple factor and f is the frequency.

$$V_{DC} = \left\{\frac{1}{12 \times f \times R_D}\right\}\left\{\left(1 + \frac{1}{\sqrt{2} \times R.F}\right)\right\} \tag{7}$$

The main components of the ELC can be designed by using (5)–(7) and control the load power through ELC in isolated systems.

4 Conclusion

The paper has presented an overview of various popular generators and their working phenomena in off-grid mode of operation, and experimental layout of small/micro hydro-based systems. The major features of these generators are analyzed and presented in the comparative study. Simplified equivalent circuit models of the generators are highlighted for further steady-state performance analysis. Further, the study focused on the effective utilization of constant generated power through the electronic load controllers for load balance phenomena. The schematic, operation and design of the ELC are discussed. The small/micro hydro system with ELC feature in isolated load applications is a cost-effective, eco-friendly, techno-friendly and required less maintenance in rural electrification.

References

1. P.K. Goel, B. Singh, S.S. Murthy, N. Kishore, Isolated Wind–hydro hybrid system using cage generators and battery storage. IEEE Trans. Ind. Electron. **58**(4), 1141–1153 (2011)
2. S.S. Murthy, Renewable energy generators and control. Chapter. 12. Electr. Renew. Energy Syst. 238–289 (2016)
3. A.B. Kanase-Patil, R.P. Saini, M.P. Sharma, Integrated renewable energy systems for off grid rural electrification of remote area. Renew. Energy **35**(6), 1342–1349 (2010)
4. Online. Available: https://www.cea.nic.in/reports. Last accessed 11 July 2020
5. R.C. Bansal, T.S. Bhatti, D.P. Kothari, Bibliography on the application of induction generators in nonconventional energy systems. IEEE Trans. Energy Convers. **18**(3), 433–439 (2003)
6. R.A. Ofosu, K.K. Kaberere, J.N. Nderu, S.I. Kamau, Design of BFA-optimized fuzzy electronic load controller for micro hydro power plants. Energy Sustain. Dev. **51**, 13–20 (2019)
7. S.S. Murthy, R. Jose, B. Singh, Experience in the development of micro hydel grid independent power generation scheme using induction generators for Indian conditions. in *Proceedings IEEE International Conference on Global Connectivity in Energy, Computer, Communication and Control*, vol. 2 (IEEE, New Delhi, 1998), pp. 461–465
8. G.K Singh, Self-Excited induction generator for renewable applications. Encyclopedia Sustain. Technol. (4), 239–256 (2017)
9. V.P. Chandran, S. Murshid, B. Singh, Voltage and frequency controller with power quality improvement for PMSG based pico-hydro system, in *2018 IEEE International Conference on Power Electronics, Drives and Energy Systems (PEDES)*, (Chennai IEEE, 2018), pp.1–6
10. P. Krause, O. Wasynczuk, S. Sudhoff, S. Pekarek (eds.), *Analysis of Electric Machinery and Drive Systems* (Wiley, New York, 2013)
11. T.-F. Chan, W. Wang, L.-L. Lai, Permanent-magnet synchronous generator supplying an isolated load. IEEE Trans. Mag. **46**(8), 3353–3356 (2010)
12. T.F. Chan, Steady state analysis of self-excited reluctance generator. IEEE Trans. Energy Conversion **7**(1), 223–230 (1992)
13. T. Fukami, T. Kondo, T. Miyamoto, Performance analysis of a self-regulated, self-excited, brushless three-phase synchronous generator, in *IEEE International Electric Machines and Drives Conference. IEMDC'99. Proceedings* (Seattle, USA, 1999), pp. 89–91
14. Y. Rahim, A. Mohamadien and A. Al Khalaf.: Comparison between the steady-state performance of self-excited reluctance and induction generators. IEEE Transaction on Energy Conversion 5(3), 519–525, (1990).
15. Nikolay Radimov, Natan Ben-Hail, Raul Rabinovici.: Switched Reluctance Machines as Three-Phase AC Autonomous Generator. IEEE Transactions on Magnetics 42(11), 3760–3764 (2006).
16. R. Raja Singh, B. Anil Kumar, D. Shruthi, and C. Thanga Raj.: Review and experimental illustration of electronic load controller used in standalone Micro-Hydro generating plants. Engineering Science and Technology, an International Journal 21(5), 886–900 (2018).
17. S. Mishra, S.K. Singal, D.K. Khatod, Cost Analysis for Electromechanical Equipment in Small Hydropower Projects. Int. J. Green Energy **10**, 835–847 (2013)
18. S. Mishra, S.K. Singal, D.K. Khatod, A review on electromechanical equipment applicable to small hydropower plants. Int. J. Energy Res. **36**(5), 553–571 (2012)

19. Okot, D.K.: Review of small hydropower technology. Renewable and Sustainable Energy Reviews 26 (C), 515–520 (2013).
20. N. Kishor, R.P. Saini, S.P. Singh, A review on hydropower plant models and control. Renew. Sustain. Energy Rev. **11**(15), 776–796 (2007)
21. Ankita Gupta.: Simulation of Advanced ELC with Synchronous Generator for Micro Hydropower Station. International Journal of Advanced Electrical Electronics Engineering 2 (1), 55–60 (2013).
22. Bhim Singh, S. S. Murthy and Sushma Gupta.: Transient Analysis of Self-Excited Induction Generator with Electronic Load Controller (ELC) Supplying Static and Dynamic Loads. IEEE Transactions on Industry Applications 41(5), 1194–1204 (2005).
23. R. Bonert and S. Rajakaruna.: Self-excited induction generator with excellent voltage and frequency control. Proceedings IEEE—Generation, Transmission, Distribution 145(1), 33–39 (1998).
24. P. Janardhan Reddy, S.P.: Voltage and frequency control of parallel operated synchronous and induction generators in micro hydro scheme. In: International Conference on Computation of Power, Energy, Information and Communication (ICCPEIC), pp.16–17. Chennai IEEE (2014).
25. A. Khodabakhshian, R. Hooshmand.: A new PID controller design for automatic generation control of hydro power systems. International Journal of Electrical Power & Energy Systems 32(5), 375–382 (2010).
26. Qazi, A., Hussain, F., Rahim, N.ABD., Hardaker, G., Alghazzawi, D., Shaban, K., Haruna, K.: Towards Sustainable Energy: A Systematic Review of Renewable Energy Sources, Technologies, and Public Opinions. IEEE Access. 7, 63837–63851 (2019).
27. E.D. Bassett, F.M. Potter, Capacitive Excitation for Induction Generators. Trans. Am. Inst. Electr. Eng. **54**(10), 540–545 (1935)
28. V. R. Sahu, R. Kesarwani, V. P. Chandran, S. Pandey, V. Kumar and S. Vadhera.: Steady state analysis of standalone SEIG for different operating conditions with interactive MATLAB Graphical User Interface. In: 2012 International Conference on Emerging Trends in Electrical Engineering and Energy Management (ICETEEEM), pp. 210–215. Chennai IEEE (2012).
29. N.H. Malik and S.E. Hague.: Steady-State Analysis and Performance of an Isolated Induction Generator. IEEE Transactions on Energy Conversion EC-1(3), 134–140 (1986).
30. B. Singh, R. Niwas and S. K. Dube.: Load Leveling and Voltage Control of Permanent Magnet Synchronous Generator-Based DG Set for Standalone Supply System. IEEE Transactions on Industrial Informatics, 10(4), 2034–2043 (2014).
31. Bhim Singh, Ram Niwas.: Performance of synchronous generators for DG set based standalone supply system. Electric Power Systems Research 113, 93–103 (2016).
32. Sakutaro Nonaka and Katsumi Kesamaru.: Analysis of Brushless Three-Phase Synchronous Generator Without Exciter. Electrical Engineering in Japan 113(7), 135–144 (1993).
33. Yawei Wang, Nicola Bianchi.: Investigation of Self-Excited Synchronous Reluctance Generators. IEEE Transactions on Industry Applications 54(2), 1360–1369 (2017).
34. Y. Wang and N. Bianchi.: Investigation of self-excitation in reluctance generators. 2017 IEEE International Electric Machines and Drives Conference (IEMDC), pp. 1–8. Miami IEEE (2017).
35. Sreenivas S. Murthy and Sriram Hegde.: Hydroelectricity. Electric Renewable Energy Systems, pp. 78–91. (2016).
36. B. S. Pali and S. Vadhera.: Renewable energy systems for generating electric power: A review. 2016 IEEE 1st International Conference on Power Electronics, Intelligent Control and Energy Systems (ICPEICES) pp. 1–6. New Delhi IEEE (2016).
37. R.S. Bhatia, S.P. Jain, B. Singh, D.K. Jain, Power conditioning of an uncontrolled micro-hydro turbine-driven induction generator for distributed generation using a battery energy storage system. International Journal of Energy Technology and Policy **5**(5), 604–618 (2007)
38. S. Mishra, S.K. Singal, D.K. Khatod, Optimal installation of small hydropower plant—A review. Renew. Sustain. Energy Rev. **15**, 3862–3869 (2011)
39. M. H. Nehrir, C. Wang, K. Strunz, H. Aki, R. Ramakumar, J. Bing, Z. Miao, and Z. Salameh.: A Review of Hybrid Renewable/Alternative Energy Systems for Electric Power Generation: Configurations, Control, and Applications. IEEE Transactions on Sustainable Energy 2(4), 392–403 (2011).

40. Zhou, D., Deng, Z. (Daniel).: Ultra-low-head hydroelectric technology: A review. Renewable and Sustainable Energy Reviews 78, 23–30 (2017).
41. A.K. Akella, M.P. Sharma, R.P. Saini, Optimum utilization of renewable energy sources in a remote area. Renew. Sustain. Energy Rev. **11**(15), 894–908 (2007)
42. R.C. Bansal, Three-Phase Self-Excited Induction Generators: An Overview. IEEE Trans. On Energy Conversion **20**(2), 292–299 (2005)
43. B. Singh, Induction Generators-A Prospective. Electric Machines & Power Systems **23**(2), 163–177 (1995)
44. S.S. Murthy, O.P. Malik, and A.K. Tandon.: Analysis of self-excited Induction Generators. IEE Proc. 129(6), 260- 265(1982).
45. N.H. Malik and A.A. Mazi.: Capacitance Requirements for Isolated Self -Excited Induction Generators. IEEE Transactions on Energy Conversion EC-2(1), 62–69 (1987).
46. S.S. Murthy, B.P. Singh C. Nagamani and K.V.V. Satyanarayana.: Studies on the use of conventional Induction Motors as Self- Excited Induction Generators. IEEE Transactions on Energy Conversion 3(4), 842–848 (1988).
47. T.F. Chan.: Capacitance requirements of Self-Excited Induction Generators. IEEE Transactions on Energy Conversion 8(2), 304–311 (1993).
48. T. F. Chan and Loi Lei Lai.: Steady-State Analysis and Performance of a Stand-Alone Three-Phase Induction Generator with Asymmetrically Connected Load Impedances and Excitation Capacitances. IEEE Transactions on Energy Conversion 16(4), 327–333 (2001).
49. Sandhu, K.S.: Analysis of Induction Generators for Renewable Energy Applications. In: Handbook of Renewable Energy Technology, pp.717–756. World Scientific Publishing Co., Singapore (2011).
50. S.R. Arya, A. Patel, A. Giri, Isolated Power Generation System Using Permanent Magnet Synchronous Generator with Improved Power Quality. J. Inst. Eng. India Ser. B. **99**, 281–292 (2018)
51. B. Singh, R. Niwas, Power Quality Improvement of PMSG-Based DG Set Feeding Three-Phase Loads. IEEE Trans. Ind. Appl. **52**(1), 466–471 (2016)
52. R. Krishnan.: Permanent Magnet Synchronous and Brushless DC Motor Drives. 1st Edition, CRC Press, Boca Raton (2010).
53. L. L. Lai and T. F. Chan.: Distributed Generation: Induction and Permanent Magnet Generators. West Sussex, U.K. Wiley (2007).
54. C. N. Bhende, S. Mishra, Siva Ganesh Malla.: Permanent Magnet Synchronous Generator-Based Standalone Wind Energy Supply System. IEEE Transactions on Sustainable Energy 2(4), 361–373 (2011).
55. T. J. E. Miller.: Brushless Permanent-Magnet and Reluctance Motor Drives. Electronic Engineering No. 21, Oxford Science Publications (1989).
56. V. P. Chandran, S. Murshid and B. Singh.: Power Quality Improvement in PMSG Based Hydro-BES System Operating in Isolated Remote Areas Using CF-FLL Control. IEEE Energy Conversion Congress and Exposition (ECCE), pp. 960–967. Baltimore, USA IEEE (2019).
57. S. S. Maroufian, P. Pillay.: Self-Excitation Criteria of the Synchronous Reluctance Generator in Stand-Alone Mode of Operation. IEEE Transactions on Industry Applications 54(2), 1245–1253 (2018).
58. M. Ibrahim, P. Pillay, The Loss of Self-Excitation Capability in Stand-Alone Synchronous Reluctance Generators. IEEE Trans. Ind. Appl. **54**(6), 6290–6298 (2018)
59. David A. Torrey.: Switched Reluctance Generators and Their Control. IEEE Transactions on Industrial Electronics 49(1), 3–13 (2002).
60. Singh, B., Niwas, R., Chandra, A., Miloud, R.: Voltage control and load leveling of synchronous reluctance generator based DG set. In: 2014 6th IEEE Power India International Conference (PIICON), pp. 1–6. Bikaner, India IEEE (2014).
61. Martinez, A., Vicuna, J., Perez, F., Laloya, E., Martin, B., Pollan, T., Juan Llado, B.S. y: Steady-state behaviour of an ac autonomous switched reluctance generator. In: 2007 European Conference on Power Electronics and Applications, pp. 1–8. Aalborg, Denmark IEEE (2007).

62. Ogunjuyigbe, A.S.O., Ayodele, T.R., Adetokun, B.B.: Steady state analysis of wind-driven self-excited reluctance generator for isolated applications. Renewable Energy 114 (B), 984–1004 (2017).
63. Barros, T.A. dos S., Neto, P.J. dos S., Filho, P.S.N., Moreira, A.B., Filho, E.R.: An Approach for Switched Reluctance Generator in a Wind Generation System With a Wide Range of Operation Speed. IEEE Transactions Power Electrons 32(11), 8277–8292 (2017).
64. Siyang Yu, Fengge Zhang, Dong-Hee Lee, Jin-Woo Ahn.: High Efficiency Operation of a Switched Reluctance Generator over a Wide Speed Range. Journal of Power Electronics 15(1), 123–130 (2015).
65. Rana, S.-D., Kar, N.C.: Steady-state analysis of self-excited synchronous reluctance generator. In: 2008 Canadian Conference on Electrical and Computer Engineering, pp. 001617–001620. Niagara Falls IEEE (2008).
66. Nagrial, M.H., Rahman, M.A.: Operation and characteristics of self-excited reluctance generator. In: Conference Record of the 1988 IEEE Industry Applications Society Annual Meeting, pp. 55–58. Pittsburgh, USA IEEE (1988).
67. B Murali Krishna V, Sandeep V.: An Investigative Study on Electric Generators for Isolated Operation. 33rd Indian Engineering Congress, pp.301–306. Udaipur IE(I) 2018.
68. B. Singh, S.S. Murthy, S. Gupta.: Analysis and implementation of an electronic load controller for a self excited induction generator. IEE Proceedings - Generation, Transmission and Distribution 151(1), 51–60 (2004).
69. B. Singh, S. S. Murthy, and S. Gupta.: A voltage and frequency controller for self-excited induction generator. Electric Power Components and Systems Journal 34(2), 141–157 (2006).
70. Gaurav Kumar Kasal and Bhim Singh.: Decoupled Voltage and Frequency Controller for Isolated Asynchronous Generators Feeding Three-Phase Four-Wire Loads. IEEE Transactions on Power Delivery 23(2), 966–973 (2008).
71. G. Castillo, L. Ortega, M. Pozo, X. Domínguez.: Control of an island Micro Hydro power Plant with Self-excited AVR and combined ballast load frequency regulator. In: IEEE Ecuador Technical Chapters Meeting (ETCM), pp. 1–6. Guayaquil IEE (2016).
72. E. Torres, F. Chan, J. Ramirez, A. Cowo.: A PWM control for electronic load controller for self-excited induction generator based in IGBT series-inverted switch. In: 12th International Power Electronics Congress (CIEP), pp 61–66. San Luis Potosi IEEE (2010).
73. J.M. Ramirez, E.M. Torres.: An electronic load controller for the self-excited induction generator. IEEE Transactions on Energy Conservations 22(2), 546–548 (2007).
74. S.S. Murthy, Rini Jose, B. Singh.: A practical load controller for standalone small hydro systems using self-excited induction generator. In: International Conference on Power Electronic Drives and Energy Systems for Industrial Growth, pp. 359–364. Perth, Australia IEEE (1998).
75. R. Panda, R.R. Singh, T.R. Chelliah.: Enforcement of ELC using reduced dump load for micro hydropower plant with the interpretation of switching transients and vibrations. In: IEEE 11th International Conference on Power Electronics and Drive Systems pp. 352–357. Sydney (2015).
76. B.N. Roodsari, E.P. Nowicki, P. Freere.: The distributed electronic load controller: a new concept for voltage regulation in microhydro systems with transfer of excess power to households. Energy Procedia 57, 1465–1474 (2014).
77. B Murali Krishna. V, V. Sandeep, Ruparani.: Design and Simulation of Voltage Sensor-based Electronic Load Balance Controller for SEIG based Isolated Load Applications. Journal of Advanced Research in Dynamical & Control Systems 12(3), 345- 352 (2020).
78. K.T.K. Teo, H.H. Goh, B.L. Chua, S.K. Tang, M.K. Tan.: Modelling and optimization of stand-alone power generation at rural area. In: IEEE International Conference on Consumer Electronics (ICCE) pp.51–56. China (2013).
79. Dipesh Shrestha, Ankit Babu Rajbanshi, Kushal Shrestha, Indraman Tamrakar.: Advance electronic load controller for micro hydro power plant. Journal of Energy and Power Engineering 8, 1802–1810 (2014).
80. M.P. Sruthi, C. Nagamani, G.S. Ilango.: An improved algorithm for direct computation of optimal voltage and frequency for induction motors. Engineering Science and Technology, an International Journal 20(5), 1439–1449 (2017).

81. R.R. Chilipi, B. Singh, S.S. Murthy.: Performance of a self-excited induction generator with DSTATCOM-DTC drive-based voltage and frequency controller. IEEE Transaction on Energy Conversation 29 (3), 545–557 (2014).
82. G. Nel, W. Doorsamy, Development of an intelligent electronic load controller for stand-alone micro-hydropower systems, in *2018 IEEE PES/IAS Power Africa* (Cape Town, South Africa IEEE, 2018), pp. 366–371
83. B. Singh, V. Rajagopal, Power balance theory-based control of an electronic load controller for an isolated asynchronous generator driven by uncontrolled pico hydro turbine, in *2009 Annual IEEE India Conference (INDICON)* (Gandhinagar IEEE, 2009), pp.1–5
84. V.I. Grigor'ev, Methods of load-frequency control for generating units of small and micro hydropower plants. Power Technol. Eng. **39**(1), 7–10 (2005)
85. B.N. Roodsari, E.P. Nowicki, Analysis and experimental investigation of the improved distributed electronic load controller. IEEE Trans. Energy Convers. **33**(3), 905–914 (2018)
86. A. Ali, M.U.R. Siddiqi, R.M. Arshad, Design and simulation of an electro-mechanical control system for mini hydro power plants, in *2018 International Conference on Power Generation Systems and Renewable Energy Technologies (PGSRET)* (Islamabad, Pakistan IEEE, 2018), pp. 6–12
87. R.R. Singh, B.A. Kumar, D. Shruthi, R. Panda, C.T. Raj, Review and experimental illustrations of electronic load controller used in standalone Micro-Hydro generating plants. Eng. Sci. Technol. Int. J. **21**, 886–900 (2018)
88. A. Ali, M.U.R. Siddiqi, R. Muhammad, S.M. Arshad, N. Ullah, Design and implementation of an electromechanical control system for micro-hydropower plants. Electr Eng. **102**, 891–898 (2020)
89. S. Nojeng, S.R Murniati, Design of the electronic load controller using micro controller based zero crossing detector for pico-hydro power plant. Electri. Electron. Eng. Int. J. (ELELIJ) **8**(3/4), 1–7 (2019)

Inexpensive Techniques to Design an Automated Home Using NodeMCU

Anthony Minj, Harsh Tank, Shipra Gautam, Yuvraj Singh Kahlon, and Shelly Vadhera

1 Introduction

It is smart devices that separate a traditional home from an automated one. These are the devices that can interact with the user and also with other smart devices in the house. The origin of home automation can be traced back to 1975 when the communication protocol X10 came into picture. X10 devices used radio waves for the purpose of communication. The method of communication wasn't much efficient either. The X10 devices communicated with switches capable of processing data sent by the X10 device in the form of 120 kHz radio bursts. A transmitter is a good example of an X10 device. For a while, the communication was a one-way communication meaning that only the X10 devices could send information and control the end devices, leaving no scope for resource monitoring. With the development of a two-way communication, the analog method of communication became more of a hindrance than progress. Today the method of communication is the internet allowing for a structured mode of communication, thereby proving to be more reliable and efficient. Also, with the internet, devices can be accessed and controlled from anywhere around the planet providing the capability of remote access [1]. The market for smart devices is increasing with giants like Amazon and Google with their devices like home assistants. Also, with the push for IoT from the industry, the smart devices becoming a home common is not a sight of the distant future [2]. This paper is aimed at providing techniques for modeling the same experience of an expensive smart home with all smart devices, through the use of traditional devices of the present. This is achieved using sensors and relays. The basic idea of a smart home or an automated home is to provide a method of granting control to a central hub which can act as the interface between the user and all the home devices at one place, be it resource monitoring, surveillance or remote access. The function of this

A. Minj · H. Tank · S. Gautam · Y. S. Kahlon (✉) · S. Vadhera
National Institute of Technology, Kurukshetra 136119, Haryana, India

S. Vadhera et al. (eds.), *Latest Trends in Renewable Energy Technologies*, Lecture Notes in Electrical Engineering 760, https://doi.org/10.1007/978-981-16-1186-5_10

central hub is incorporated through the Wi-Fi-enabled controller, NodeMCU. The interface used is an application that has to be run on smartphones called 'Blynk' [3].

2 System Design

2.1 Input and Output

The input side of the system consists of sensors, namely the ultrasonic sensor, PIR motion sensor and a photosensor. The ultrasonic sensor used is the HC-SR04 which is an electromagnetic-type sensor operating at 5 V. It is used for the purpose of the measurement of distance. This sensor consists of a trigger circuit and an echo circuit. The trigger circuit sends sound waves in the form of eight pulse burst while the echo circuit listens to the bounced back sound wave sent by the trigger circuit. The distance can be easily calculated as the speed of the sound waves is known (330 m/s) and the echo circuit is capable of calculating the time.

The PIR motion sensor stands for 'Passive InfraRed' motion sensor. The PIR sensor used is EC-0142. This sensor consists of two materials which are sensitive to infra-red, separated by a little distance. The scanning area of the sensor is created using a frenal lens. When the sensor is idle, that is to say that there is no activity, both the materials receive the same amount of infra-red from the ambient surrounding. But when a person or an object crosses the sensor detection area, a differential signal is produced in the sensor, one positive while the other is negative. This signal is then used to take further actions such as in the case of this project has been used to turn on lights in rooms when a person enters.

A photosensor is basically a photoresistor which produces signals depending on the amount of light transmitted by the transmitter and that received by the receiver. The photosensor used is EC-2387. There are three main types of photosensors. One works on the diffused beam principle, while others through beam and retro reflective principles.

The output side consists of relays, LEDs and the motor drive circuit.

2.2 System Controller

For the purpose of providing central control hub to the user and the devices, a NodeMCU is used. NodeMCU is an open source IoT platform using the ESP32 Wi-Fi SoC by Espressif Systems. This incorporates a dual-core Tensilica Xtensa LX106 core CPU with a maximum clock speed of 240 MHz. It also comes equipped Wi-Fi and dual Bluetooth-enabled capabilities. The firmware is programmed in Lua Scripting language but is also compatible with the Arduino IDE which has been used here.

2.3 *Blynk Application*

In order to communicate with the devices at homes, a middle hub called NodeMCU, as described in the previous section, has been used. NodeMCU has been chosen because it offers advantage of internet connectivity, thus providing an opportunity of remote access and control. In order to interact with NodeMCU, an interface called Blynk, which is a third-party smartphone application, capable of IoT has been used.

3 Implemented Codes

3.1 *Controlling Home Appliances from Remote Location Using Blynk App*

Using this application, we can easily control and operate appliances in our home with just a click of a button and hence actual human effort is reduced which could be highly useful for homes of differently abled and elderly persons (see Fig. 1). Here we have selected three buttons that will control the air conditioner, water tanker pump and geaser.

Arduino IDE is programmed with the desired program providing the authentication key, Wi-Fi hotspot name and password for the respective house. The instruction from our smartphone reaches the NodeMCU through internet and then NodeMCU instructs the respective relay to operate the appliance.

The flowchart of the code on Arduino IDE application provides the authentication key, Wi-Fi hotspot name and password for the respective house (see Fig. 2).

Description of code. Different home devices are connected through relays to different pins on NodeMCU. The pins are first configured to work as output pins. This is followed by connection authorization. After authorization, Blynk app is operated on a smartphone to set the output pins as high/low.

Code 1. Figure 3 shows how the program for controlling home appliances from remote location using Blynk application.

3.2 *Ultrasonic Sensor for Garage Parking*

The connection for ultrasonic sensor with NodeMCU for the garage parking was done (see Fig. 3). The code we have written basically gives various lights and various distances, and the buzzer gives sound in extreme conditions alerting the driver if he approaches too close to the door (Fig. 4).

The flowchart of the code for connection of ultrasonic sensor with NodeMCU and different color LEDs is shown in Fig. 5.

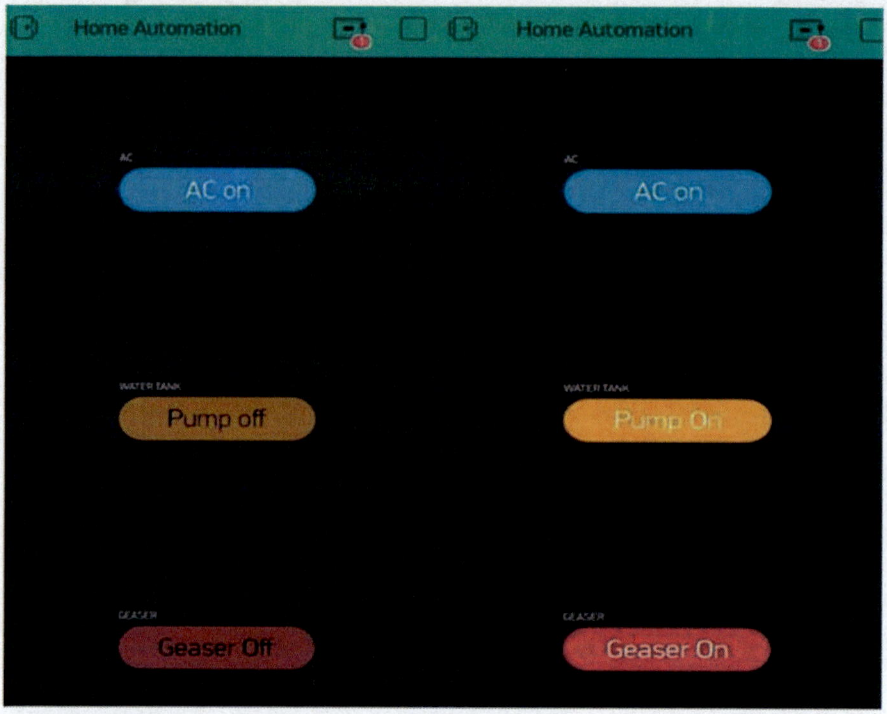

Fig. 1 A graphic interface is created in this Blynk application digital dashboard with three selected buttons for air conditioner, water tanker pump and geaser

Description of code. All the pins assigned from Pin_blue_led to Pin_red_led are set as 0 initially, making use of the 'for' loop. Next, respective pins are set high or low depending upon the distance calculated through the get distance function. This function takes time as it requires input from the sensor and converts that into distance.

Code 2. Figures 6 and 7 show the program for controlling the ultrasonic sensor for the purpose of garage parking.

3.3 Use of PIR Motion Detector Sensors

For the connection of PIR motion sensor with NodeMCU, see Fig. 5.

Also, the program code for the lights control in rooms and bathroom and switch off them when not in use is done on the Arduino IDE. And this was done to save a lot of energy (Fig. 8).

The flowchart of the code for the control of PIR sensor and detecting the motion from it is shown in Fig. 9.

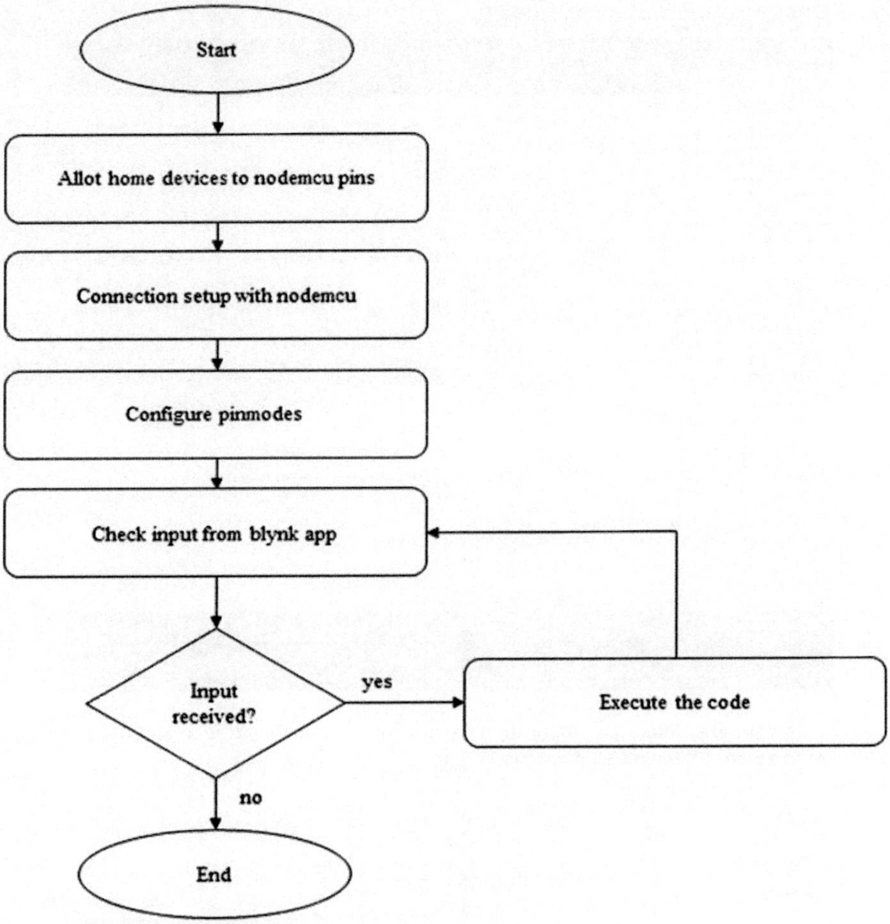

Fig. 2 Flowchart of the program on Arduino IDE application for providing the authentication key, Wi-Fi hotspot name and password for the respective house

Description of code. Two pins are chosen. One pin is set as input for NodeMCU, while the other is set as output pin. The input from the sensor is read at the configured input pin and depending on whether the input is high/low, the output pin is set as high/low.

Code 3. Figure 10 shows the program for motion detection by using PIR sensor.

Fig. 3 Program on Arduino IDE application for providing the authentication key, Wi-Fi hotspot name and password for the respective house

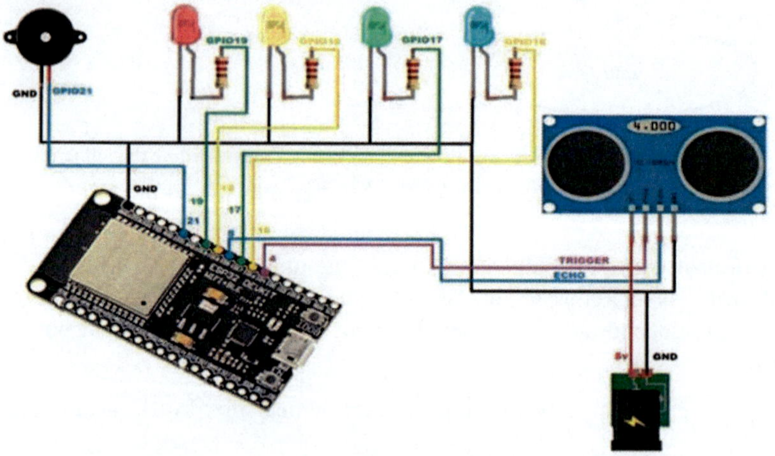

Fig. 4 Connection of ultrasonic sensor with NodeMCU and different color LEDs

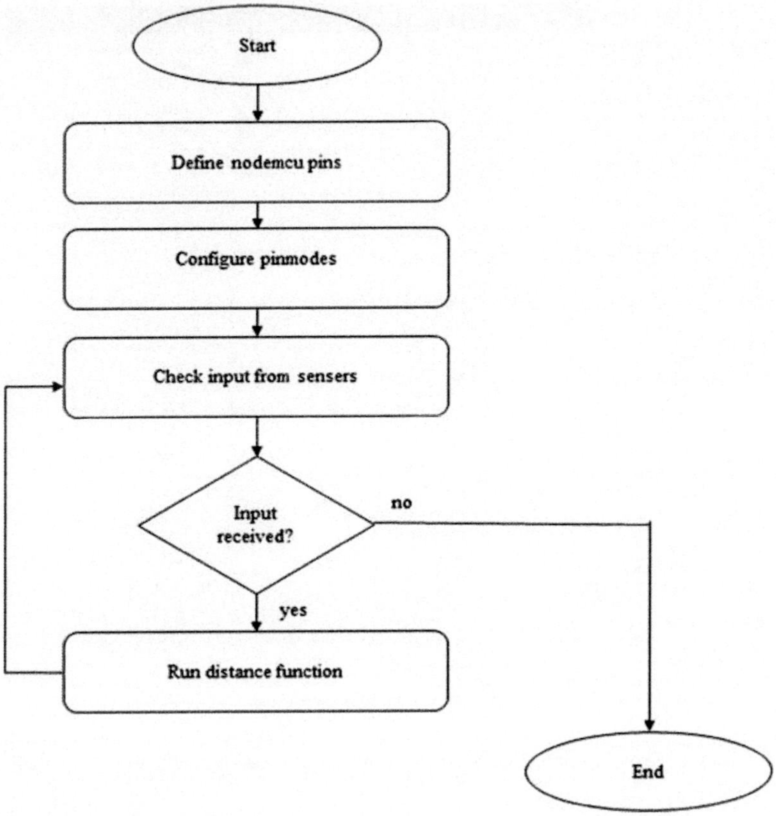

Fig. 5 Flowchart of the code in Arduino IDE for working of LEDs according to the distance calculated by ultrasonic sensor

3.4 *Use of Photosensors*

This sensor detects the presence of light and give different signal during the presence of different intensity of lights. Connection diagram for photosensor with NodeMCU, two external resistors of 100 Ω and LED is shown in Fig. 7.

In the program this photoresistor is used to turn on and off a LED when the intensity of light around the resistor changes (Fig. 11).

The flowchart of the code for photosensor use to turn off and on the LED is shown in Fig. 12.

Description of code. Two pins are chosen. One pin is set as input for NodeMCU, while the other is set as output pin. A reading is taken from the sensor under well-lit conditions. This value is then stored as threshold in the lightInit variable. During operation, the reading collected from the sensor is read and stored in lightVal variable.

```
homeauto_ultrasonic §
#include < ultrasonic.h>
#define INTERVAL_LEITUR 250 //(ms)

#define PIN_BLUE_LED     16
#define PIN_GREEN_LED    17
#define PIN_YELLOW_LED   18
#define PIN_RED_LED      19
#define PIN_BUZZER       21

unsigned int distancia = 0;

#define PIN_TRIGGER   4
#define PIN_ECHO      5

Ultrasonic ultrasonic(PIN_TRIGGER, PIN_ECHO);

#define CLK 14
#define DIO 13
 void configurePins()
{
  pinMode(PIN_BLUE_LED,    OUTPUT);
  pinMode(PIN_GREEN_LED,   OUTPUT);
  pinMode(PIN_YELLOW_LED,  OUTPUT);
  pinMode(PIN_RED_LED,     OUTPUT);
  pinMode(PIN_BUZZER,      OUTPUT);
}
void loop()
{
  verificarDistance();
  delay(INTERVAL_LEITUR);
}
int getDistance()
{
  int distance;
    long microsec = ultrasonic.timing();
      distance = ultrasonic.convert(microsec, Ultrasonic::CM);
    return distance;
```

Fig. 6 Program code in Arduino IDE for working of LEDs according to the distance calculated by ultrasonic sensor

This value is then compared with the lightInit value. If lightVal value is lower than lightInit, then the output pin is set high, else output pin is set low.

Code 4. Figure 13 shows the program for controlling the LEDs by using photosensor.

3.5 Program to Control Motors

For connection, see Fig. 14, and for program, see Fig. 16.

The flowchart of the code for the control of motors with motor driver circuit is shown in Fig. 15.

Description of code. Four pins are chosen: leftForward, leftBackward, rightForward and rightBackward. All the four pins are set as output pins. For motor operation functions, refer Table 1.

Code 5. Figure 16 shows the program for controlling DC motor using motor driver circuit.

```
homeauto_ultrasonic §
int getDistance()
{
  int distance;
    long microsec = ultrasonic.timing();
      distance = ultrasonic.convert(microsec, Ultrasonic::CM);
    return distance;
}
void verificarDistance()
{
     distancie = getDistance();
  display.showNumberDec(distance);

for(int i=PIN_BLUE_LED; i<=PIN_RED_LED; i++)
  {
    digitalWrite(i, LOW);
  }
  digitalWrite(PIN_BUZZER, LOW);

 if( distance <= 5 )
  {
    digitalWrite(PIN_RED_LED, HIGH);
    digitalWrite(PIN_BUZZER, HIGH);
  }
 else if(distance <=20)
  {
    digitalWrite(PIN_YELLOW_LED, HIGH);
  }

  else if(distance <= 40)
  {
    digitalWrite(PIN_GREEN_LED, HIGH);
  }
 else
  {
    digitalWrite(PIN_BLUE_LED, HIGH);
  }
}
```

Fig. 7 Program code in Arduino IDE for working of LEDs according to the distance calculated by ultrasonic sensor

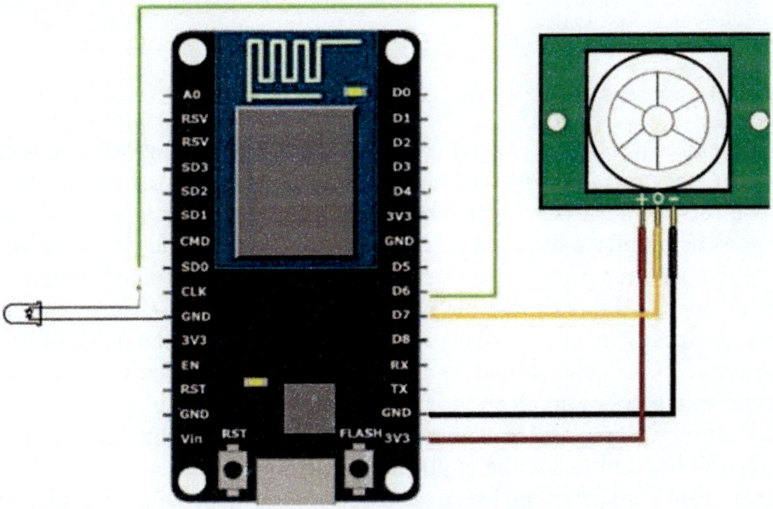

Fig. 8 Connection diagram of PIR motion sensor with NodeMCU

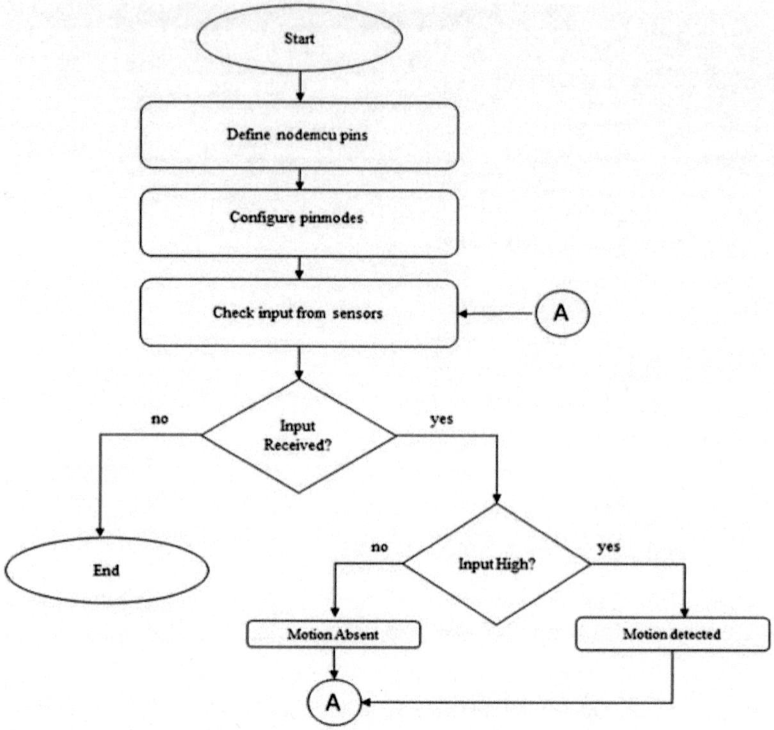

Fig. 9 Flowchart of code for the control of PIR sensor and detecting the motion from it

4 Conclusion

This paper proposes an inexpensive approach toward providing home automation experience to the users [4]. With the help of the proposed methods, traditional devices can be controlled, regulated as well as monitored as the project makes use of physical connections using relays to communicate control signals to the devices giving the ability to the user to perform actions such as resource management, remote access and surveillance. All this is possible with the help of various sensors connected to the home devices. The internet plays a great role in acting as the mode of communication between the control module and the user. The commands are sent using a third-party smartphones application, and thus provides scope of customization for communication with the device. The methods make use of a controller enabled with Wi-Fi capabilities and hence the applications of the device also include its ability to communicate with smart devices over wireless networks. With a microcontroller onboard, the project can be programmed as per the needs, paving a way for future technologies such as artificial intelligence and cloud computing to be incorporated into the project for an even more effective home automation experience.

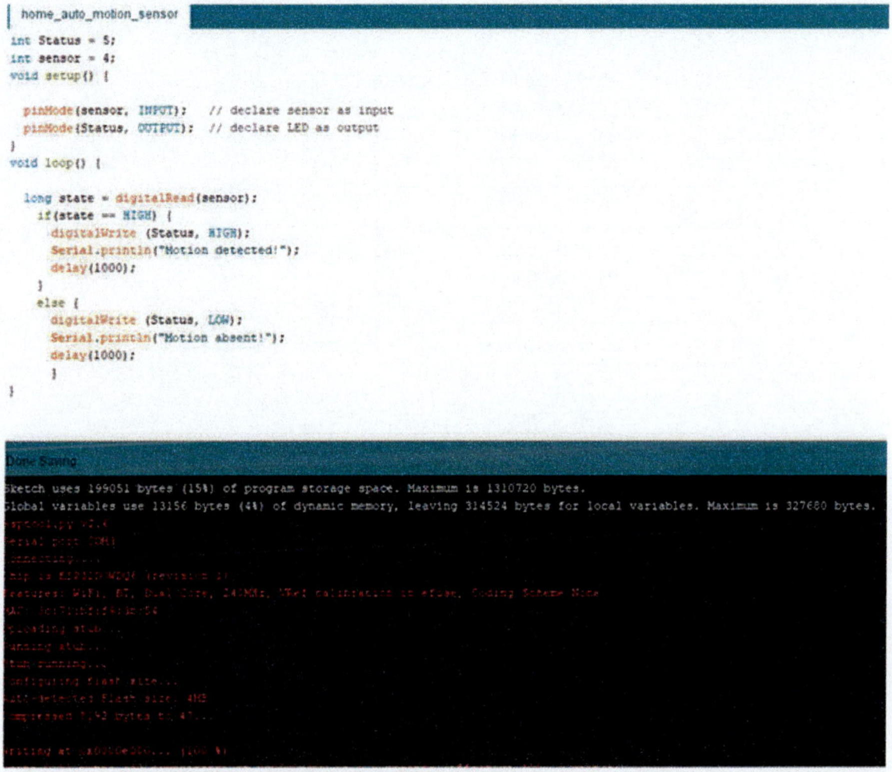

```
home_auto_motion_sensor
int Status = 5;
int sensor = 4;
void setup() {

  pinMode(sensor, INPUT);   // declare sensor as input
  pinMode(Status, OUTPUT);  // declare LED as output
}
void loop() {

  long state = digitalRead(sensor);
    if(state == HIGH) {
      digitalWrite (Status, HIGH);
      Serial.println("Motion detected!");
      delay(1000);
    }
    else {
      digitalWrite (Status, LOW);
      Serial.println("Motion absent!");
      delay(1000);
      }
}
```


Fig. 10 Code for the control of PIR sensor and detecting the motion from it

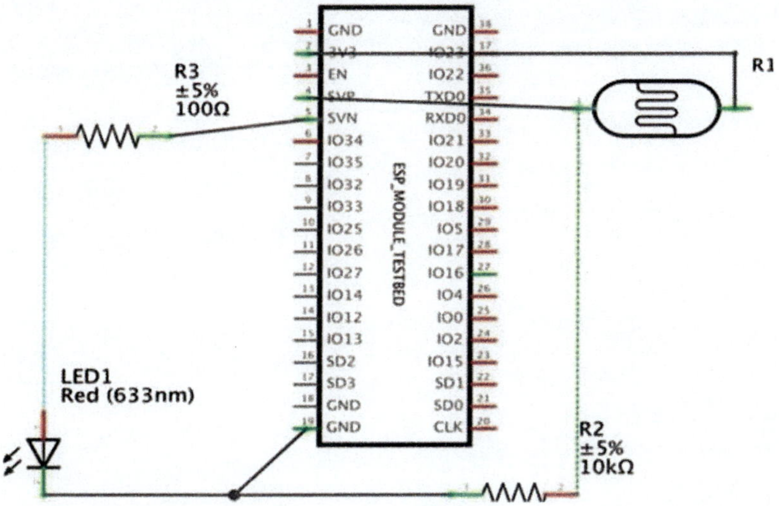

Fig. 11 Connection diagram of photosensor with NodeMCU

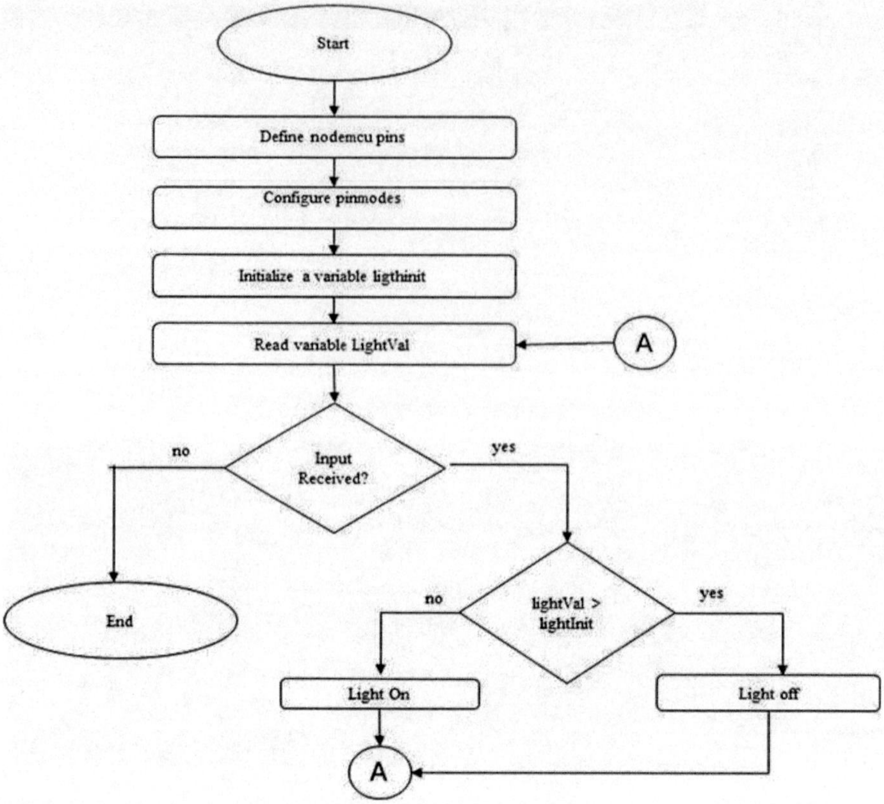

Fig. 12 Flowchart of program for photosensor use to turn off and on the LED

home_auto_lightsensor§

```
const int sensorPin = 2;
const int ledPin = 5;

//We Set up some global variables for the light level an initial value.
int lightInit;  // initial value
int lightVal;   // light reading

void setup()
{
  // We'll set up the LED pin to be an output.
  pinMode(ledPin, OUTPUT);
  lightInit = analogRead(sensorPin);
  //we will take a single reading from the light sensor and store it in the lightCal
}


void loop()
{

  lightVal = analogRead(sensorPin); // read the current light levels

  //if lightVal is less than our initial reading withing a threshold then it is dark.
  if(lightVal - lightInit <  50)
  {
      digitalWrite (ledPin, HIGH); // turn on light
  }

  //otherwise, it is bright
  else
  {
    digitalWrite (ledPin, LOW); // turn off light
  }

}
```

Done Saving

Fig. 13 Program for photosensor used to turn off and on the LED

Fig. 14 Connection of motor with motor driver circuit board

Fig. 15 Flowchart of program for the control of motors with motor driver circuit

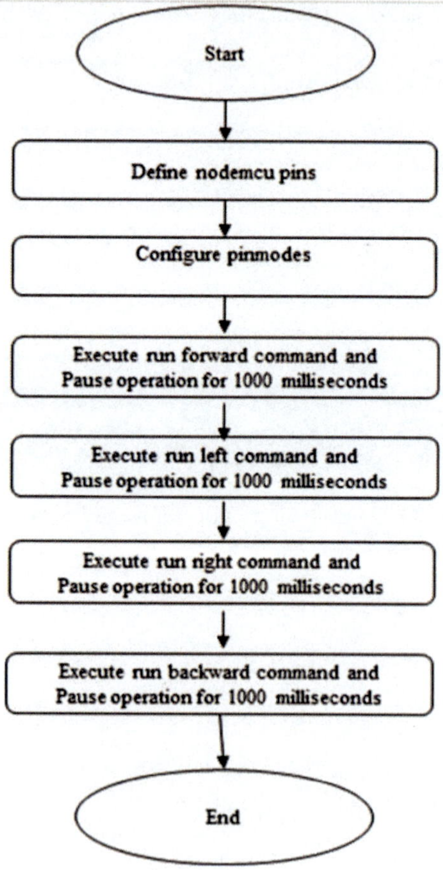

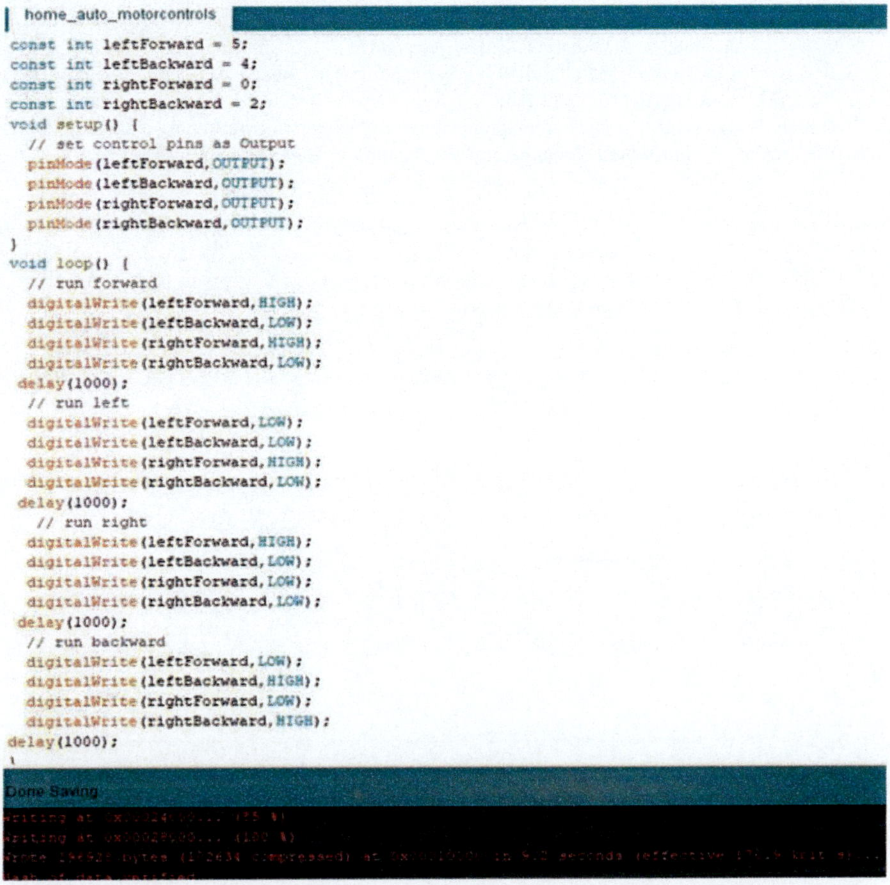

Fig. 16 Program for the control of motors with motor driver circuit

Table 1 Motor operation functions

Operation name	leftForward pin	leftBackward pin	rightForward pin	rightBackward pin
Run forward	HIGH	LOW	HIGH	LOW
Run left	LOW	LOW	HIGH	LOW
Run right	HIGH	LOW	LOW	LOW
Run backward	LOW	HIGH	LOW	HIGH

References

1. A. Adriansyah, A.W. Dani, Design of small smart home system based on Arduino, in *2014 Electrical Power, Electronics, Communications, Controls, and Informatics Seminar (EECCIS)*
2. S.-Y. Chen, C.-F. Lai, Y.-M. Huang, Y.-L. Jeng, Intelligent home-appliance recognition over IoT cloud network, in *2013 9th International Wireless Communications and Mobile Compu- ting*

Conference (IWCMC) (2013), pp.639–643. Accessed 1–5 July 2013
3. R. Mahindar, M. Prakash, S. Ghosh, S. Mukherjee and Dr. R. Ghosh, IoT-based home appliances control system using NodeMCU and Blynk server. Int. Adv. Res. J. Sci. Eng. Technol. (ISO 3297:2007 Certified) **5**(6), 16–22 (2018)
4. P. Mtshali, F. Khubisa, A smart home appliance control system for physically disabled people, in *Conference on Information Communications Technology and Society (ICTAS)* (2019)

Improving Energy Efficiency and Reducing CO_2 Emission of Institutional Building: An Energy Audit Case Study

Arjun Deo and Saurabh K. Rajput

1 Introduction

Availability of power plays a crucial role in the overall development of any country. For this purpose, electricity is one of the core requirements. The conventional methods of energy generation are limited and thus need some form of substitution. Renewable resources being "Clean" and "Green" not only produce minimal amounts of waste but are also available in surplus. As it is said, "Energy saved is energy generated". Thus, instead of increasing the energy generation at much higher cost, one should opt for energy conservation at much lesser cost. Conserving energy not only ensures financial savings but also reduces net carbon emission, thus protecting nature. IIT, Kanpur conducted a project named Samadhan, where they conducted energy auditing in their hostel block to make it energy-efficient. They suggested replacement of rheostat regulator with electronic ones and installation of solar water heater [1].

Institute of Chemical Technology, Akhleshwar, India conducted energy auditing through utility bill analysis of SRICT Institute. They focused on reducing the total electricity bill by reducing the maximum power demand [2]. In India, solar-based energy is the most easily available resource and installing a rooftop PV system can aid in extracting and utilizing the solar energy [3]. The energy analysis of the photovoltaic module was conducted for the first time and they found that in 40 years. The PV module will recover the energy used in its construction and manufacturing. The EPBT (Energy Payback Time) for PV modules was taken as the ratio of total

A. Deo (✉)
Energy Institute Bangalore, A Centre of Rajiv Gandhi Institute of Petroleum Technology, Amethi, India
e-mail: arjun.iitd86@gmail.com

S. K. Rajput
Madhav Institute of Technology Science, Gwalior, India
e-mail: saurabh9march@gmail.com

S. Vadhera et al. (eds.), *Latest Trends in Renewable Energy Technologies*, Lecture Notes in Electrical Engineering 760, https://doi.org/10.1007/978-981-16-1186-5_11

energy required in construction of PV modules to the yearly energy output of the module produced by using sun energy [4].

Chauray and Kandpal have evaluated the CO_2 emission for photovoltaic systems in India. During the study, they found that carbon credit not only reduces pollutant emissions but also the financial burden on the user is reduced by 19% [5]. According to analysis of an off-grid photovoltaic (PV) system for the climate of Varanasi Uttar Pradesh India, it is found that when using the photovoltaic system against the coal-based thermal plant for the generation, reduction in CO_2 emission is 2.525 tCO_2e and the monetary savings due to carbon credit is 143162.19 for a lifetime of 35 years [6]. The proposed work is with the aim to make the institutional building energy-efficient. Conserving energy will not only result in financial savings but also enable to discharge the moral duty of preserving the scarcely available resource.

1.1 Objectives of the Energy Audit Study

1. The first objective is to perform utility bill analysis of the focus institutional building and analyse the results to give suggestions for reducing the annual bill and thus saving energy.
2. The second objective is to conduct an energy audit in class rooms of building for lighting loads.
3. The third objective is to suggest changes to ensure reduction in energy consumed and CO_2 emitted.
4. The final objective is to analyse the rooftop solar PV plant and calculate energy savings and CO_2 emission reduction.

2 Types of Energy Auditing

2.1 Walkthrough Audit

In this audit, we have a walkthrough inspection to check maintenance and operations. It is also used to know the area of further analysis. This audit includes recognizing the energy-saving opportunity. Benchmarking is most important for walkthrough audit; it is a preliminary step by which we analyse the previous consumption and cost and the juxtaposition of the pursuance of the building to that of other buildings.

2.2 *Detailed Audit*

On inspecting the pre-audit results, this audit has an energy consumption survey and energy conservation measures and alternate resources. It involves modification of sources that are capital intensive which requires study [7].

3 Electricity Bill Analysis

To improve efficiency of the building, analysis of electricity bills is done so as to study the energy that is wasted. Tables 1 and 2 and Fig. 1 show various quantities obtained while doing one-year bill analysis.

3.1 *Maximum Demand Control*

A preventive control method to control maximum demand is through maximum demand control meter. By using this, the loads get automatically detached if the demand connected exceeds a present value.

Savings anticipated: Recent MD meter set value is 350 kVA and the average maximum demand is 304.38 kVA. If we take maximum demand as 320 kVA, then monthly monitory saving through maximum demand control will be Rs 12900.

Table 1 Monthly analysis

Month & Year	Total Units Consumed (kWh)	Energy Consumed In Peak Hours (kWh)	Average Power Factor	Power Factor Penalty Rebate (Rs)	Total bill Rs	Average Cost Rs/KWh
Feb-19	70,197	12,084	0.96	−9919.8	680,142	9.69
Mar-19	53,754	8880	0.96	−3798.1	569,191	10.59
Apr-19	94,899	13,821	0.97	−13,430	866,713	9.13
May-19	120,900	16,428	0.96	−16,687.3	1,099,398	9.09
Jun-19	102,957	13,479	0.96	−6661.23	1,000,373	9.72
Jul-19	83,193	11,235	0.95	−5405.07	876,386	10.53
Aug-19	104,349	14,142	0.96	−10,861.6	1,022,016	9.79
Sep-19	114,849	15,552	0.97	−24,790.2	1,123,507	9.78
Oct-19	85,416	12,846	0.97	−19,502.4	848,413	9.93
Nov-19	60,315	17,676	0.97	−13,623.4	632,535	10.49
Dec-19	54,724	10,220	0.96	−8329.8	608,271	11.12

Table 2 Maximum demand charge analysis

Month & Year	Maximum demand (kVA)	Demand Charge (kVA)	Demand Charge Rs	Penalty due to MD
Feb-19	209	360	159,637	0
Mar-19	168	360	159,637	0
Apr-19	365	365	161,787	0
May-19	405	405	181,137	6450
Jun-19	432	432	198,552	23,865
Jul-19	435	435	200,487	25,800
Aug-19	420	420	24,187	0
Sep-19	415	415	200,675	13,800
Oct-19	294	360	170,775	0
Nov-19	260	360	170,775	0
Dec-19	169	360	170,775	0
Jan-20	219	315	144,900	0
Feb-20	166	315	144,900	0
Average	304.38	377.0769	160,632.62	

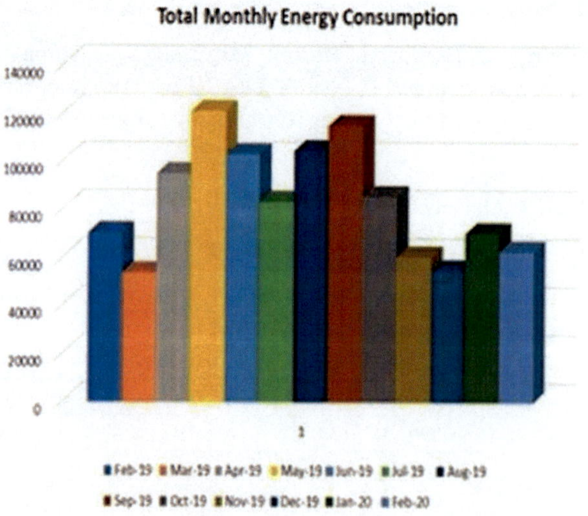

Fig. 1 Total monthly energy consumption

Considering the investment price of MD meter Rs 70000, the payback period is 6 months (approx.).

Also, if maximum demand does not exceed the set maximum demand, all the penalties can be saved here. Therefore, along with the load shedding procedure it is recommended to install a maximum demand meter.

3.2 *Improving Power Factor*

For the equal amount of useful energy transferred, a load which has a lower power factor always draws more current when compared with a load which has a high power factor. A lower power factor is more expensive and comparatively inefficient. Also, linear loads that have lower power factors (example, induction motors) can be improved through with a passive network of few capacitors.

Installing capacitors to increase KVAR ratings:

Power factor = kWh/kVAh

Required capacitance = (Power) × (Multiplying Factor)

Monthly Contracted Demand = 350 KVA

Average Power Factor = 0.96

Targeted Power Factor = 0.99

Power(KW) = 350×0.96 = 336

Power factor correction multiplying factor for desired 0.99 PF is 0.149 [8].

Hence, capacitor rating = 336 × 0.149 = 50.064 KVAR

Thus, capacitors of rating approximately 50 KVAR must be used in order to get power factor of approximately 0.99.

4 Analysis of Class Rooms

During a walkthrough audit, it was found that one old academic block uses fluorescent tube light (FTL) and old ceiling fans in many class rooms and corridor area. FTL is rated for 40 watt and illuminates the same lux level which is produced by 12 watt rated LED light.

We found a total of 180 FTLs and 100 old fans, and observed that wattage of both FTLs and fans was huge resulting in more energy consumption.

4.1 *Analysis of Replacing Conventional FTL with LED Lights*

Rating of each FTL = 40 W

At an average light used for 330 days in a year

Power consumption of each FTL per year = per consumption per day × total days for which FTL is used in a year = (40 × 12 × 330)/1000 = 158.4 kWh

Annual expenditure while using FTL = power consumption per year × energy cost per unit

= 158.4 7.1 = Rs 1,124.64

Rating of each LED = 12 W

Power consumption per year = (1212330)/ 1000 = 47.52 kWh

Annual expenditure = 47.52 × 7.1 = Rs 337.392

Annual energy saved = 158.4 - 47.52 = 110.88units

Annual savings in electricity bills = 110.88 × 7.1 = Rs 787.248

Total units saved for 180 LED bulbs = 110.88 × 180 = 19,958 units/year

Total annual savings on electricity bills after replacing all the 180 FTLs with LED bulbs =19,958 × 7.1 = Rs 1,41,704.64

1. The initial investment for installation of one LED bulb is Rs 500, so the payback period for the above-mentioned replacement becomes 8 months only.
2. Since for electricity emission factor is 0.85 kg CO_2 per unit [9], so the CO_2 emission saving is 16,964.3 kg

4.2 *Analysis of Replacing Pre-existing Fans by Modern Energy Efficient Fans*

Rating of old fans = 75 W

Rating of new fans = 50 W

Average use of fan per day = 12 h

Total power consumption by 100 old fan per month= (75 × 100 × 12 × 30)/1000 = 2,700 kWh

Total power consumed by 100 new fans per month = 900 kWh

Saving in Rs per month = 900 × 7.1 = Rs 6,390

Considering fans are used for 8 months in a year

Total energy saved 8 months in a year is,

1. Annual savings = 6,390 × 8 = Rs 51,120
2. Since for electricity emission factor is 0.85 kg CO_2 per unit [9], so CO_2 Emission saved = 7200 × 0.85 = 6,120 kg

Thus, it is clearly visible that the replacement can result in huge savings.

5 Analysis of Rooftop Solar PV Plant

The rooftop area of a section of building is used for the generation of electricity by installation of photovoltaic (PV) systems. This generation of electrical energy (kWh) partly fulfils the requirement of electrical energy of the mill and also the penalty of exceeding maximum demand as over connected load is reduced.

Table 3 Energy generated by the PV system

Month & Year	Maximum Demand (kVA)	Energy Consumption by load (kWh)	Energy taken by grid (kWh)	Energy generated by PV (kWh)
Nov, 19	260	68,429.4	60,315.00	8114.4
Dec, 19	160	61,483.04	54,724.00	6759.04
Jan, 20	219	79,448.59	70,094.00	9354.59
Feb, 20	166	73,874.9	62,688.00	11,186.90

The on-grid rooftop solar PV system was installed in the month of October 2019. About a 10 square meter area is required to set up 1 kWh grid connected rooftop solar system. The open area available inside is 57785.0 m^2, out of which 800.0 m^2 area is used for the generation of 80 kWh electricity by solar photovoltaic (PV) systems. Table 3 shows the energy generated by the PV system and its effect on energy taken by the grid.

It has been observed from the last two years bills that the maximum demand of the plant is very high in a few months when the connected load is high. The maximum demand is near to the contracted demand, and sometimes it overshoots the contracted demand. But after the installation of PV plants, the maximum demand has reduced. This is because a part of the energy requirement of load is fulfilled by the PV plant. This reduction in maximum demand not only conserves the energy but also provides the savings in the electricity bills. Analysing using data given in Table 3, average per month energy generated by PV plant is 8,853.73 kWh.

6 Reduction in CO_2 Emission

If electricity is generated from thermal plants, an average carbon dioxide equivalent is 0.98 kg of carbon dioxide per unit. However, in coal-based power plant, losses in transmission and distribution are approx. 40%. So, for a coal-based power plant total carbon dioxide equivalent will be increased and this intensity for generation of electricity will now become 1.63 kg of carbon dioxide per unit at source side [6].

Based on the estimates, approx. 50% of individual carbon footprint is caused by electricity and 17% is caused by lighting only [10].

Table 4 shows the comparison of CO_2 production for 1 kWh of electricity production through different generating sources.

Therefore, to decrease CO_2 emission, the following steps can be taken.

Table 4 Amount of CO_2 emitted while generating 1 kWh of electricity

Source	Grams of CO_2 produced
Coal	955
Oil	893
Natural gas	650
Nuclear energy	60
Hydropower	15
Solar energy	40

6.1 Reduction in CO_2 Emission by Controlling Maximum Demand and Improving Power Factor

Almost 830 g of CO_2 equivalents are released into the atmosphere in the usage of 1 unit of electricity in the country. It can be said that carbon footprint can be minimised significantly if energy consumption is reduced. Based on the availability of time and financial resources, several measures can be used to reduce energy consumption. Out of these measures, one important is reduction in maximum demand and improvement of power factor.

6.2 Reduction in CO_2 Emission by Replacing Incandescent Bulbs with LED Bulbs and Old Fans with New Efficient Fans

LED bulbs absorb very small amounts of energy as compared to incandescent bulbs and they have the added benefit of long lasting and also, they do not contain mercury. The replacement with new LED lighting seems to be expensive based on the initial investment; however, the cost–benefit analysis shows that LED lights easily give a minimum of 6% yearly return in the form of running costs alone. The comparative study between incandescent lamp and LED bubs is shown in Table 5.

Table 5. Comparative Study.

6.3 Reduction in CO_2 Emission by Using Solar Energy

According to the latest data available, daily CO_2 emission global levels are at 406.47 ppm (parts per million). During 2019, growth in India's net CO_2 emissions slowed down drastically. The main reasons for this include slowdown in the expansion of thermal electricity generation and surging of renewable energy. Also, an announcement of a target of 450 GW of non-fossil, i.e. renewable energy was made by Prime Minister of the country, Mr. Narendra Modi. The present scenario of the country

Table 5 Amount of CO_2 emitted while generating 1 kWh of electricity

	Incandescent	LED
Power consumption (W)	50	6
kWh (units of electricity used per hour)	0.05	0.006
Hours of operation per day	10	10
Carbon emissions (tons) per year per lamp	0.152083	0.01825
Reduction in carbon footprint (tons) per lamp		0.133833
Lighting carbon emission (tons) per year per household	6.84375	0.82125
Reduction in carbon		6.0225

has renewable energy generation capacity of around 74 GW compared to a previous target of 175 GW by year 2022 [11].

Annual reduction in carbon dioxide emission is found to be 173.179 tCO_2e

7 Conclusion

After conducting walkthrough auditing, the following conclusions have been drawn:

(1) Using a maximum demand control meter would result in an annual saving of Rs 154,800. The average power factor at present is 0.96. Installing a capacitor of approx. 50 KVAR rating is suggested to bring the power factor approximately to 0.99.
(2) Replacing FTL with LED bulbs will give monetary savings of Rs 1, 41,704.64 per year. And replacing old fans gives savings of Rs 2,556 per year. And also this replacement will result in a reduction of 16.964 tCO_2 and 6.120 tCO_2 respectively.
(3) Annual energy generated by solar panels is 106244.79 units. Reduction in CO_2 emission per year is 173.179 tCO_2e
(4) In our study, it is observed that there is a large scope of improving energy efficiency in the ring frame machine section of textile mills.
(5) Following suggestions are made to reduce carbon footprint:
    (a) Save energy and reduce wastage.
    (b) Renew the lighting system.
    (c) Install solar panels to reduce energy generation by thermal plants.

Applying all the recommendations given will result in total monetary saving of Rs 1,053,398.65 and reduction of 196.263 tCO_2e emission.

Acknowledgements The authors are thankful to Er. Vikas Sharma (accredited energy auditor) and energy audit team of NITRA Ghaziabad (linked to MoT, Govt. of India) for their valuable guidance.

References

1. IIT Kanpur Halls of Residence energy audit Project Samadhan, Group of Environmental and Energy Engineering.
2. Patel, P., Patel, V., Patel, A., Solanki D., Patel, D. (2017) "Energy audit: utility bill analysis in an institute," *International Journal of Electronics,Electrical and Computational System.*
3. Kumari, A., Patra, J., Pal, N. et al. (2020) Impact of solar panel on the transformer performance: A case study. *Iranian Journal of Science and Technology, Transactions of Electrical Engineering, 44*, 1197–1206 (2020).
4. Slesser, M., Hounam, I. (1976) "Solar energy breeders," *Nature* 262, 244–245 1976. J. Clerk Maxwell, A Treatise on Electricity and Magnetism, 3rd ed., vol. 2. Oxford: Claren-don, 1892, pp.68–73.
5. Chaurey, A.A., Kandpal, T.C., (2009) Carbon abatement potential of solar home systems in India and their cost reduction due to carbon finance. *Energy Policy 37*, 115–125.
6. Kumar Rajput, S., Singh, O., (2017) "Reduction in CO_2 emission through PV system: A case study," In: *3rd IEEE International Conference on Nanotechnology for Instrumentation and Measurement*, pp. 1–5.
7. General aspects of energy management and audit. (2005) *Guide book 1 for National Certification Examination for Energy Managers and Auditors*, bureau of energy efficiency, India, Available on www.bee-india.nic.in.
8. Eaton.com. (2020) [online] Available:https://www.eaton.com/ecm/groups/public/%40pub/%40electrical/documents/content/sa02607001e.pdf.
9. Green clean Guide. (2011) Carbon footprint calculation—A small introduction of ISO 14064 - Green Clean Guide. [online] Available : https://greencleanguide.com/calculate-your-carbon-footprint/.
10. Apo-tokyo.org. (2012) [online] Available at: https://www.apo-tokyo.org/publications/wp-content/uploads/sites/5/2012_Sep-Oct_p4-5.pdf.
11. Analysis Carbon Brief. (2019) India'S CO2 emissions growth poised to slow sharply in 2019. [online] Available: https://www.carbonbrief.org/analysis-indias-co2-emissions-growth-poised-to-slow-sharply-in-2019.

Relationship Between Renewable Energy and CO_2 Emissions in BRICS Countries

Totakura Bangar Raju, Astha Sharma, and Vasundhara Sen

Abstract This paper investigates the nexus of renewable energy, non-renewable energy, GDP and CO_2 emissions in a 1990–2017 sample of BRICS countries to analyse the need for better policy framework and future action needed to tackle climate change. Panel Unit Root tests, Panel co-integration tests and long-run estimates that there exists a long-run relationship between CO_2 emissions, GDP, renewable energy and non-renewable energy for the BRICS countries. It was then seen in the panel OLS results that increased renewable energy consumption can significantly reduce CO_2 emissions, whereas the panel FMOLS and panel DOLS results do not show any significant relationship between these two variables. All the three estimation results show that increase in non-renewable fuel consumption would result in CO_2 emission, thus confirming the positive relationship between non-renewable energy consumption and CO_2 emissions. The analysis also suggests that increased presence of renewable energy in the BRICS nations have avoided emissions, but the extent has not been significantly large. Thus, BRICS countries' policymakers should encourage more consumption of renewable energy for better results.

Keywords Renewable energy · BRICS · GDP · CO_2

T. B. Raju (✉)
University of Petroleum and Energy Studies, Dehradun, India
e-mail: bangarraju@gmail.com

A. Sharma
Spectra Fuels, Ahmedabad, India
e-mail: asthasharma2609@gmail.com

V. Sen
Symbiosis International (Deemed University), Pune, India
e-mail: vasundhara_sen@scmhrd.edu

S. Vadhera et al. (eds.), *Latest Trends in Renewable Energy Technologies*, Lecture Notes in Electrical Engineering 760, https://doi.org/10.1007/978-981-16-1186-5_12

1 Introduction

BRICS countries, consisting of Brazil, Russia, India, China and South Africa, are a group having high gross domestic product (GDP) rates and significant energy consumers. These countries also have useful technological innovations, low labour cost and high energy demand [1]. These countries are top CO_2 emitters of the world and among top 10 energy consumers. Therefore, BRICS countries have a responsibility towards climate change and to mitigate GHG emissions. The Koyoto Protocol fixes the responsibility on developed countries, but BRICS has been a significant source of emissions over a decade [2]. The Paris Agreement [3] has brought forward the need for countries to tackle green house gas emissions and BRICS have affirmed their commitment for the same. The energy sector is the primary source of emissions in BRICS except for Brazil, where land-use change and agriculture are the primary sources [4]. In order to support climate change, the BRICS countries need to check the emissions from the energy sector. Coal is the primary source of energy for India, China and South Africa. Oil and gas remain to be a significant source of energy for Russia and Brazil. Thus, fossil fuels in various forms remain to be a significant source of energy for BRICS, and thereby major source of emissions. In the eight BRICS summit held in India on 16 October 2016, the countries brought forward the need for usage of nuclear energy and natural gas in order to meet Paris support agreement. However, BRICS failed to outperform the OECD and EU members in terms of sustainable development climate goals and climate change [5]. Keeping in view the above, there is a need for the development of renewable energy sources in BRICS countries to meet the climate change goals. There are many factors influencing renewable energy in BRICS countries [6]. The factors are geographical location and area, availability of natural resources, prices of resources like coal, oil, electricity and natural gas, economic growth, GHG's emissions and carbon footprint, infrastructure and regulatory framework. Several financial models have been introduced in BRICS for development of renewables [7]. Growth of solar, wind and hydropower has been phenomenal and there has been an increase in investment of funds for the same. However, there is an absence of study between renewable energy, non-renewable energy, GDP and CO_2 emissions from the BRICS countries to analyse the need for better policy framework, and future action is needed to tackle climate change.

2 Literature Review

The literature review studies are related to renewables and CO_2 emissions in BRICS countries. In the early studies, the relationship between economic growth and renewable energy consumption in China for the period 1977–2011 was analysed [8]. From the results, it was found that there is a presence of bidirectional causality relationship between renewable energy and economic growth. The causal relationship between renewable energy and economic growth was also investigated in BRICS countries during the period 1971–2010 [9]. From the empirical results, there was a long-run relationship between renewable energy and economic growth.

The interactions between renewable energy, output and carbon emissions have been analysed for China and India using VECM analysis for the period 1972–2011 [10]. The results suggest unidirectional causality between CO_2 emissions and renewable generation and output for India. For China, the results show a unidirectional relationship between output and renewable energy. The relationship between environment, health, energy consumption and influence on BRICS's economic growth has been analysed in [1]. The study brings forward the strong relationship between energy, economy, health and environment. Need for agricultural technology to be accessible to landowners and tenants at subsidised rates has been brought forward. Panel data analysis was used to examine the factors affecting BRICS countries for CO_2 emissions [11]. The interactions among economic growth, FDI, CO_2 emissions and energy consumption were investigated, and indirect relationship between CO_2 and energy consumption was established. A comparative study of Turkey's energy performance between BRICS and OECD countries has been done using multiple criteria [12]. DEA analysis has been used along with the Malmquist productivity index for making differences between technical improvements and efficiencies over a period. Encouragement of renewables was analysed by examining the relationship between renewable energy, agriculture and CO_2 along with non-renewable energy and output during the period 1992–2013 [13]. Results show a negative relationship between CO_2 emissions and renewables and output. The nexus between GDP, CO_2 emissions, renewable energy and natural gas study is attempted in the BRICS countries during the period 1985–2016 [14]. The effects of financial development, economic growth, trade openness and capital on energy consumption for BRICS countries have been investigated for the period 1991–2013 [15]. From the empirical evidence, there is a strong presence of co-integration between the variables; presence of unidirectional causality between energy consumption and economic growth; and bidirectional causality between financial development and energy demand. The relationship between CO_2 emission volumes, lag of emissions and GDP for the period 1980–2011 of BRICS countries is analysed [16]. The environmental consequences due to economic activity were found to be mixed, and examination on case-to-case basis was envisaged. NARDL bounds approach was used to investigate the asymmetric impact of economic growth and globalisation on the consumption of energy in BRICS countries [17]. Results indicate that positive shocks stimulate the consumption of energy and adverse shocks to diminish the consumption of energy. The EKC curve was estimated for CO_2 in India and the role of renewable energy generation in India was investigated [18]. Renewable energy was found to give a negative impact on CO_2 emissions, and trade volume was also uncovered to have an adverse impact on CO_2 emissions. From the above literature review, it is found that in most of the studies, renewable energy was found to be negatively impacting CO_2 emissions, and non-renewable was impacting positively on CO_2 emissions. However, for studies related to BRICS, absence of model exploring the relationship between renewable energy, non-renewable energy, GDP and CO_2 emissions was found to be the gap in the literature review.

3 Methodology, Model and Data

The objective of the study is for the BRICS region, to understand the impact of renewable energy, non-renewable energy and GDP on CO_2 emissions. For the empirical analysis, the multivariate equation is given as

$$CO_{it} = (RE_{it}, NRE_{it}, GDP_{it}) \tag{1}$$

After converting all variables into logarithmic structures, the equation can be taken as

$$CO_{it} = \alpha_i + \partial_i t + \beta_{1i} re_{it} + \beta_{2i} nre_{it} + \beta_{3i} gdp_{itv} + \epsilon_{it} \tag{2}$$

where CO is the CO_2 emissions, RE is the renewable energy consumption, NRE is the non-renewable energy consumption, and GDP is the gross domestic product of the country based on the five BRICS countries. Note that t is the time period from 1990 to 2017. α is the constant term, ∂ is the deterministic trend, ε is the error term and β is the elasticity impact on CO_2 by other variables.

3.1 *Long-Run Estimates, Panel Unit Root Tests, Panel Co-integration Tests*

In order to confirm the order of time series data integration, the variables are investigated by unit root tests. The article would use unit roots tests by [19], W-statistics [20], Phillips-Perron and the Fisher augmented Dicky fuller test [21]. To understand the long-run equilibrium relationship among variables, panel co-integration tests are deployed. After testing for panel unit root tests, the co-integration is examined by using Pedroni residual co-integration test [22, 23] and residual co-integration test by Kao [24]. This is based on the residual strategy of Engle and Granger [25]. Even though long-run co-integration relationships could be explored using panel co-integration tests, but long-run estimates could not be calculated. Therefore, in order to estimate the relationship from Eq. (2) among the variables, the article uses the FMOLS and DOLS methods to validate long-run equilibrium relationships. The approaches of DOLS and FMOLS can control serial correlation of the long run and endogeneity of the regressors [26].

3.2 *VECM Granger causality*

In order to check the directional causalities among the variables, the VECM Granger causality test is employed. The said test [25] is used to explore the short-run and long-run causality among the variables in the time series data. The first one is to find the short-run causalities using the Wald test in dynamic VECM and long-run relationships established by examining the residuals in Eq. (2).

3.3 Data

See Table 1.

4 Analysis

From the above descriptive statistics (Table 2), it is evident that China has the highest growth rate of CO_2 emissions at 5.353% followed by India at 5.173% and Brazil at 3.33%. It is interesting to note that Russia has negative growth rate of −1.337% of CO_2. China has the highest growth rate of renewables of 11.328%, followed by India at 6.502%, and South Africa has the least growth rate at 0.45%. Russia is having a negative growth rate of non-renewables at −0.765 and China leading at 5.353%, followed by India at 5.162. China leads at high GDP rate at around 11%, followed by India at 4.9%, and South Africa is least at 0.009%.

Prior to estimating the models, we estimate the overall correlation structure in the given data (Table 3). It can be seen that the variables are significantly correlated with each other. However, the existence of correlation alone cannot be treated as proof of any long-run relationship among the variables. Hence, we proceed towards estimating the possible co-integration within the given dataset for BRICS countries. We implement both Pedroni's [22, 23] and Kao's co-integration test [24]. However, both these tests require the variables to be I(1) to test for possible co-integration. Towards this, we employ four-panel unit root tests. The results are shown in Table 4. From the results, it can be seen that the null of non-stationarity cannot be rejected at the level. Further, all the variables are found to be I(0) at the first difference. As all the variables are found to be non-stationary of the same order, we proceed towards testing for possible co-integration in the given panel data (Table 5).

From the Pedroni co-integration test results, it is evident that with the alternative of common AR coefficients, both panel PP statistic and panel ADF statistic reject the null of no-co-integration. Similarly, with the alternative of individual AR coefficients, both the group PP statistic and group ADF statistic reject the null of no-co-integration. The Pedroni test results conclusively prove the presence of co-integrating relationships in the given panel data. Further, the Kao's T statistic also confirms the co-integrating relationship among the variables. It can be stated that there exists a long-run relationship between CO_2 emissions, GDP, renewable energy and non-renewable energy for the BRICS countries. After confirming the existence of the co-integrating relationship, the long-run influence of these variables on CO_2

Table 1 Data and sources

Variable	Unit	Symbol	Measuring method	Data source
Renewable energy	Terawatt-hours	RE	From wind, solar, hydro, biomass	BP statistical review
Non-renewable energy	Terawatt-hours	NRE	From fossil fuels such as crude oil, coal and gas	BP statistical review
Grow domestic Product	US$	GDP	Real GDP (constant 2010)	World development indicator
Carbon dioxide emission	Million tonnes	CO		BP statistical review

Table 2 Descriptive statistics

Countries	Variables	Mean	Median	Minimum	Maximum	Std deviation	Growth rate in %
Brazil	GDP	9641.256	9117.394	7797.827	11915.417	1358.812	1.187
	Renewable	83.049	76.412	48.519	127.914	24.150	3.734
	Non-renewable	134.931	125.664	77.743	204.069	38.692	3.416
	CO_2	339.179	317.943	197.242	511.871	92.697	3.330
Russia	GDP	8804.740	8980.488	5505.627	11803.708	2244.923	0.903
	Renewable	38.488	38.764	34.942	42.305	1.938	0.537
	Non-renewable	639.705	631.640	559.564	821.855	68.098	−0.765
	CO_2	1598.256	1524.844	1446.970	2234.669	219.748	−1.337
India	GDP	1029.486	876.599	530.894	1963.546	441.239	4.949
	Renewable	31.792	21.437	15.054	74.329	18.042	6.502
	Non-renewable	385.962	337.501	180.807	701.154	161.359	5.162
	CO_2	1271.583	1089.429	603.218	2344.243	540.622	5.173
China	GDP	3102.022	2365.749	730.772	7329.089	2069.994	8.934
	Renewable	139.485	74.234	28.247	474.983	131.814	11.328
	Non-renewable	1599.989	1398.528	655.225	2763.923	783.525	5.574
	CO_2	5579.228	4918.447	2325.992	9232.581	2663.175	5.353
South Africa	GDP	6487.342	6242.985	5423.587	7582.552	830.516	0.009
	Renewable	0.842	0.492	0.033	4.145	1.041	0.450
	Non-renewable	108.243	109.418	87.232	126.122	13.629	0.010
	CO_2	378.313	381.633	303.469	449.337	52.602	1.168

Table 3 Correlation for the panel data

	LnGDP	LnRenew	LnNonRen	LnCO2
LnGDP	1			
LnRenew	−0.032	1		
LnNonRen	−0.221**	0.613**	1	
LnCO2	−0.303**	0.534**	0.989**	1

Table 4 Panel unit root tests

Variable	LLC test (Levin, Lin and Chu test)		LPS test (Im– Pesaran–Shin test)		Fisher ADF test		Fisher PP test	
Level	Statistic	Prob	Statistic	Prob	Statistic	Prob	Statistic	Prob
GDP	−0.8980	0.1846	1.9633	0.9752	4.1618	0.9395	13.5194	0.1961
Renewable	2.0511	0.9799	3.1417	0.9970	7.8601	0.6425	14.8425	0.1379
Non-renewable	−0.9177	0.1422	0.0552	0.5220	9.8755	0.4515	3.7829	0.9566
CO_2	−0.5051	0.3046	0.0792	0.5315	17.3712	0.0665	3.6364	0.9623
First.diff								
GDP	−3.1460	0.0000	−3.2905	0.0050	32.0408	0.0002	39.3505	0.0000
Renewable	−5.6190	0.0008	−6.4552	0.0000	75.0283	0.0000	139.787	0.0000
Non-renewable	−2.0482	0.0203	−2.6570	0.0039	27.2836	0.0023	78.3902	0.0000
CO_2	−1.9513	0.0255	−1.7781	0.0377	23.8705	0.0080	69.8456	0.0000

Table 5 Panel co-integration tests

Pedroni co-integration test: Alternative hypothesis: common AR coefficients				
	Statistic	Prob	Weighted statistic	Prob
v-Statistic	1.590512	0.0559	1.344365	0.0894
rho-Statistic)	−1.303161	0.0963	−1.121267	0.1311
PP-Statistic	−2.698636	0.0035	−2.411809	0.0079
ADF-Statistic	−3.208560	0.0007	−3.193788	0.0007
Pedroni co-integration test: Alternative hypothesis: individual AR coeffs. (between-dimension)				
	Statistic	Prob		
rho-Statistic (group)	−0.398319	0.3452		
PP-Statistic (group)	−2.445250	0.0072		
ADF-Statistic (group)	−3.249396	0.0006		
Kao's residual co-integration test				
T statistic	−3.3653	0.0004		

Note The optimal lag length was estimated using AIC criterion

emissions is estimated using panel OLS, panel FMOLS and panel DOLS methods [26]. The results are given in Table 6.

The R^2 and adjusted R^2 values for all the three models are found to be quite high, indicating that the models were a good fit. The GDP is found to have a significant long-run negative impact on CO_2 emissions as per the OLS and FMOLS results. It can be said that increasing per-capita output would reduce CO_2 emissions and thereby improve environmental quality. This outcome supports the argument that increased economic growth would result in inefficient ways of production and thereby reduce

Table 6 Panel, FMOLS, DOLS and OLS where dependent variable is CO_2

Variables	GDP	Non-renewable	Renewable	R^2	Adjusted R^2
OLS	−0.0847	1.0662	-0.0542	0.9921	0.9919
	(0.000)	(0.000)	(0.000)		
FMOLS	−0.0928	1.068	0.012	0.9996	0.9995
	(0.000)	(0.000)	(0.070)		
DOLS	−0.0518	1.017	0.015	0.9997	0.9996
	(0.107)	(0.000)	(0.213)		

Note p-values in the ()

environmental degradation. All the three estimation results show that 1% increase in non-renewable fuel consumption would result in an almost equal amount of CO_2 emission. This confirms that there is a positive relationship between non-renewable energy and CO_2 emissions. The panel OLS results show that enhanced renewable energy consumption can substantially bring down CO_2 emissions, whereas the panel FMOLS and panel DOLS results do not show any significant relationship between these two variables.

After confirming the long-run co-integration relationship, the next step would be to model the short and long-run causal relationship present among the variables. Towards this, a panel VECM model is estimated [25]. The results are shown in Table 7. The ECT column values are shown to be either positive and significant or negative and non-significant. Hence it can be said that there is no long-run causal relationship among these variables. However, we can identify short-run unidirectional causal relationships between emissions of CO_2 and GDP, non-renewable energy utilization and GDP. It can be said that the reduction of non-renewable energy consumption would result in reduced output. Further, any strategies to reduce CO_2 emissions

Table 7 Panel VECM granger causality tests

Dependent variable	Short run				Long run
Error correction:	D (CO_2)	D (GDP)	D (Non-renewable)	D (Agency)	ECT
D(CO_2)		−0.050649	0.177313	−0.423269	0.037643
		(0.8913)	(0.6632)	(0.9112)	(0.001)
D(GDP)	0.326674		0.325135	1.033281	−0.002262
	(0.0002)		(0.0001)	(0.1892)	(0.8247)
D(Non-renewable)	0.116912	0.111530		0.383492	0.033906
	(0.7916)	(0.777)		(0.9245)	(0.0026)
D(Agency)	−0.005725	−0.004545	−0.004372		−0.027757
	(0.553)	(0.5891)	(0.6364)		(0.7907)

Note p-values in the ()

would result in reduced output. As BRICS countries majorly employ non-renewable energy sources, this outcome is logical and somewhat expected.

5 Conclusion

This study was undertaken to test the presence of a long-run nexus between the key variables of interest, namely, carbon emissions, renewable energy consumption, non-renewable energy consumption and GDP growth, with a specific focus on BRICS nations. We find that a 1% increase in non-renewable energy consumption leads to 1% increase in carbon emissions and any efforts taken to reduce carbon emissions will lead to reduced output for the BRICS countries. The analysis also suggests that increased presence of renewable energy in the BRICS nations have avoided emissions, but the extent has not been significantly large. This could be attributed to the fact that the share of renewable energy sources-based electricity in the total electricity supply mix for these countries is still relatively small. Amongst the BRICS nations, only the two rapidly developing economies of India and China record renewable energy share (as a percentage of total electricity generation) of 26–29%, but for the others (that is Russia, South Africa and Brazil) it stands between 10% and 16% [27]. Thus, the renewable energy generation portfolio in all BRICS countries should expand for results to really show. The high tariffs offered to renewable energy generators in the past have meant financial stress for power purchasing utilities. Hence a move is now seen to adopt power purchase auctions. In an auction system, renewable energy generators compete to sell the power generated at least price to power utilities. India, Brazil and China are amongst the BRICS countries that have moved to an auction system, from a Feed-in-tariffs system. Moreover, the lower realized power purchase rates have given rise to financial concerns for generators. Projects are getting delayed and have, at times, been even declared financially not feasible. Thus, it can be said that while all the BRICS countries have independent renewable energy policies and mandates, there are significant deterrents to high sectoral growth as well. For the BRICS countries like Brazil and Russia, the potential of wind and solar energy has long submitted to the dominance of hydropower resources in Brazil [28] and ample available fossil fuel resources (in Russia). In Brazil, utility-scale wind and solar projects took off after the dedicated programs launched in the early 2000s in an effort to explore non-hydro renewable resources. However, with Brazil turning to auctions, and distributed renewables through solar rooftops picking up pace (supported by a robust net metering policy), the green energy transition is well under way for the country but will remain dominated by hydropower resources. For Russia only 0.03% of total installed capacity came from wind and solar energy as of 2016. The country now stands committed to increase the renewable energy footprint by announcing innovative incentive schemes, not only are feed-in-tariffs being offered to renewable generators, additional payments are also being promised to renewable generator that cover capital expenditure and guarantee a 12% return on their investment over a 15-year horizon [29]. However, very high local content requirements (65% for solar projects from 2019 onwards) restrict the pace of development.

South Africa's Renewable Energy Independent Power Producers Procurement Program has met with tremendous success. The auction process, through which renewable power is being procured in the county has witnessed sharp fall in wind and solar power prices, owing to the falling technology costs. On-shore wind and solar photovoltaic have recorded some of the lowest levellized cost of electricity lending success to renewable energy auctions [30]. South Africa also has a very high footprint of distributed renewables, notably solar rooftops, which has drastically reduced energy poverty and feeds the lighting needs in rural areas [31]. The BRICS countries experience lend to the following conclusions and points to several suggestions on how to increase the renewable energy footprints in these set of nations:

Given the lagging investments in transmission capacities in many BRICS nations, renewable capacities should be augmented through off-grids, mini-grids and micro-grids, especially in the remote and rural areas. However, as of now, funding for such off-grid systems has mainly come from multi-lateral funding agencies and donor institutions. It is essential that a sustainable financing model is designed in order to take off-grid renewable systems to a commercial scale. This remains a key focus area for future research. Public financing of renewable projects can be helpful, where community development of projects can be the model of development. Local community members can participate in project development and can be assured returns on equity investments made from the project returns.

References

1. K. Zaman, A. Bin Abdullah, A. Khan, M.R. Bin Mohd Nasir, T.A.A.T. Hamzah, S. Hussain, Dynamic linkages among energy consumption, environment, health and wealth in BRICS countries: Green growth key to sustainable development. Renew. Sustain. Energy Rev. 56, 1263–1271 (2016)https://doi.org/10.1016/j.rser.2015.12.010
2. R. Leal-Arcas, The BRICS and climate change. Int. Aff. Forum **4**, 22–26 (2013) https://doi.org/10.1080/23258020.2013.824246
3. United Nations, Framework Convention on Climate Change (France, Paris) (2015)
4. C. Downie, M. Williams, After the Paris agreement: what role for the BRICS in global climate governance? global policy **9**, 398–407 (2018). https://doi.org/10.1111/1758-5899.12550
5. M.N. Rahman, A.M. Turay, Climate change issues in BRICS countries. Manage. Econ. Res. J. **4**, 174 (2018). https://doi.org/10.18639/merj.2018.04.678933
6. L. Pathak, K. Shah, Renewable energy resources, policies and gaps in BRICS countries and the global impact. Front. Energ. **13**, 506–521 (2019). https://doi.org/10.1007/s11708-018-0601-z
7. S. Zeng, Y. Liu, C. Liu, X. Nan, A review of renewable energy investment in the BRICS countries: history, models, problems and solutions. Renew. Sustain. Energy Rev. **74**, 860–872 (2017). https://doi.org/10.1016/j.rser.2017.03.016
8. B. Lin, M. Moubarak, Renewable energy consumption—economic growth nexus for China. Renew. Sustain. Energy Rev. **40**, 111–117 (2014). https://doi.org/10.1016/j.rser.2014.07.128.
9. M. Sebri, O. Ben-Salha, On the causal dynamics between economic growth, renewable energy consumption, CO_2 emissions and trade openness: fresh evidence from BRICS countries. Renew. Sustain. Energy Rev. **39**, 14–23 (2014).https://doi.org/10.1016/j.rser.2014.07.033

10. S. Rafiq, H. Bloch, R. Salim, Determinants of renewable energy adoption in China and India: a comparative analysis. Appl. Econ. **46**, 2700–2710 (2014). https://doi.org/10.1080/00036846.2014.909577
11. G.Y. Zakarya, B. Mostefa, S.M. Abbes, G.M. Seghir, Factors affecting CO_2 emissions in the BRICS Countries: a panel data analysis. Procedia Econ. Finan. **26**, 114–125 (2015). https://doi.org/10.1016/s2212-5671(15)00890-4
12. A. Sözen, F. Karık, Comparison of Turkey's renewable energy performance with OECD and BRICS countries by multiple criteria. Energy Sources, Part B: Economics, Planning and Policy **12**, 487–494 (2017). https://doi.org/10.1080/15567249.2016.1200163
13. X. Liu, S. Zhang, J. Bae, The nexus of renewable energy-agriculture-environment in BRICS. Appl. Energ. **204**, 489–496 (2017). https://doi.org/10.1016/j.apenergy.2017.07.077
14. K. Dong, R. Sun, G. Hochman, X. Zeng, H. Li, H. Jiang, Impact of natural gas consumption on CO_2 emissions: Panel data evidence from China's provinces. J. Cleaner Prod. **162**, 400–410 (2017). https://doi.org/10.1016/j.jclepro.2017.06.100
15. K. Ahmed, Revisiting the role of financial development for energy-growth-trade nexus in BRICS economies. Energy **128**, 487–495 (2017). https://doi.org/10.1016/j.energy.2017.04.055
16. V.G. Azevedo, S. Sartori, L.M.S. Campos, CO_2 emissions: a quantitative analysis among the BRICS nations. Renewable and Sustainable Energy Reviews **81**, 107–115 (2018). https://doi.org/10.1016/j.rser.2017.07.027
17. M. Shahbaz, S.J.H. Shahzad, S. Alam, N. Apergis, Globalisation, economic growth and energy consumption in the BRICS region: the importance of asymmetries. J. Int. Trade Econ. Dev. **27**, 985–1009 (2018). https://doi.org/10.1080/09638199.2018.1481991
18. A. Sinha, M. Shahbaz, Estimation of environmental Kuznets curve for CO2 emission: role of renewable energy generation in India. Renewable Energy **119**, 703–711 (2018). https://doi.org/10.1016/j.renene.2017.12.058.
19. A. Levin, C.F. Lin, C.S.J. Chu, Unit root tests in panel data: asymptotic and finite-sample properties. J. Econometrics **108**, 1–24 (2002). https://doi.org/10.1016/S0304-4076(01)00098-7
20. K.S. Im, M. Hashem Pesaran, Y. Shin, Testing for unit roots in heterogeneous panels. J. Econometrics **115**, 53–74 (2003). https://doi.org/10.1016/S0304-4076(03)00092-7
21. G.S. Maddala, Wu. Shaowen, A comparative study of unit root tests with panel data and a new simple test. Oxford Bull. Econ. Stat. **61**, 631–652 (1999). https://doi.org/10.1111/1468-0084.61.s1.13
22. P. Pedroni, Panel cointegration: asymptotic and finite sample properties of pooled time series tests with an application to the PPP hypothesis. Econometric Theory **20**, 597–625. (2004) https://doi.org/10.1017/S0266466604203073
23. P. Pedroni, Critical values for cointegration tests in Heterogeneous panels with multiple regressors. Oxford Bull. Econ. Stat. **61**, 653–670 (1999). https://doi.org/10.1111/1468-0084.61.s1.14
24. C. Kao, Spurious regression and residual-based tests for cointegration in panel data. J Econometrics **90**, 1–44 (1999). https://doi.org/10.1016/S0304-4076(98)00023-2
25. R.F. Engle, C.W.J. Granger, Co-integration and error correction: representation, estimation and testing. Econometrica **55**, 251–276 (1987). https://doi.org/10.2307/1913236
26. A. Banerjee, Panel data unit roots and cointegration: an overview. Oxford Bull. Econ. Stat. **61**, 607–629 (1999). https://doi.org/10.1111/1468-0084.61.s1.12
27. Secretariat, REN21. Renewables 2019: Global Status Report. REN21.2019. Paris, France. (2019) https://doi.org/10.3390/resources8030139
28. A. Bradshaw, Regulatory change and innovation in Latin America: the case of renewable energy in Brazil. Utilities Policy **49**, 156–164 (2017). https://doi.org/10.1016/j.jup.2017.01.006
29. T.A. Lanshina, A. John, V.Y. Potashnikov, V.A. Barinova, The slow expansion of renewable energy in Russia: competitiveness and regulation issues. Energy Policy **120**, 600–609 (2018). https://doi.org/10.1016/j.enpol.2018.05.052
30. D.R. Walwyn, A.C. Brent, Renewable energy gathers steam in South Africa. Renew. Sustain. Energy Rev. **41**, 390–401 (2015). https://doi.org/10.1016/j.rser.2014.08.049

31. A.K. Aliyu, B. Modu, C.W. Tan, A review of renewable energy development in Africa: a focus in South Africa, Egypt and Nigeria. Renew. Sustain. Energy Rev. **81**, 2502–2518 (2018). https://doi.org/10.1016/j.rser.2017.06.055

Proposed Framework for Sustainable Village Strategy in the Semi-Arid Region of Maharashtra, India

Hemraj R. Kumavat, Rohan V. Kumavat, and Hemal V. Bhangale

Abstract Urbanization is a trend that is unavoidable and is in fact a result of a developed economy. In every field, India is witnessing tremendous growth. India has recently launched a smart city program, but when cities in the developing world lack long-term strategic planning, management skills, struggle with infrastructure provision, and high concentration of poverty and slums, it does not seem to be a realistic goal to think of smart cities. Developing countries should concentrate more on their villages rather than being obsessed with the vision of smart cities and try to curb the growth of urbanization. Yet smart village has not made in roads as the cost of developing physical infrastructure in rural areas of low density has been too high, and Panchayats village has little income to support such infrastructure even if it is built through state or federal funding. Sustainable villages can therefore be another good option. The main objective of this paper is to establish a sustainable perspective strategy for a Maharashtra community. The current village was analyzed in the present study in terms of productive resource and energy usage. A questionnaire was distributed to all villagers to determine the overall consumption of electricity, water, and other energy sources. Detailed analysis of the rainfall data and groundwater level survey of the last 10 years has been performed. In addition, the program was also introduced for renewable energy (solar power), toilets with soak pit. A perspective strategy for the sustainable development of that village was established after a detailed analysis of the survey data. Such results will help the government strategize its efforts to achieve prosperity in the village rather than in smart cities.

Keywords Urbanization · Smart villages · Sustainable development · Perspective plan

H. R. Kumavat (✉)
Assistant Professor, Civil Engineering Department, R C Patel Institute of Technology, Shirpur, India
e-mail: kumavathr1981@gmail.com

R. V. Kumavat · H. V. Bhangale
Civil Engineering Department, Veermata Jijabai Technological Institute, Mumbai, India
e-mail: rohankumavat30@gmail.com

H. V. Bhangale
e-mail: c1820002@spce.ac.in

S. Vadhera et al. (eds.), *Latest Trends in Renewable Energy Technologies*, Lecture Notes in Electrical Engineering 760, https://doi.org/10.1007/978-981-16-1186-5_13

1 Introduction

1.1 Engineering for Sustainable Development

Civil engineers have fundamentally changed society's growth throughout history. They have contributed over time to incremental changes in living standards from the Roman viaducts to the highest structures, to the Hoover Dam; civil engineers have undeniably influenced society and the planet. Nevertheless, this development has helped create many issues of the twentieth and twenty-first centuries: poor land use that encourages urban sprawl; industrial processes that lead to excessive energy use and pollution; and destroying or diverting natural rivers that endanger biodiversity and habitats [6, 8, 9].

The Code of Ethics of the American Society of Civil Engineers (ASCE) was established to create a framework on how engineers can work professionally. This Code of Ethics sets out seven basic canons that an engineer will always control their conduct, the first of which introduces us to the idea of sustainable development [10].

"Engineers shall hold paramount the safety, health and welfare of the public and shall strive to comply with the principles of sustainable development in the performance of their professional duties" [10].

1.2 Smart or Sustainable Village

The word "smart" would involve creating more effective institutional and governance systems that aim for self-sustainability, both in economic as well as in the production and consumption of energy and resources. Smart villages can be a better option for controlling environmental degradation and climate change, especially in developing and developed countries. One of the main causes of urban growth or urbanization is migration from rural to urban areas for better job and job prospects and better facilities, as can be seen in India, in particular, over the past two decades [1]. So sustainable villages can be another good option. Any society with less use of resources and sustainable behavioral practices can help control environmental deterioration, achieve a lower carbon footprint, and thus help to minimize greenhouse emissions and climate change. Villages are more robust and prosperous than cities [7].

1.3 Need for Sustainable Development

Over the past decade, the world has woken up to rapidly depleting non-renewable resources, biodiversity loss, land destruction, rising air pollution, ozone depletion, rapidly disappearing glaciers, contaminated freshwater supplies, sea land erosion, nuclear waste, electronic waste, increasing deforestation, uncontrolled /

unplanned growth, and broader, unexpected disasters [2]. The CIB (International Building Research and Innovation Council) described sustainability in construction through the Sustainable Building Agenda 21 (CIB, 1999) as achieving sustainable development through environmental, socio-economic, and cultural aspects [5].

Brand and construction concerns relate to optimizing the building process based on local factors such as environment, culture, building practices, and technology level. These also include material quality concerns for a healthy and productive indoor environment, such as energy content, recycling capacity, and reducing emissions. Consumption of resources means maximizing the use of energy and resource utilization, i.e., making the most productive use of actual resources.

1.4 Sustainable Development in India

India listed in the Sustainable Development Index (SDI) a low of 110 out of 149 nations. This rates countries across the 17 global goals that are a collection of ambitious goals across the three dimensions of sustainable development, based on their performance. Economic development, social inclusion, and protection of the environment, backed by good governance [3].

In India, states such as Sikkim, Arunachal Pradesh, Mizoram, Manipur, Meghalaya, Uttarakhand, Assam, Tripura, Meghalaya, and Odisha have made good progress toward sustainable development. However, the remaining states are far from sustainable development, and the environmental degradation is rising.

As the sector is growing rapidly, it poses a host of challenges to preserve the environment. There is now an imminent need in India to adopt green concepts and techniques that can sustainably support growth.

1.5 Objective

The paper focuses on the need for sustainable building in relation to the current world scenario and the role of civil engineer in sustainable design. The paper aims to define person position in sustainability, method of introducing to sustainability, and the importance of critical thinking in achieving the goal.

2 Ease of Case Studıes: A Village in Semi-Arid Region, Maharashtra, India

2.1 Hiware Bazaar

Hiware Bazaar in the state of Maharashtra was a typical semi-arid village from the late 1970s to the early 1990s. It had run out of most of its natural assets-cutting trees, drying water sources, and unproductive land. The village was facing an acute water crisis and was in ruins with its conventional water storage facilities. Only 12% of the land was cultivable in 1989–1990. There was widespread deprivation and there were no job opportunities for the youth. This resulted in large-scale migration to cities, much of which was permanent in the village's total geographic area of 976.84 hectares, in which 795.23 hectares was cultivable. The district's average annual rainfall is 579 mm, though sporadic as well as irregular. In addition, social issues such as alcohol addiction and gambling also plagued the area. According to a study in 1995, 168 out of 180 families lived below the poverty line.

2.2 Ralegan Siddhi: A Village Transformed

Ralegan Siddhi's village was one of India's poorest in the 1970s. It had only one acre per family of irrigated land. Food production was only 30% of the village requirements, and most people migrated to seek work each year. A groundbreaking program of participatory land–water-forest management was undertaken by the Ralegan Siddhi people in the late 1970s. In the early 1970s, Ralegan Siddhi was a poor village with high levels of migration of distress and extreme poverty. Anna Hazare, now a well-known environmental crusader, returned from the army to his home, gradually involving the villagers in regenerating their climate. The excellent transition that followed in Ralegan Siddhi had begun to become evident by the mid-80 s in a short span of only 10 to 15 years.

2.3 Govardhan Eco Village

Govardhan Eco Village (GEV) is a 70-acre sustainable farming village on the Sahyadri Mountains foothills, 108 km north of Mumbai, India. It is located in the Wada district of Maharashtra on 50 acres of pristine farmland. GEV is a modest attempt to stress the importance of living in harmony with nature. The goal is to develop peasant communities as they are the only viable way to lead a physically, mentally, socially, economically, spiritually, and environmentally friendly existence.

Table 1 Village Survey

Name of village	Population parameters			
	Area (Hect.)	Population	House hold	Problems
Vadbare	506.65	846	162	Waste disposal
Panhale	754.82	1387	262	Waste disposal, power cuts
Talwade	398	864	151	Water shortage, frequent, power cuts, waste disposal, sanitation, poverty, unemployment, basic education facilities, etc.
Malsane	368	842	156	
Ganur	953.21	2274	461	
Nanhaw	473	754	139	
Dahiwad	316.46	597	156	
Nimgavhan	221.61	939	189	

3 Methodology

Three villages—Hiware Bazaar, Ralegan Siddhi, Govardhan Eco Village—have introduced various measures such as rainwater harvesting, groundwater recharge, recharge pit development, nallah bonds, water quality contour trenches, use of solar power and biogas to meet energy needs, environment friendly building techniques, waste management techniques, etc. All these measures eventually helped the villagers to grow their economy by increasing yields and milk production, preserving the environment and established ecosystems, while reducing pollution and maintaining good health conditions. Therefore, for the rest of the country, these are the model sustainable villages. They provide a sustainable solution to the ecological issues that we are currently facing worldwide. Until selecting a village for development, various villages will be visited. It has been found that almost all villages face problems such as water shortage, shedding of loads, lack of waste disposal and sanitation facilities, unemployment, deprivation, etc., as shown in Table 1.

Nimgavhan has been chosen for the project after visiting eight villages, such as Ganur, Wadbare, Dahiwad, Nanhaw, Malsane, Panhale, Talwade, and Nimgavhan, as this village faces more problems than others such as frequent power cuts, water crisis, irrigation water shortage, unemployment, and garbage. Also, the village is small as compared to others villages, making it easier to carry out a detailed survey, evaluate it, and identify the small-scale mitigation initiatives.

4 Questionnaire Analysis

In order to find out the problems in Nimgavhan village, a questionnaire was designed. This questionnaire was divided into five parts:

Part A records the responses related to background of respondent including their family details, source of income, etc.;

Part B records the responses related to water supply and consumption, whether it satisfies the requirements of villagers for household as well as irrigation purpose;

Part C attributed for sources of energy used, electricity supply, and load shedding;

Part D and E record the information related to farming and waste management, respectively. This questionnaire was circulated among the villagers to collect the relevant data and 60 responses were recorded. The responses obtained from the villagers are analyzed.

After extensive questionnaire study, it was found that only 20% of families are living in the village and all the families are still living in their own farms. Agriculture is more than 80% of the families' main source of income. Below are the various problems faced by the villagers.

4.1 Water Availability

A minimum Under 44 Gaon water supply system, there is a water tank with a capacity of 25,000 L that is filled once a week or once in 15 days, and then only those villagers who live in the village are supplied with water. According to the review of the questionnaire, only 200–250 of remaining families who live on their farms have their own water sources such as bore wells and open wells. They have plenty of water for farming until the end of December, but after that the wells are dry. Since December, water sources ran dry and land remains unproductive due to a shortage of water. Since January to June, the villagers are facing acute water crisis.

4.2 Sanitation Facilities

Approximately 70% of households do not have toilet facilities, public toilets do not have water, and therefore they are not maintained and cleaned; people are not willing to use public toilets. Therefore, people use the sides of the road for the disposal of excreta and create unhygienic conditions. The waste water from the kitchen and bathroom is allowed to run directly on the ground. Some of the villagers have created soak pits for waste disposal in their fields.

4.3 Load Shedding

In the villages, electricity supply is the biggest issue, as electricity is supplied for only 12 hrs. in both residential and farming areas. For cooking and water heating, villagers use wood, kerosene, and coal as sources of energy that generate emissions.

4.4 *Waste Disposal*

The village does not have a waste collection and disposal system, so villagers dump the garbage on the side of the road and open spaces that create unhygienic conditions. Agricultural waste combustion is another waste disposal practice practiced by villagers that causes a lot of pollution.

5 Results and Discussion

Since there is a lot of water scarcity in Nimgavhan, measures need to be taken to increase the ground water level in a village, as water plays a major role in irrigation. Roof and non-roof rainwater harvesting are proposed as solutions for this, as well as measures to increase groundwater depletion are proposed so that the recharge activities should be undertaken. In this area, the construction of recharge pits for artificial recharge of wells is considered to be appropriate measure for groundwater recharge. Building farm ponds is a good water storage choice. Recharge or deep percolation of groundwater is a hydrological process in which water passes down from surface water to groundwater. Among the various factors affecting a site's rainwater harvesting capacity, the most important are climatic conditions, especially rainfall and catchment characteristics.

Rainfall is the most unpredictable variable in the calculation and therefore, to calculate the potential supply of rainwater for a given catchment, accurate rainfall data is required, ideally for a period of at least 10 years. It would also be much easier to use rainfall data with equivalent conditions from the nearest station. Estimating average annual rainfall in the village: data from the Maharashtra Engineering Research Institute (MERI), Department of Water Resources, Nashik have been collected over the past 10 years, as shown in Fig. 1.

The details of the last 10 years of groundwater level (GWL) were obtained from the Ground Water Survey and Development Agency, Nashik, as shown in Fig. 2.

It was found that groundwater level falls well below monsoon in May, and after monsoon it rises again to 4 to 6 m. Figures 2 and 3 offer detailed analysis of rainfall and groundwater. It is clear that in summer there is a lot of water scarcity as all water sources get dried. Land remains unproductive in summer because water is not sufficient for irrigation, so measures need to be taken to improve energy efficiency and increase the level of groundwater in a village.

The total cost of transforming existing Nimgavhan village into a sustainable village can be estimated by adding the cost of all steps to achieve water quality, energy efficiency, waste management, sanitation facilities, and outdoor climate.

Thus, total cost of converting existing Nimgavhan village into sustainable village is 2,35,73,554.4 (INR), and cost for sustainable development of per hectare area is 1,06,374.05 (INR) shown in Table 2. In order to convert the current Nimgavhan

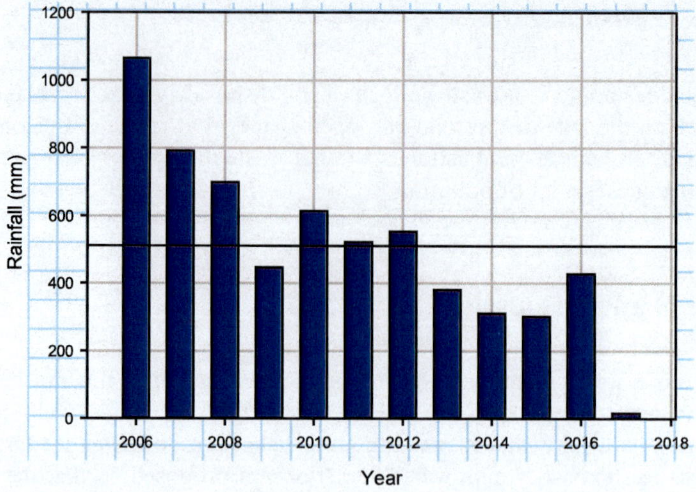

Fig. 1 Year wise rainfall (Nimgavhan Village)

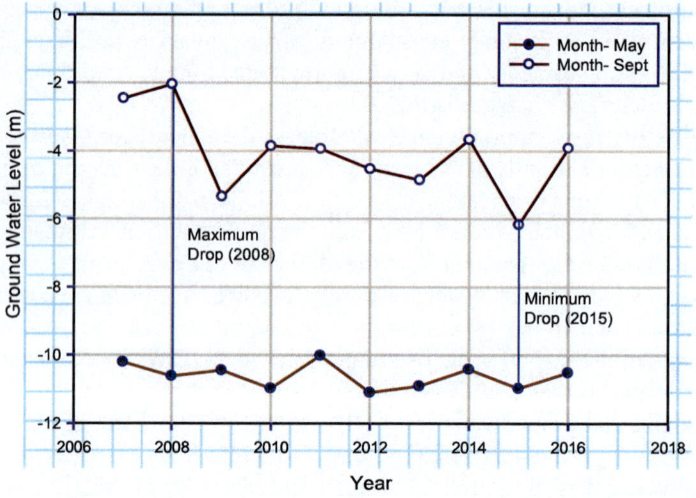

Fig. 2 Average GWL (Meters) of observation Wells in Nimgavhan (Pre and Post Monsoon)

village into a sustainable village, suggested steps along with cost analysis, payback period, and benefits obtained after these measures have been implemented as shown in Table 3.

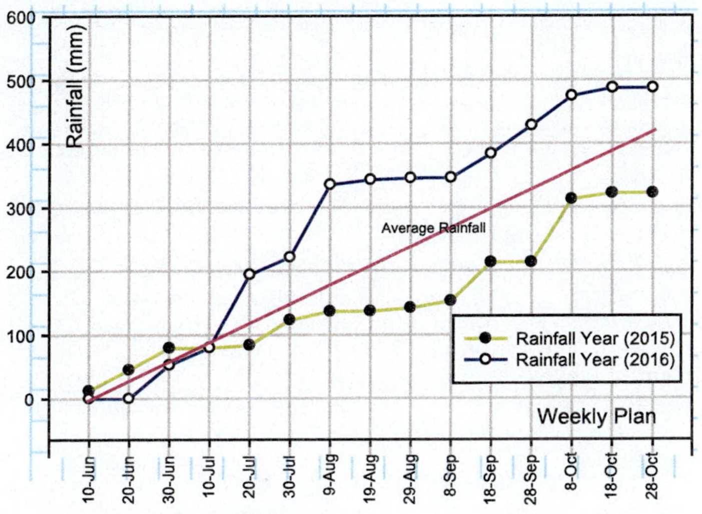

Fig. 3 Cumulative rainfall (Nimgavhan village) (Reference—Government of India Guidelines for Swachha Bharat Mission, Dec 2014) [4]

Table 2 Cost analysis for sustainable village

Suggested measures	Cost (Rs.)	Period (years)
Water efficiency	11,60,504.4	45
Energy efficiency	1,62,35,550	10
Waste management	9,62,500	12
Sanitation facility	12,15,000	12
Outdoor environment	40,00,000	12
Total cost = INR 2,35,73,554.4		
Cost per hectare = INR 1,06,374.05		

6 Conclusion

1. The total cost of transforming the current village into a sustainable village is INR 2,35,73,554.4 and the cost per hectare is INR 1,06,374.05 after a detailed survey and design of Nimgavhan.
2. The main cost was to supply electricity using solar panels of INR 1,62,35,550 and a payback period of 10 years.
3. The cost of implementing water conservation measures in a village is INR 11,60,504.4 and the payback period is approximately 4–5 years.
4. The cost for provision of biogas plants in village is INR 9,62,500 and the payback period is 2 years. The cost for community toilets is INR 12,15,000. The cost for improving outdoor environment quality is INR 40,00,000.
5. The sustainable cost is always reduced than existing cost of village measures.

Table 3 Benefit for sustainable village after suggested measure

Suggested measures	Benefit
Water efficiency – Roof top rain water harvesting for all houses, 10 storage tanks of 10,000 L capacity for storage – Recharge pits –6 – Artificial recharge of existing 10 bore wells	– Increase in groundwater level, water availability for irrigation, increase in yield – Reduced dependency on public water supply scheme – Jobs are created
Energy efficiency – Grid tied solar system – Combined 1 kW system for 2 house (for 60 houses = 30 kW) -2.5 kW system for 129 houses in farms	24 h electricity supply – Reduced load on MSEB Reduced dependency – Jobs are created
Waste management and energy Construction of biogas plants –100 biogas plants of 1 cum. Capacity	Provides cooking and heating fuel – Cheaper source of energy – Problem of biodegradable waste disposal gets solved Reduced dependency
Sanitation facility – Provision of community toilets – Toilet blocks with 5 cubicles for women, 4 cubicles for men and 2 urinals for men (with septic tank)	– Open defecation free village – Prevents unhygienic conditions – maintains clean, healthy environment Jobs are created
Outdoor environment	Maintains clean, healthy environment

6. The use of the proposed measures can provide the much-needed means of improving water quality, energy efficiency, waste management, and sanitation in Nimgavhan. Implementing these steps would create jobs for villagers and therefore lead to economic development. Clean and healthy environmental conditions, which are part of social growth, are preserved by avoiding unhygienic conditions due to open defecation and waste disposal.

Acknowledgements Thanks to MERI (Maharashtra Engineering board), Government of India Guidelines for Swachha Bharat Mission for more valuable help to collection this type of material.

References

1. M. Bilec, R. Ries, H. Scott Matthews, "Sustainable Development and Green Design-Who Is Leading the Green Initiative?", Department of Civil and Environmental Engineering, University of Pittsburgh, Pittsburgh
2. R. Dodds, R. Venables, *Engineering for Sustainable Development: Guiding Principles* (The Royal Academy of Engineering, London, Sept. 2005)
3. P.K. Gautam, Dr. A.S. Jethoo, Dr. S. Shrivastava, S.K. Gupta, "Sustainability in civil construction and the role of civil engineers", Int J Eng Technol. Manag. Appl. Sci. **3**, 101–103 (2015)

4. Government of India Guidelines for Swachha Bharat Mission (2014)
5. Indian Green Building Council (2001)
6. U. Jain, M.I. Faraz, S. Singh, K. Jain, Analysis to convert traditional building to green building. Int J Eng Trends Technol **23**(9), 432–438 (2015)
7. P.W. Jowitt, "Systems and Sustainability: Sustainable Development", Civil Engineering and the formation of the Civil Engineer, The Scottish Institute of Sustainable Technology
8. N. Kavani, F. Pathak, "Retrofitting of An Existing Building Into A Green Building" Int. J. Res. Eng. Technol. **3**(6), 339-341
9. J.D. Rodgers, "Sustainability and Civil Engineering", Ohio Valley Regional Student Conference, Western Kentucky University, (2009)
10. The American Society of Civil Engineers (ASCE) Code of Ethics (1996)

Fuzzy-Based Modeling of Photovoltaic Power Loss Due to Soiling Based on Ecological Parameters

Sujit Kumar, Neel Kamal, Kumar Shri Nivas, Anuj Banshwar, and Naveen Kumar Sharma

1 Introduction

Nowadays, people are mainly focusing on renewable energy resources rather than non-renewable resources for energy consumption due to its scarcity in the coming future. Of all the major renewable resources, solar power is drawing a lot of attention. The solar panel made up of semiconductors restrict its maximum efficiency to 15%–16% only under optimum conditions [1]. Furthermore, the efficiency of solar panels is drastically affected due to environmental factors such as temperature, irradiance, wind speed, and soiling. A report from CSIR-CSIO Chennai suggests that there may be a loss of up to 49.42% in the power if the solar panel is not cleaned for two months [2]. Due to this, soiling on the surface of the panel becomes a major concern. The traditional method of cleaning the panel through automation processes are a bit costly. The soiling problem on the surface of the panel is quite severe which

S. Kumar (✉)
Department of Electrical and Electronics Engineering, Jain (Deemed-to-be University), Bangalore, Karnataka, India
e-mail: sujitvj.kumar@gmail.com

N. Kamal · K. S. Nivas
Department of Electrical and Electronics Engineering, Noida Institute of Engineering and Technology, Greater Noida, India
e-mail: drsaini25aug@gmail.com

K. S. Nivas
e-mail: shrinivas2108@gmail.com

A. Banshwar
Government Polytechnic Tada Gulabrai, U.P, Madhotanda Road, Puranpur (Pilibhit), India
e-mail: anujbanshwar@gmail.com

N. K. Sharma
IKG Punjab Technical University, Main Campus, Jalandhar, Punjab, India
e-mail: naveen31.sharma@gmail.com

S. Vadhera et al. (eds.), *Latest Trends in Renewable Energy Technologies*, Lecture Notes in Electrical Engineering 760, https://doi.org/10.1007/978-981-16-1186-5_14

needs to be taken care properly. The processes used in traditional methods are quite complicated and in comparison with that fuzzy logic provides ease of control. The fuzzy logic will be used as a feedback system that will give information about the soiling on the panel surface [3]. The fuzzy logic controller will provide the signal to the motor driving circuit which drives the wiper attached with the jet sprayer to wipe the panel. After that hardware simulation will be done for the working module and simulation of fuzzy logic for controlling of the removal process. In hardware analysis, the analysis of physical components for designing the cleaning system will be done. In hardware designing, all the physical components will be assembled and tested.

2 Solar Structure

2.1 Panel Description

Solar panel, also named as photovoltaic panel (PV), utilizes semiconductor diode that assimilate solar energy and transduces it into electrical energy (DC) that can also be transformed into AC. The working of the solar panel is based on the principle of the photoelectric effect. Solar panels are formed by connecting several cells. The solar cells are fabricated on a silicon wafer. One part of the wafer is fabricated with an N-Type semiconductor which has an excess of electrons and the other part is fabricated with P-Type semiconductor having holes in excess. The silicon wafer will now act as a diode having a P–N junction [3].

As shown in Fig. 1, the energy of Sun excites electrons in N-doped region and if the energy from the sun is more than the bandgap energy then there will be a bouncing of electrons from the valence band to the conduction band and they move to the P-doped region through the load.

Fig. 1 Solar panel working

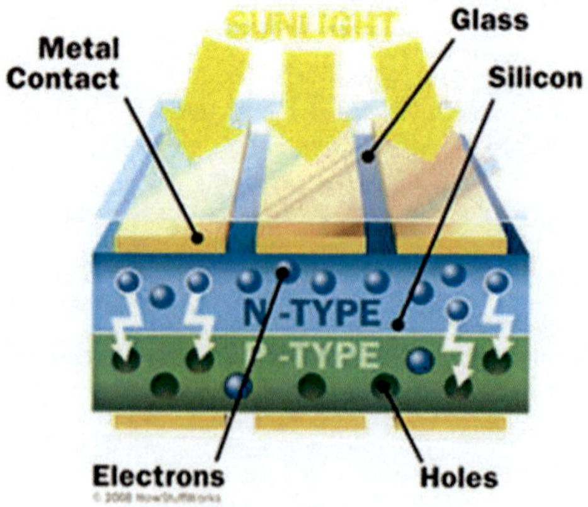

Fig. 2 Solar panel working

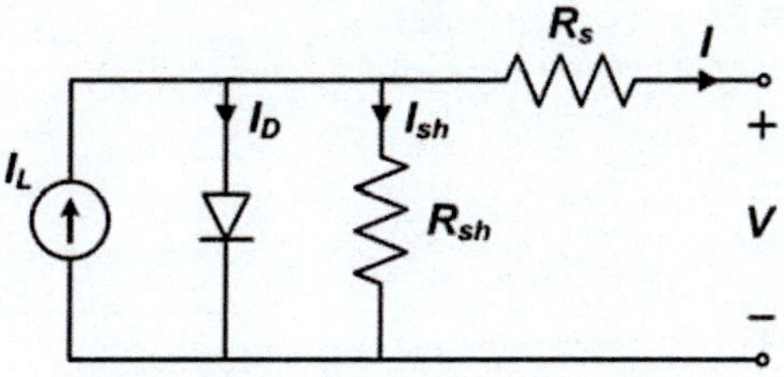

Since the wafer acts as diode, some reverse current will also flow from P to N in the form of diode current. So, the overall current after diode current will flow through the load. The output current value highly depends upon the series and parallel resistances of the cell. Series and parallel resistances are internal parameters of a cell. Ideally, the value of series resistance should be zero and the value of parallel resistance should be infinity. The panel is formed by combining several cells. A metal strip interconnects those cells to each other. Currents are collected from each cell through that metal strip and are sent to the load. The solar panels are also covered headings with glass to prevent the metal strip from corrosion.

The circuit of a solar panel consists of a current source reverse biased with diode. The current source represents the irradiance that is obtained from the sun. The PN junction is represented by diode that is formed in the wafer. From Fig. 2, it can be observed that there are a series and a parallel resistance connected to the circuit. Series and parallel resistances are internal property of the cell. These are developed during manufacturing or transportation. Ideally, the value of series resistance should be zero and that of parallel resistance should be infinity. The output current that will be generated from the cell will consider the effect of these resistances also [4].

2.2 *Panel Simulation*

We have chosen ELDORA 150P (36 cells) panel with a maximum power of 150 watts, the specifications of which have been shown in Table 1. The simulation of the above-mentioned panel is done on MATLAB (Simulink) on Nov 9, 2019 by 3:05 PM in Greater Noida, Uttar Pradesh. The temperature and irradiance at that time were 29 °C and 450 W/m^2. Total output current was given by Eq. 1.

$$I = I_{ph} - I_0\left[e^{(q*(V+IR_s)/n*K*N_s*T)} - 1\right] - I_{sh} \tag{1}$$

where Ns is the number of cells in series, n is ideality factor of the diode, V is the diode thermal voltage, T is the nominal temperature(K), G represents solar ırradiance, q is the Electron Charge, K is the Boltzmann Constant, R_s is the series resistance, R_{sh}-shunt resistance, Ish is the current through shunt resistor which is calculated in Eq. 2

Table 1 Parameters used to solve the equation of PV cells

Parameters used	Value
k_i	0.0032
q	1.6e-19
K	1.3e-23
n	0.5659
E_{g0}	1.1
R_s	0.363
R_{sh}	609.67
T_n	298
V_{oc}	22.58
I_{sc}	8.70
N_s	36
N_p	1

$$I_{sh} = \frac{V + I * R_s}{R_{sh}} \tag{2}$$

The main purpose of developing a simulation model was to find out that what will be the output power under general environment condition. If the output power will be less than that of the power that we find out through simulation then it will be said that there will be soiling on the surface of the panel. The mathematical model has been referred for simulating the panel on MATLAB R2014a as shown in Fig. 3 and the subsystem is shown in Fig. 4.

The output power 150 W power panel will be 69.1935 W. The short-circuit current is 3.9185 A. There is a decrement in the power as of standard test condition (STC) when the system is simulated based on temperature and irradiance of Nov 9, 2019. The output current and voltage characteristics of the mentioned solar panel under the above condition are shown in Figs. 5 and 6.

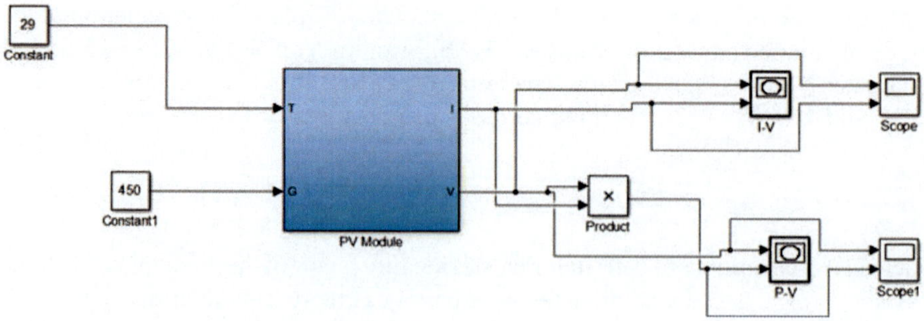

Fig. 3 MATLAB model of simulation

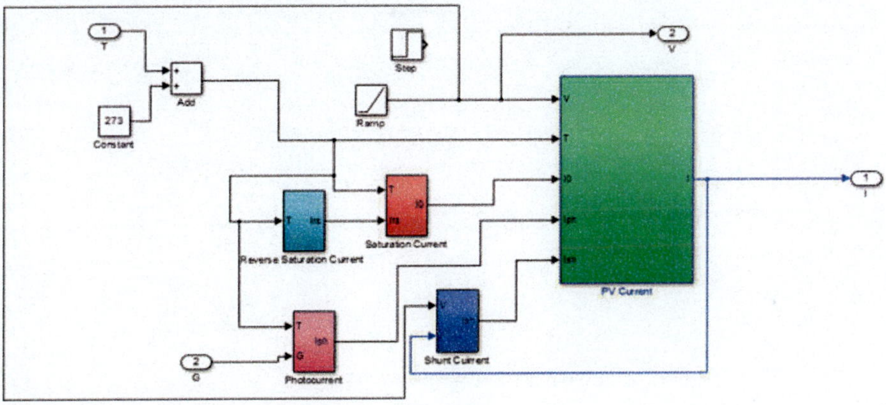

Fig. 4 MATLAB mathematical subsystem

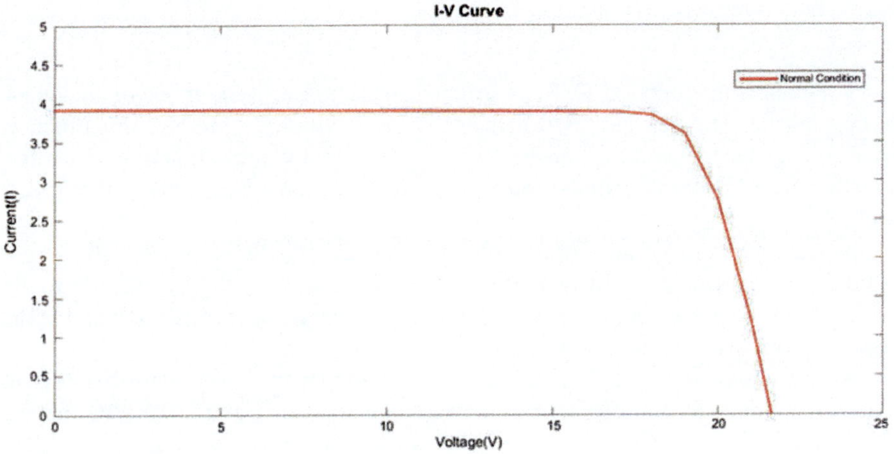

Fig. 5 I–V characteristic of the system under STC

3 Fuzzy Implementation

The benefit of the fuzzy logic control (FLC) is that it tends to be applied in conditions in which the controlled body is excessively unpredictable or hard for the utilization of scientific models [5]. The structure of the controller is communicated to guarantee that it can mirror a framework with a couple of fuzzy standards [6]. However, FLC can be partitioned into four sectors:

(1) Fuzzification
(2) Decision-Making Logic
(3) Fuzzy Knowledge Base
(4) Defuzzification.

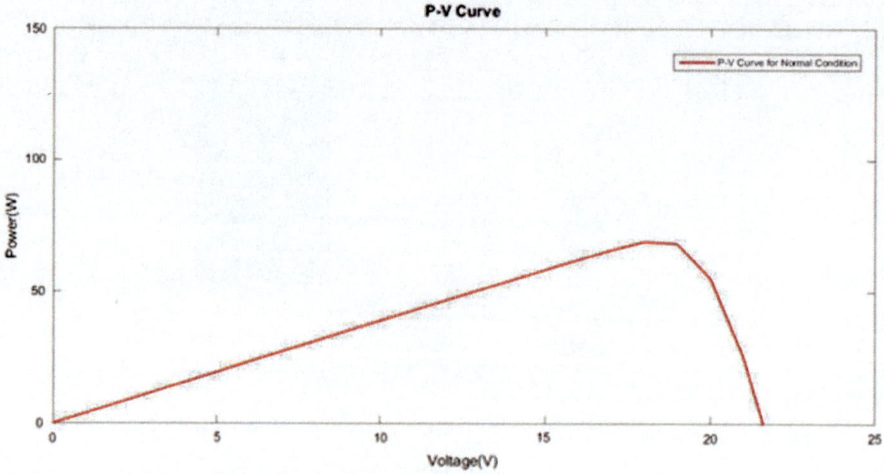

Fig. 6 P–V characteristic of a solar panel under STC

A membership function (MF) is utilized to decrease the number of mistakes, portraying the fuzzy set, and perform quantitative activities [7]. Membership function of different environmental parameters was created with the continuous trigonometric function. The MF for the input–output parts of the system are dilated in Figs. 7, 8, 9 and 10.

Output current and voltage of the solar panel act as input to the system. The output is utilized for washing the surface of panels. Each input variable corresponds to the ownership function value. Collection element represents the horizontal axis for the input variable whose size is represented by its degree of ownership.

Determination of the fuzzy rules is the second phase in the fuzzy controller design. These rules refer to the relationship between the solar cell output current and the illumination from the Sun. Figure 11 shows the rule which was made using data from the previous data.

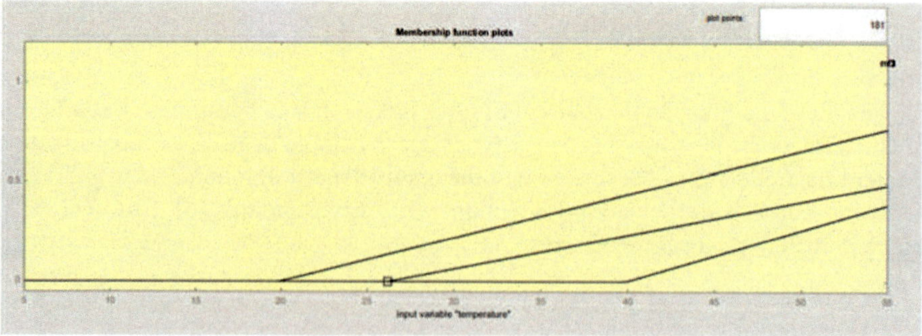

Fig. 7 Membership function of temperature

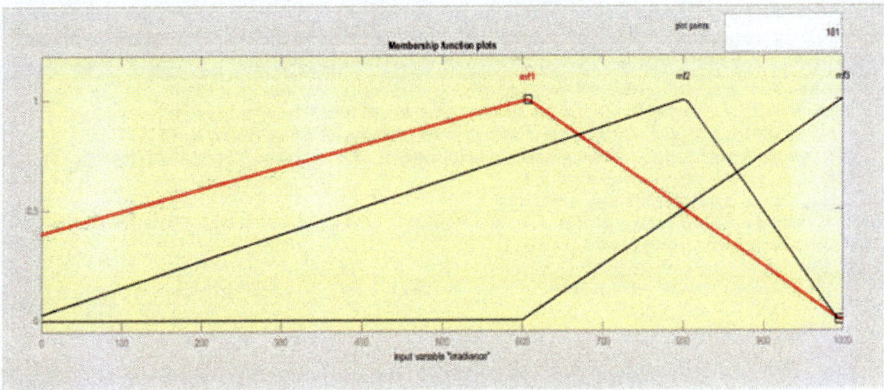

Fig. 8 Membership function of irradiance

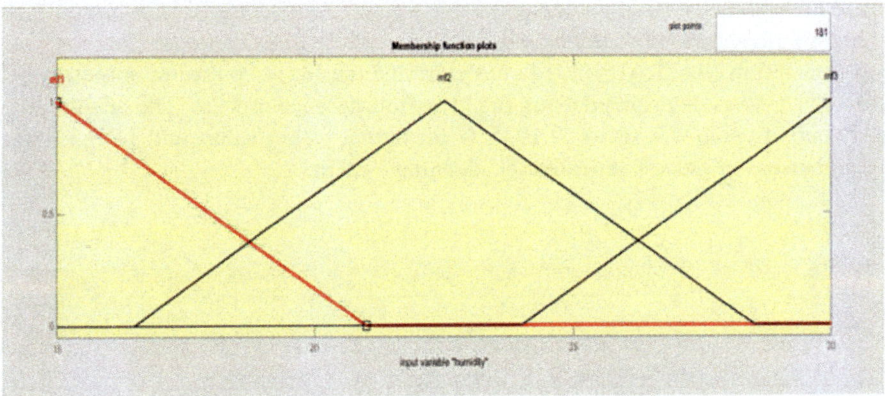

Fig. 9 Membership function of humidity

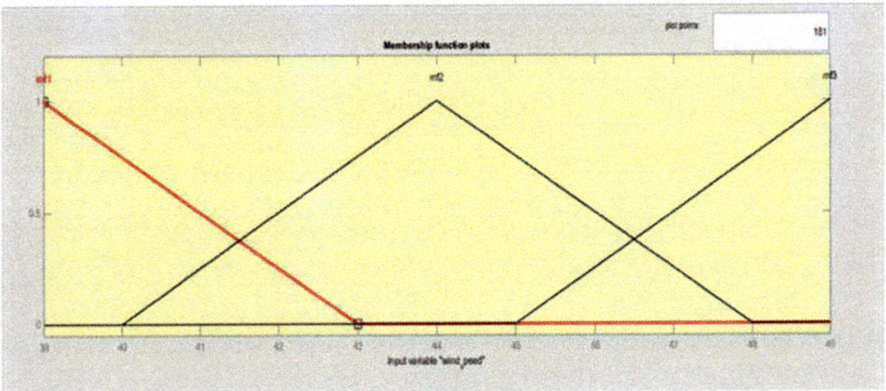

Fig. 10 Membership function of wind speed

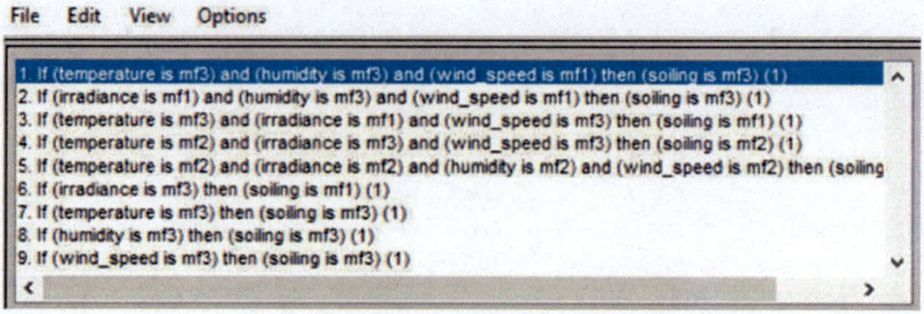

Fig. 11 Fuzzy interference system rules

The type of fuzzy inference engine to be selected is decided in the third step of the fuzzy controller design. Here, Max–Min synthetic fuzzy inference method was utilized.

Figure 12 shows the defuzzification. In this study, Fuzzy Logic Toolbox was implemented in MATLAB R2014a to determine the fuzzy controller output value. Output value was determined using the fuzzy interference process. The offset of the fuzzy output value was set at 69.1935 W for output power value, and 3.9185 A for output current or greater activated the cleaning system.

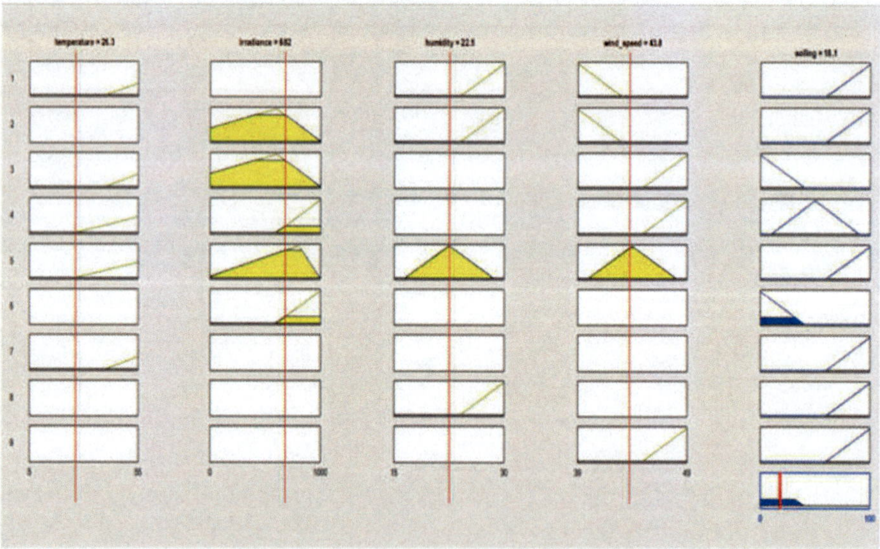

Fig. 12 Fuzzy interference process

4 Conclusion

The proposed prediction of solar panels and washing the surface of the panels is based on FLC. As per the current scenario, no solar panels in the market are furnished with washing system. With changing atmospheric events, soiling can accumulate on surface panels, results in shading, and reduced efficiency. Nevertheless, few particulates are erosive, decreasing the lifespan of the cells.

In this study, the effect of temperature on the changeover competancy is also investigated. In addition to the washing process, a water spraying system has been attached to reduce the ambient temperature of modules. Hence it ensures maximum efficiency of the solar cells. The proposed system can predict the soiling of the panel which was given by a fuzzy inference system. Results also concluded that the fuzzy inference system can effectively predict the dust data with good accuracy and successfully operated the wiper which wipes the panel. This verifies the system has high stoutness when compared to other techniques.

References

1. S. Dhar, R. Sridhar, V. Avasthy, Modeling and simulation of photovoltaic arrays. Proceedings NPSC, pp. 1–5. Kanpur, India (2012)
2. M.A. Hasan, S.K. Parida, An overview of solar photovoltaic panel modeling based on analytical and experimental viewpoint. Renew. Sustain. Energy Rev. **60**, 75–83 (2016)
3. Ibrahim, Haider., Anani Nader.: Variations of PV module parameters with irradiance and temperature. Energy Procedia. 134, 276–285 (2017)
4. K.J. Sauer, T. Roessler, C.W. Hansen, Modeling the irradiance and temperature dependence of photovoltaic modules in PV syst. IEEE J. Photovoltaics **5**, 152–158 (2014)
5. M. Seyedmahmoudian, S. Mekhilef, R. Rahmani, R. Yusof, E. Renani, Analytical modeling of partially shaded photovoltaic systems, Energies **6**, 128–144 (2013)
6. H. Patel, V. Agarwal, MATLAB-based modeling to study the effects of partial shading on PV array characteristics. IEEE Trans. Energy Convers. **23**, 302–310 (2008)
7. J. Carpenter, Y. Ramadass, "Fundamentals of ambient energy transducers in energy harvesting and monthly global solar radiation on inclined surfaces: Models re-visited." Energies **10**, 134–141 (2017)

Application of Plasma Gasification Technology in Handling Medical Waste as an Approach to Handle the Waste Generated by COVID-19 Pandemic

Rohit, Rajneesh Kaushal, and Amit Kumar Dhaka

Abbreviations

GDP	Gross Domestic Product
MSW	Municipal solid waste
PPE	Personal protective equipment
DC	Direct Current
RF	Radio Frequency
CAGR	Compounded Annual Growth Rate
UNEP	United Nation Environment Program
WHO	World Health Organization

1 Introduction

A rapid increase in urbanization is giving rise to more consumeristic behavior in people. The consumeristic behavior is further stimulating the inefficient usage of resources, overconsumption, and generation of waste at a rapid pace [51]. Indian economy is growing very fast and the percentage of urbanization in India has grown from a meager 17.29% in 1951 to 31.16% in the year 2011. Further, it is projected

Rohit (✉) · A. K. Dhaka
School of Renewable Energy and Efficiency, NIT Kurukshetra, Kurukshetra, Haryana, India
e-mail: Rohit_31910110@nitkkr.ac.in

A. K. Dhaka
e-mail: amit_31910120@nitkkr.ac.in

R. Kaushal (✉)
Department of Mechanical Engineering, NIT Kurukshetra, Kurukshetra, Haryana, India
e-mail: Rajneesh41@gmail.com

S. Vadhera et al. (eds.), *Latest Trends in Renewable Energy Technologies*, Lecture Notes in Electrical Engineering 760, https://doi.org/10.1007/978-981-16-1186-5_15

that the majority of the population will be living in urban areas and is set to breach the 69% mark by the year 2050 [34].

Solid waste is undesired substances or materials that are generated after use which has no direct usage value until treated or recycled properly. They are discarded by the people as they may possess hazards to the society or they do not have any usage. Types of solid waste generated in India are listed as Domestic waste or household waste, Municipal waste, biomedical waste or hospital waste, Agricultural waste, Hazardous waste or Industrial waste, radioactive waste. These wastes have been classified under different categories and various methods of disposal and treatment are put in place to handle them effectively [39].

The waste generation in India has seen a rapid increase due to the advance of urbanization in the country. It has been a really difficult task to manage this humongous amount of waste that is being generated on daily basis. Indian cities have seen an exponential increase in waste generation from 6 million tons in 1947 to around 48 million tons in 1997. This is further projected to increase up to 300 million tons by 2047. The annual growth rate of waste generation is seen to be 4.25% [46]. India stands fifth in the list of countries with an overall MSW generation of 56 million tons in 2017–18. This translates to approximate 0.12 million tons per day as per 2016 data [37].

As the GDP of India is increasing along with the rapid increase in population, the generation of MSW has seen a gradual increase. Indian cities are so unplanned and faced haphazard development that around half of MSW remains unattended. This leads to filthy and unhygienic conditions which will further deteriorate the situation and give rise to health and environmental mismanagement [16]. Municipal authorities are held accountable under Municipal Solid Waste (Management and Handling) Rules, 2000 for disposal of waste and its treatment. But these municipal bodies have financial constraints which restrains them from designing suitable strategies and building infrastructure for sustainable waste disposal. This leads to the disposal of waste in dumping grounds that are uncovered, mismanaged landfills, and areas outside city limits that are low lying. This unhygienic disposal is posing a significant hazard for the environment and people [10].

During the last 2–3 decades there has been a phenomenal increase in the generation of medical waste which has raised a serious concern for its disposal and treatment. This rise can be attributed to the use of disposable such as gloves, syringes, PPE kits, containers, etc. Increased usage for these disposables is mostly seen due to precautions against diseases such as AIDS, SARS, MERS, and COVID-19, etc. Mostly the medical waste is generated by hospitals, clinics, and other facilities used for medical purposes and research. As a thumb rule, it has been estimated that the production of medical waste amounts to around 1 kg per bed per day during a shift of 8 h [32] Hospital waste is managed by landfilling, recycling, reduction at source, and incineration i.e. burning under controlled conditions. A study done by Sharma and Kaushal [43] has also discussed gasification as a feasible and economical technology for waste management. There are very high cost associated with the disposal of medical waste which can go up to 100–500$ for disposal of every single ton of medical waste. The market is set to grow $1 billion annually in the US itself [32].

ASSOCHAM study has estimated that India will be generating 775.5 tons of medical waste per day in 2022. The current level of medical waste generation stands out to be 550.9 tons daily. A compounded annual growth rate of 7% is observed. United Nations Environment Program has estimated an increase of 0.5 kg per bed per day during the COVID-19 crisis [24]. China has seen an increase of 200 tons per day of extra medical waste due to the Covid-19 pandemic in Wuhan city only. Similarly, an increase of 300% in medical waste generation is observed in the rural UK [52].

Plasma Gasification offers an environment-friendly solution to treat almost all types of waste [54]. Temperature as high as around 10,000–15,000 °C can be achieved by Plasma in comparison to around 3000 °C in the conventional gasification process. The high temperature in the process accelerates the decomposition by enhancing the chemical reactions. This fast speed of degradation of medical waste can't be achieved by any conventional processes [31].

Research done by the author [29] has revealed in their experiments that processing of biomedical waste with Plasma Gasification can produce synthesis gas with volume percentage ranging from 53.4 to 84.9%. Maximum yield was seen at a temperature not more than 1320 °C.

2 Introduction to Technology: Plasma Gasification

2.1 Principle and Operation

Plasma Gasification is a process that involves the decomposition of biomass into its basic ingredients like hydrogen, carbon monoxide, carbon dioxide, methane, etc. The reaction takes place in a limited supply of oxygen at a very fast speed due to the high temperature involved in the reaction.

Plasma Gasification was first used in the 1960s with carbon electrodes. These carbon electrodes were short-lived as they get decomposed due to the high temperature of plasma involved in the reaction. As the technology developed with time, plasma gasification evolved to be game-changer due to its property of handling almost any kind of wastes. Due to the abundance of ultraviolet radiation present in the thermal plasma, organic chlorine can also be dehydrogenated [32].

Plasma is produced by the ionization of gases which snatches the electrons from the atoms. The molecules dissociate into atoms at 2000 °C while the atoms further dissociate into plasma at temperatures higher than 3000 °C [6].

The electric arc is produced in the Plasma Torch setup which is like a tubular device. The setup shown in Fig. 1 consists of two electrodes that are separated from each other. Plasma provides a heat source that is neither dependent on a feed of materials nor on the supply of air or oxygen. Gas is ionized by feeding electricity to the electrodes. Plasma is thus created which can transfer the heat to the waste material which is fed into the system. The temperature of the feedstock is raised to

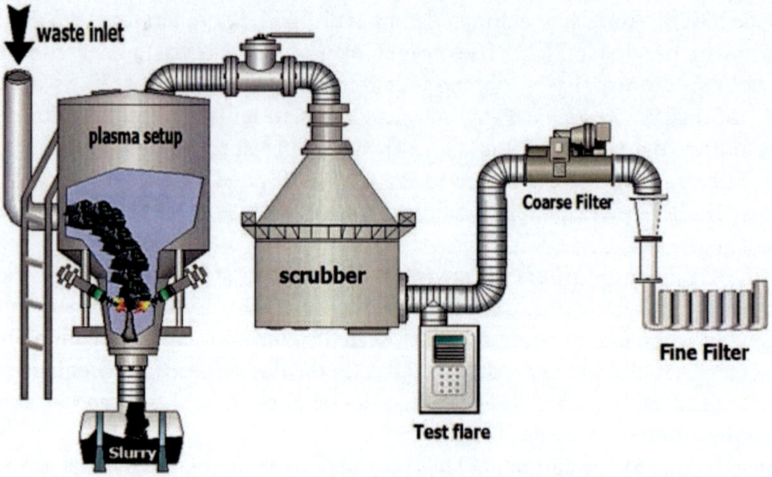

Fig. 1 Plasma gasification setup

3000 °C or higher depending on the power source used. The feedstock consisting of waste is gasified within a fraction of seconds at such extremely high temperatures [41]. The organic compounds are immediately converted into syngas, methane, and carbon dioxide while inorganic compounds and polymers are converted into a glass-like structure. The slag can be easily taken out and further processed and recycled for usage in various industries like road construction and landscaping. Metals can also be recovered from the setup in the molten state which can be cooled and utilized. Volume reduction of waste is as high as 95%, takes place within the system [23].

2.2 *Plasma Generation*

Generation of Plasma takes place in plasma torch setup when electrical energy is supplied to a gas at a particular pressure and temperature. The ionization and excitation of gas take place which means the electrons generated further collides with the atoms of the gas to produce more ions. This process continues till the atoms are excited to the level that conversion in plasma initiates. The resistance of the system helps in the generation of high temperatures which are in the range of 3000–15,000 °C [42].

Various methods have been used to generate the thermal plasma. These methods have been discussed by Bogaerts et al. [2] in their research paper. Commonly two methods have been used extensively to generate plasma as discussed in several research articles, these are the generation of plasma using Direct Current (DC) and Radio Frequency (RF) inductively coupled discharges [15, 33].

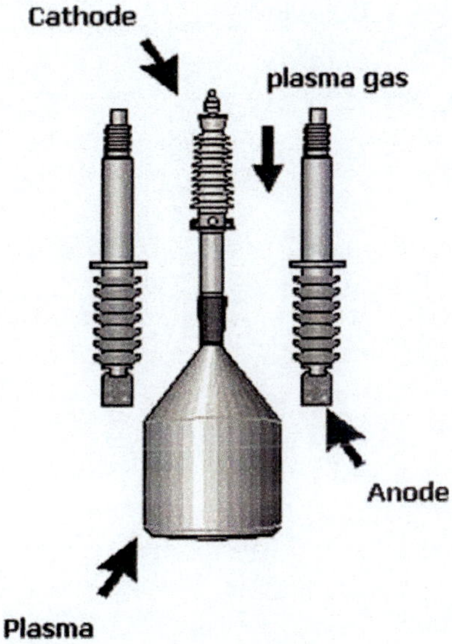

Fig. 2 DC non transferred plasma torch

2.2.1 DC Non-transferred Arc Plasma Torches

In non-transferred arc plasma torches, the torch itself contains both the electrodes i.e. anode and cathode. Direct Current as high as 10^5 A is required to generate plasma in between the two electrodes. The electrical breakdown of gas passing through these two electrodes takes place due to the electrical energy supplied. The gaseous atoms break down into ions and electrons thus leading to plasma generation at high temperatures. The plasma comes out via an opening in one of the electrodes and is generally unstable. The arc needs to be stabilized by using an external magnetic field or by controlling the flow rate of the gas by rotameters [42].

Figure 2 represents the schematic sketch of a DC non-transferred arc plasma torch. It is clearly seen that both the electrodes are located inside the torch itself. Arc is non-transferred as the two electrodes have the sole function of plasma generation only and they don't participate in the processing of the feedstock materials.

DC non transferred torches have several disadvantages associated with the processing of feedstock materials. As the arc is generated in between the two electrodes, the electrodes erode away slowly with the passage of time. This slow erosion of electrodes also contaminates the product. The efficiency can be as low as 50% [3].

2.2.2 DC Transferred Arc Plasma Torches

In transferred arc plasma torches, the body contains only a single electrode as opposed to two electrodes in the case of non-transferred plasma torches. The two electrodes are separated from each other by a distance ranging from a few cm to 1 m [3]. A

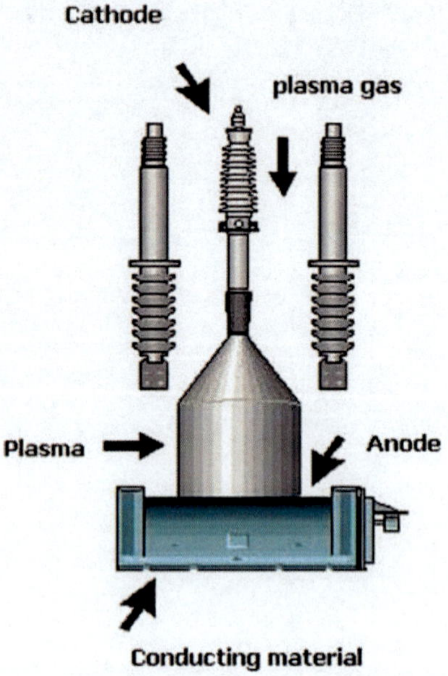

Fig. 3 DC transferred plasma torch

torch is made cathode or anode and is placed concentrically with the axis of the jet. The workpiece is made cathode/anode depending on the electrode configuration of a torch. Therefore to process the waste material, the feedstock is placed in a metallic vessel which acts as the anode. High thermal fluxes are obtained in transferred arc torches as the plasma arc is generated outside the torch body which is water-cooled [13]. This makes the system more efficient than its counterpart as the heat transfer between plasma and water is minimized.

Figure 3 represents the schematic sketch of a DC non-transferred arc plasma torch. It is clearly seen that only a single electrode is located inside the torch itself. Graphite electrodes possess good properties to be used in plasma arc torches. Due to its refractory properties, water cooling is generally not required [38]. Also, it can be used with feed materials where carbon contamination is not a big problem to handle such as for waste materials processing.

DC transferred arc torches with graphite electrodes can be used with diatomic gases like nitrogen. Nitrogen is cost-efficient as compared to argon torches [11]. DC transferred arc torches have the advantage that series of electrodes can be used within the setup to generate plasma [49].

2.2.3 Radio Frequency Inductively Coupled Discharge

In RF Plasma torches, the energy required to convert gas into plasma is provided by an RF power source consisting of inductive or capacitive coils. The feed is injected

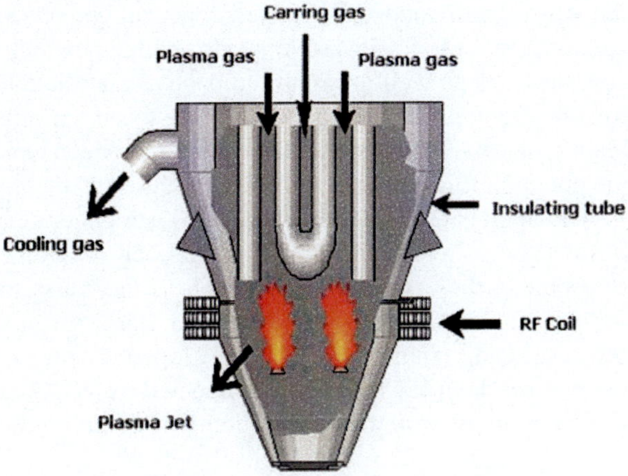

Fig. 4 Radio frequency plasma torch

directly into the plasma region. The electromagnetic field is generated by induction coils which transfer the energy to the input gas [19].

As the electrodes and plasma gas are situated separately i.e. contactless arrangement as shown in Fig. 4, contamination of workpiece materials is avoided, making the setup suitable to operate in almost any conditions ranging from corrosive, reducing, oxidizing, etc. Due to this contactless arrangement of electrodes, various possibilities of investigations are opened which were not available due to rapid corrosion of electrodes at high temperature [48].

RF plasma torches have a very long life as the electrodes are not subjected to high temperatures of plasma. A 120 KW setup can be made commonly available for technology demonstration as well as localized setup of Plasma Torch Gasification. Plant as high as 1 MW has been scaled also [4]. The main disadvantage associated with RF setup is that due to the utilization of oscillator electronics, low efficiencies are observed [49].

3 Medical Waste

3.1 Medical Waste Pattern

Medical waste is waste that comes out from hospitals, veterinary centers, nursing and maternity clinics and centers, medical research facilities, neonatal care units, etc. This waste is generated during treatment and diagnosis as well as immunization of human beings as well as animals. They can also be generated during some research work or production of medicines. Anatomical waste such as body tissues, organs, blood and human body fluids, pathological waste and rejected medicinal waste, cotton dressings, linen is included in these wastes [32]. Medical waste also includes the waste from food used, paper, logs, rubber products, different plastics, glass, ceramics, etc. Medical waste may also include medicinal and chemical reagents and compositions as well as some radioactive materials like those used in oncological treatments, microbiological wastes like tissue culture, waste sharps like syringes, blades, lancets, scalpels, etc. Therefore it has been concluded that any waste that has the potential to spread infections is classified as medical waste [45].

The aggregate amount of medical waste generated per day by a single bed in a hospital varies a lot depending on a number of factors including the type of disease or problem, the care provided as well as the waste management practices that are followed by the hospital. It is observed that it varies up to 1–2 kg per day per bed for developing countries but for developed countries, this value goes up to 4.5 kg per bed per day [45]. The study conducted by WHO for waste management patterns across different countries and found that globally 18–64% of health care facilities don't use appropriate waste disposal methods. This value is 56% for southeast Asian region countries [5].

The amount of medical waste that is generated in India varies greatly with the occupancy ratio of beds in the hospital. A rough estimation has suggested that this data is around 1–2 kg per bed per day same as seen in the case of a typical developing country. In a general physician's clinic, this data is 600 grams per bed per day. Going by this data a typical hospital consisting of 100 beds in India will generate around 100–200 kg per day of medical waste [39]. In general, the view is that of this medical waste produced in developed countries only 10–15% is contaminous but for India 45–50% of waste comes in this category due to poor waste management practices [45].

As per the latest data from the Government of India around 484 tons per day of medical waste is generated in the country. This amount is generated by 1.7 lakhs of healthcare facilities situated all over the country. Of this 484 tons of waste only 447 tons are treated and 37 tons are left untreated as garbage dumped open [8].

International Clinical Epidemiology Network conducted a survey during 2002-04 to assess the BMW situation in the country across different states. Results indicated that 82% of primary, 60% of secondary, and 54% of tertiary healthcare facilities do not have any trustworthy Medical Waste Management System in place [20]. The medical waste management guidelines were adhered to by only 67.2% of public healthcare facilities while a mere 40.4% adhered to it in the private sector at the primary level. But for the secondary level, both have seen compliance of 100% for these rules [9].

New Biomedical waste management rules 2016 has been notified by the Ministry of Environment, Forest, and Climate. These rules are much broader in the sense as they have more coverage, simplified the authorization as well as categorization of waste. These rules have improved the segregation of waste, its transportation, and disposal.

3.2 *Classification*

Medical waste has been categorized as General waste as well as Hazardous Waste. General waste include those waste that does not possess any threat, therefore these type of waste do not require some special equipment and methods for its handling as well as disposal. This waste includes paper, pulp, leather, wood, cloth, food waste generated from the hospital that is non-contaminated.

Hazardous waste is those kinds of waste that have significant risks associated with it during handling and disposal. These waste are handled and treated under specific guidelines of the government such as Biomedical waste management rules notified in India. Mishandling or mismanagement may lead to significant health and environmental hazards.

Hazardous waste is further classified into Chemical waste, radioactive waste, infectious waste. Chemical waste is waste from chemical and biological preparations like medicinal and active ingredients. Radioactive waste has radioactive elements included in that has significant health and environment risk. This waste is generated during the treatment of various diseases and has to handle very carefully. These need to be dumped in landfills by filling them in a lead container. Infectious waste on the other hand has contamination risk associated with it. These include microorganism culture, blood, body fluids, pathogens, etc. They are mostly generated from pathological labs and clinics.

3.3 *Medical Waste Generation Due to COVID-19 Pandemic*

The sudden change in the medical environment due to Covid-19 has increased the demand for personal protective equipment, plastic hand gloves, masks, ventilators, sanitizers, containers, etc. Before this pandemic, the usage of plastic was increasing by a very moderate rate every year. Global plastic production has exceeded 400 million Mt per year [50]. An increase in the plastic medical waste generation of 0.5 kg per bed per day has already been predicted by United Nations Environment Program during the Covid-19 pandemic [24]. Covid-19 pandemic has thrown a unique challenge as the world be seeing a rapid increase in medical waste generation. WHO has estimated that during the Covid-19 pandemic, 8.9 crores mask, 3 crores PPE kits, 16 lakhs goggles, 7.6 crores glove will be consumed in a single month. Segregation of Covid-19 related waste is also an issue in waste management and treatment. The government of India has already issued detailed guidelines for handling Covid-19 related waste.

3.4 Treatment Methods

There are several treatment and disposal methods to handle medical waste. Various studies have been conducted on the treatment of medical waste. These studies gave us a rough idea of treatment methods that are generally used for the treatment of medical waste. Due to its potential risk of contamination spread, incineration is the most common method preferred with a share of around 59–60%. Steam sterilization technology is also known as Autoclaving is used for the treatment of 20–37% of total waste. This technology is particularly useful as it converts hazardous waste into non-infectious ones. Rest other methods like microwave treatment, chemical, and biological treatment, landfilling, plasma pyrolysis has a share of 4–5% [14, 22].

In recent years, Plasma Gasification comes out to be a more promising technology for the treatment of medical waste as suggested by various authors [12, 25, 28, 47, 55].

In India data shows that only 35% of total medical waste generated has access to common waste management facilities. Most of the medical waste is dumped outside the hospital in open landfills and a lot of waste doesn't make up its way to common waste management facilities [18]. Biomedical waste possesses such hazards to humans and our environment that methods designed for its disposal or treatment need to be authentic and totally reliable. Management of biomedical waste should not be taken as only legal responsibility of government but it is a social responsibility for all.

4 Plasma Gasification Treatment of Medical Waste

During recent years a significant change has been observed in medical waste composition due to heavy usage of plastics and polymers. Disease such as AIDS, CORONA has increased the PPE kit consumption along with gloves, surgical masks, etc. The plastic content has seen a rise from 10% to 30% in the medical waste composition during the last decades. This has pushed the heating value of the medical waste generated [21]. Experiments have shown that the average calorific value of biomedical waste varies from 12,550 to 16,740 kJ/kg [27].

Plasma Gasification has evolved as a promising technology to handle the medical waste of all types. Plasma Gasification can kill all the microorganisms in the medical waste due to temperature as high as 3000 °C associated with it. At such high temperatures, ultraviolet radiation is also generated which aids the killing of bacterial and viral loads. Active ingredients of medicines are also destroyed in the process leaving no trace [15]. This technology can be utilized to destroy even cytotoxic medicines which are highly resistant to decomposition. Significant reduction in dioxins, benzopyrene, and furan emissions is also observed [12]. Heavy use of electrical power in range for hundreds of KW to few MW is the only disadvantage associated with the technology. But the syngas produced as by-product in the whole process can have various usage. It can be converted to liquid fuel by the Fischer-Tropsch process. It can also be utilized in running gas turbines or as fuel input in a fuel cell for electricity production. Ammonia, methane, and hydrogen can be produced by using syngas [29]. A lot of studies are being conducted to utilize syngas for running Compression ignition engines. An increase in mileage and reduction in operation cost has been observed [1].

Plasma Gasification technology is relatively new if its implementation for medical waste treatment is observed. Plasma Gasification reactors vitrify the waste in a glass-like substance that has potential usage in construction as well as the infrastructure industry. Scrap metals can be recovered from the waste in a molten form that can be put to reuse [30]. Due to the high temperature associated with plasma, complex structures can be broken into simple forms that are less hazardous to the health and environment. An important byproduct is the formation of syngas that can substitute the fuels used in compression ignition engines [17]. The process flow of Plasma Gasification for the treatment of medical waste is shown in Fig. 5.

A typical Plasma Gasification unit consists of various components such as a system for feeding waste, the chamber for processing waste, handling system of solid residues that also caters for its removal, the management system of the synthesis gas generating, monitoring, and control system [15].

Research conducted by Zeng et al. [53] for co-treatment of medical waste and fly ash in Plasma Gasification so as to obtain a higher value end product with zero waste discharge. Residual slag contains approx. 36% of Oxygen, 28.6% of Calcium, and 16% of Silicon if only medical waste is fed into the Plasma gasifier. Co-treatment of medical waste with fly ash in a weight ratio of 2:1 as shows a similar result. As the co-treatment ratio is increased for the fly ash along with medical waste, the slag obtained thereafter shows an increasing content of transition metals. Increasing

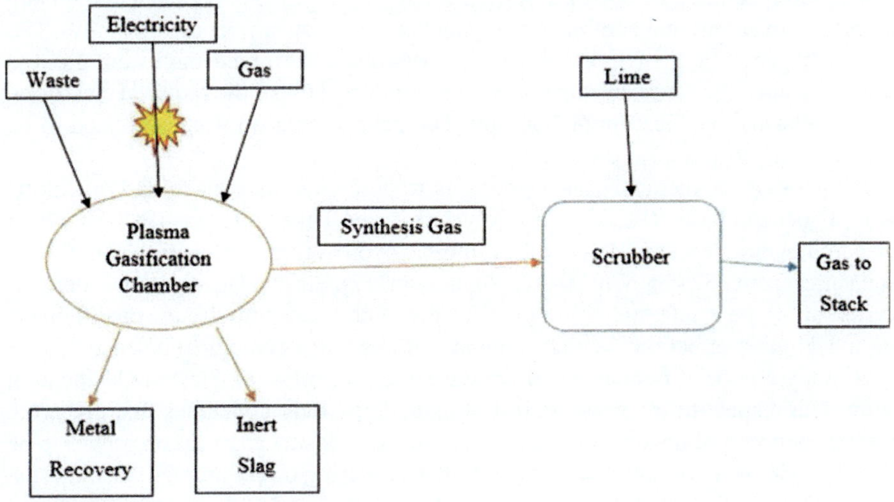

Fig. 5 Plasma gasification process flow diagram

activity is also observed for nitric oxide which goes for selective catalytic reduction. The treatment of fly ash obtained after co-treatment i.e. secondary fly ash by Plasma gasification shows an enriched amount of heavy metals.

Plasma Gasification of waste yields a different amount of synthesis gas at different temperatures. The calculations done by Messerle et al. [29] have shown that maximum yield can be observed at a temperature not greater than ~1300 °C. Bone Tissue has been put to plasma gasification in air and steam medium separately and it has been observed that the concentration of synthesis gas was not more than 54% and 85% (volume %) respectively. The heat of combustion for the synthesis gas is measured to be 3510 kJ/kg and 5664 kJ/kg respectively. X-ray analysis conducted on the byproducts of Plasma Gasification of medical waste has shown no presence of harmful impurities.

An experimental investigation conducted on solid waste which also includes medical waste has revealed that Carbon monoxide (73.4%), Hydrogen (6.2%), Nitrogen (29.6%), and Sulfur (0.8%) as volume percentage composition of synthesis gas obtained. The total concentration of synthesis gas obtained is 69.6% (vol.%) This experimental data obtained matches correctly with the calculations done by the authors [26]. The consumption of electricity for Plasma gasification has been calculated on a unit basis and that comes out to be around 1.92 Kwh/kg for air gasification and 2.44 Kwh/kg for steam gasification.

A study conducted by the author [56] has concluded that processing waste by Plasma Gasification can be a commercially viable venture based on the ecological benefits and energy efficiency of the system. Results of the experiment show that Plasma Gasification is highly efficient in treating hazardous waste with 99.99% efficiency. The technology can accept a comprehensive range of waste materials and is described as a flexible treatment system. 900 Kwh electricity can be produced with each ton of solid waste fed into the Plasma gasifier [36].

Thermodynamic analysis of a system consisting of a Plasma Gasification setup coupled with an internal combustion engine has shown an efficiency of 78.58%. The electricity generation potential for the setup was also determined and it comes out to be 31%, that is, the setup has the potential to produce 31% of the required electricity used in Plasma Gasification technology. The payback period was calculated to be approximately 6 years for the setup [35].

Incineration of medical waste produces toxic compounds, such as Polychlorinated Dibenzodioxins (PCDD) and Polychlorinated Dibenzofurans (PCDF) which are carcinogenic in nature. Polycyclic aromatic hydrocarbons generation in incineration varies from 3377 ng N/m^3 to 11,390 ng N/m^3 [7]. Plasma Gasification is demonstrated out to be a safe technology for the disposal of hospital waste considering it has a negligible effect on the environment and maintains ecological balance.

A study has been conducted to compare the feasibility of Plasma Gasification setup with respect to incineration in Portugal. This study concludes that given the overall condition of health safety and environment, Plasma gasification proves to be more beneficial as compared to incineration. Comparing on economic terms, Plasma gasification has lower or similar costs over a period of 20 years as compared to incineration [40].

5 Conclusion

Medical waste generation has seen a rapid increase during recent decades. India is also observing a CAGR of 7% and medical waste generation in the country is set to cross the 775 tons per day mark in 2022. This is quite a huge number considering the infectious nature of medical waste. COVID-19 pandemic has further worsened the situation as the plastics are increasing as percentage composition in medical waste. WHO and UNEP has further shown concern to handle such a humongous amount of waste and its treatment. In this scenario, Plasma Gasification technology proves out to be a viable alternative for the treatment of medical waste. It has the advantage that it can treat almost any kind of waste with volume reduction post-processing to be as good as 95% is observed. It can reduce the need for landfills and is termed a quite safe technology as it has negligible impact on the environment. The byproduct is vitrified slag which can be utilized for building aggregates, tiles, and brick, etc. Synthesis gas generated during the process can be further utilized to run internal combustion engines [44]. A lot of difficulties such as lack of proper standards, return on investment are hampering the growth of technology. Plasma Gasification requires intervention from the government to make it commercially successful. There is further scope of studying the cost of technology for treating medical waste in India as there is a dearth of work in this field. Further standards and design can be studied to develop a design that caters to the Indian perspective.

References

1. R.P. Bates, K. Dölle, Syngas use in internal combustion engines-a review. Adv. Res. 1–8 (2017)
2. A. Bogaerts et al., Gas discharge plasmas and their applications. Spectrochim. Acta, Part B **57**(4), 609–658 (2002)
3. G. Bonizzoni, E. Vassallo, Plasma physics and technology; industrial applications. Vacuum **64**(3), 327–336 (2002)
4. M.I. Boulos, New frontiers in thermal plasma processing. Pure Appl. Chem. **68**(5), 1007–1010 (1996)
5. M. Capoor, K. Bhowmik, Current perspectives on biomedical waste management: rules, conventions and treatment technologies. Indian J. Med. Microbiol. **35**(2), 157–164 (2017)
6. H. Conrads, M. Schmidt, Plasma generation and plasma sources. Plasma Sources Sci. Technol. **9**(4), 441–454 (2000)
7. S. Das et al., PCDD/PCDFs: a burden from hospital waste disposal plant; plasma arc gasification is the ultimate solution for its mitigation, in *Energy Recovery Processes from Wastes*, ed. by S.K. Ghosh (Springer, Singapore, 2020), pp. 9–21
8. P. Datta et al., Biomedical waste management in India: Critical appraisal. J. Lab. Physicians **10**(01), 006–014 (2018)
9. A. Devi et al., Evaluation of biomedical waste management practices in public and private sector of health care facilities in India. Environ. Sci. Pollut. Res. **26**(25), 26082–26089 (2019)
10. L.F. Diaz, Options for improving solid waste management in economically developing countries. Waste Manag. **29**(1), 1 (2009)
11. C. Ducharme, Technical and economic analysis of plasma-assisted waste-to-energy processes, in *Research Paper I*. School of Engineering and Applied Science, Columbia University (2010)

12. F. Fabry et al., Waste gasification by thermal plasma: a review. Waste Biomass Valorization **4**(3), 421–439 (2013)
13. D.L. Flamm, O. Auciello, *Plasma Deposition, Treatment, and Etching of Polymers: The Treatment and Etching of Polymers*. Elsevier (2012)
14. M.K. Ghasemi, R.B.M. Yusuff, Advantages and disadvantages of healthcare waste treatment and disposal alternatives: Malaysian scenario. Polish J. Environ. Stud. **25**(1), 17–25 (2016)
15. E. Gomez et al., Thermal plasma technology for the treatment of wastes: A critical review. J. Hazard. Mater. **161**(2), 614–626 (2009)
16. S. Gupta et al., Solid waste management in India: options and opportunities. Resour. Conserv. Recycl. **24**(2), 137–154 (1998)
17. F.Y. Hagos et al., Trends of syngas as a fuel in internal combustion engines. Adv. Mech. Eng. **6**, (2014)
18. P. Hanumantha Rao, Report: hospital waste management—awareness and practices: a study of three states in India. Waste Manag. Res. **26**(3), 297–303 (2008)
19. H. Huang, L. Tang, Treatment of organic waste using thermal plasma pyrolysis technology. Energy Convers. Manag. **48**(4), 1331–1337 (2007)
20. INCLEN Program Evaluation Network study group, N. D., India, Bio-medical waste management: situational analysis & predictors of performances in 25 districts across 20 Indian States. Indian J. Med. Res. **139**(1), 141–153 (2014)
21. H. Jianjun et al., Thermodynamic study of water-steam plasma pyrolysis of medical waste for recovery of CO and H2. Plasma Sci. Technol **7**(6), 3148–3150 (2005)
22. B.-K. Lee et al., Alternatives for treatment and disposal cost reduction of regulated medical wastes. Waste Manag. **24**(2), 143–151 (2004)
23. B. Lemmens et al., Assessment of plasma gasification of high caloric waste streams. Waste Manag. **27**(11), 1562–1569 (2007)
24. P. A. Mahanwar, M. P. Bhatnagar, Medical Plastics Waste (2020)
25. I.B. Matveev et al., New combined-cycle gas turbine system for plasma-assisted disposal of sewage sludge. IEEE Trans. Plasma Sci. **45**(12), 3100–3104 (2017)
26. V. Messerle et al., Plasma processing of model residential solid waste. J. Eng. Phys. Thermophys. **90**(5), 1192–1197 (2017)
27. V.E. Messerle et al., Plasmachemical processing of medicobiological wastes. J. Eng. Phys. Thermophys. **88**(6), 1471–1475 (2015)
28. V.E. Messerle et al., Municipal solid waste plasma processing: thermodynamic computation and experiment. IEEE Trans. Plasma Sci. **44**(12), 3017–3022 (2016)
29. V.E. Messerle et al., Processing of biomedical waste in plasma gasifier. Waste Manag. **79**, 791–799 (2018)
30. M.T. Munir et al., Plasma gasification of municipal solid waste for waste-to-value processing. Renew. Sustain. Energy Rev. **116**, (2019)
31. A.B. Murphy et al., Plasma destruction of ozone depleting substances. Plasma Chem. Plasma Process. **22**(3), 371–385 (2002)
32. S. Nema, K. Ganeshprasad, Plasma pyrolysis of medical waste. Curr. Sci. **83** (2002)
33. H. Nishikawa et al., A treatment of carbonaceous wastes using thermal plasma with steam. Vacuum **73**(3), 589–593 (2004)
34. B.C. O'Neill et al., The effect of urbanization on energy use in India and China in the iPETS model. Energy Econ. **34**, S339–S345 (2012)
35. R.F.S. Paulino et al., The use of syngas from biomedical waste plasma gasification systems for electricity production in internal combustion: thermodynamic and economic issues. Energy **199**, (2020)
36. M. Pourali, Application of plasma gasification technology in waste to energy challenges and opportunities, in *2009 IEEE PES/IAS Conference on Sustainable Alternative Energy (SAE)*. IEEE (2009)
37. Y. Pujara et al., Review on Indian Municipal Solid Waste Management practices for reduction of environmental impacts to achieve sustainable development goals. J. Environ. Manag. **248**, (2019)

38. M. Punčochář et al., Development of process for disposal of plastic waste using plasma pyrolysis technology and option for energy recovery. Proc. Eng. **42**, 420–430 (2012)
39. R. Rajput et al., Scenario of solid waste management in present Indian context **7**(1) (2009)
40. A. Ramos et al., Life cycle costing for plasma gasification of municipal solid waste: A socio-economic approach. Energy Convers. Manag. **209**, (2020)
41. J. Rezaiyan, N.P. Cheremisinoff, *Gasification technologies: a primer for engineers and scientists* (CRC Press, Boca Raton, 2005)
42. B. Ruj, S. Ghosh, Technological aspects for thermal plasma treatment of municipal solid waste—a review. Fuel Process. Technol. **126**, 298–308 (2014)
43. M. Sharma, R. Kaushal, Advances and challenges in the generation of bio-based fuels using gasifiers: a comprehensive review. Int. J. Ambient. Energy 1–19 (2018)
44. M. Sharma, R. Kaushal, Performance and emission analysis of a dual fuel variable compression ratio (VCR) CI engine utilizing producer gas derived from walnut shells. Energy **192**, (2020)
46. Z. Singh et al., An introduction to essentials of bio-medical waste management. Med. J. Armed Forces India **57**(2), 144–147 (2001)
45. R.P. Singh et al., An overview for exploring the possibilities of energy generation from municipal solid waste (MSW) in Indian scenario. Renew. Sustain. Energy Rev. **15**(9), 4797–4808 (2011)
47. A.V. Surov et al., Multi-gas AC plasma torches for gasification of organic substances. Fuel **203**, 1007–1014 (2017)
48. L. Tang, H. Huang, Biomass gasification using capacitively coupled RF plasma technology. Fuel **84**(16), 2055–2063 (2005)
49. L. Tang et al., Development of plasma pyrolysis/gasification systems for energy efficient and environmentally sound waste disposal. J. Electrostat. **71**(5), 839–847 (2013)
50. J.S.C. Viera et al., On replacing single-use plastic with so-called biodegradable ones: the case with straws. Environ. Sci. Policy **106**, 177–181 (2020)
51. D. Vij, Urbanization and solid waste management in India: present practices and future challenges. Proc. - Soc. Behav. Sci. **37**, 437–447 (2012)
52. S. You et al., COVID-19's unsustainable waste management. Science **368**(6498), 1438 (2020)
53. J. Zeng et al., Co-treatment of hazardous wastes by the thermal plasma to produce an effective catalyst. J. Clean. Prod. **208**, 243–251 (2019)
54. L. Zhang et al., Overview of recent advances in thermo-chemical conversion of biomass. Energy Convers. Manag. **51**(5), 969–982 (2010)
55. Q. Zhang et al., Gasification of municipal solid waste in the Plasma Gasification Melting process. Appl. Energy **90**(1), 106–112 (2012)
56. V. Zhovtyansky, V. Valinčius, Efficiency of plasma gasification technologies for hazardous waste treatment. Gasif. Low-Grade Feedstock 165–189 (2018)

Comprehensive Updates on Various Fast Charging Technology for Electric Vehicles

Anand Yadav and Shelly Vadhera

1 Introduction

Nowadays electric vehicles are gaining popularity worldwide in an attempt to reduce the different problems such as an increase in oil pricing, shortage of fossil fuels and gasoline and environmental pollution. In addition, the automobile sector is facing increasing pressure from community and government regulations to adopt more sustainable and clean technology such as PEVs in order to reduce carbon emissions. A large part of oil consumption is on the road for 2030 as with a report from an international energy outlook, the transportation sector will raise its share of fuel in the global market by 55%. To aim for the development of energy efficiency, there is an evolution in the field of transportation being undertaken. Huge money is spent on research to restore electronic appliances, mechanical properties and data and control of the electrical system. Recently, EVs are developing rapidly, as they strive for green energy in the world. However, the adoption of this new technology depends largely on people's economy and on the development of adequate PEV technology. The reliability of the new vehicle is considered to be a human concern that is of great importance due to distance concerns. The limited driving distance of PEV make public charging requirement for a longer trip, and, thus, the convenient and fast charging station (FCS) is key to strengthening the adoption of PEVs.

Generally, EVs can operate for a distance of 100 kilometres, hence it's necessary to have a charging station just like gasoline stations. There are three main features of the charging station: fast charging (less than 15 min.), longer battery life (lesser rise in temperature during charging), and standby. Constant current (CC) and constant voltage (CV) mode of charging the electric vehicles batteries are generally used.

A. Yadav (✉) · S. Vadhera
Department of Electrical Engineering, NIT Kurukshetra, Kurukshetra, Haryana, India
e-mail: yanand81@gmail.com

S. Vadhera
e-mail: Shelly_vadhera@rediffmail.com

S. Vadhera et al. (eds.), *Latest Trends in Renewable Energy Technologies*, Lecture Notes in Electrical Engineering 760, https://doi.org/10.1007/978-981-16-1186-5_16

During constant current (CC) mode battery is subjected to a constant high current for a shorter duration with higher temperature increases; and the CV offers low heat for long charging. Per power limit per cell (V. Icell) and max. charging current (Imax.) from battery providers, 15 min. charging time and low-temperature requirements cannot be met from the CC and CV method. For fast charging of electric vehicle at higher rated current pulse charging is considered as a better choice. With this method of charging the values measured Voltage per cell and the maximum peak current are provided. Therefore with this charging technique charging time is greatly reduced. There are basically two ways to charge the batteries of electric vehicles, that is, Onboard charging and Offboard charging.

Onboard charger/AC charging: Onboard charger is, that every electric vehicle is equipped with. It takes AC form the wall outlet and converts this charge into direct current (DC) and transfers it to batteries for storage.

Offboard charger/DC Charging: DC charging method or fast charging method is often used at charging stations which are also known as Electric vehicle service equipment (EVSE). DC regulated power is provided directly by the EVSE to charge the battery. This type of charger is fixed at the place because of its bulky nature, the power rating is as high as hundreds of kilowatts (kW) (Figs. 1 and 2).

The charging system on the board consists of an AC-DC rectifier, a DC-DC amplifier, and a high-frequency converter DC-DC. The aforementioned configuration consists of three transition steps that result in a significant power loss [1, 2]. So overall PTW operating capacity is very low, just like a regular petrol car. Road electric vehicles today are more efficient with PTW between 15 and 20%. In terms of AC input, many changes were made to reduce the size of the charger, without impacting the efficiency and cost of the charger. Another option involve a two-stage charger, which includes an AC-DC converter and a PCC power amplifier, followed by zero voltage switch DC-DC converters.

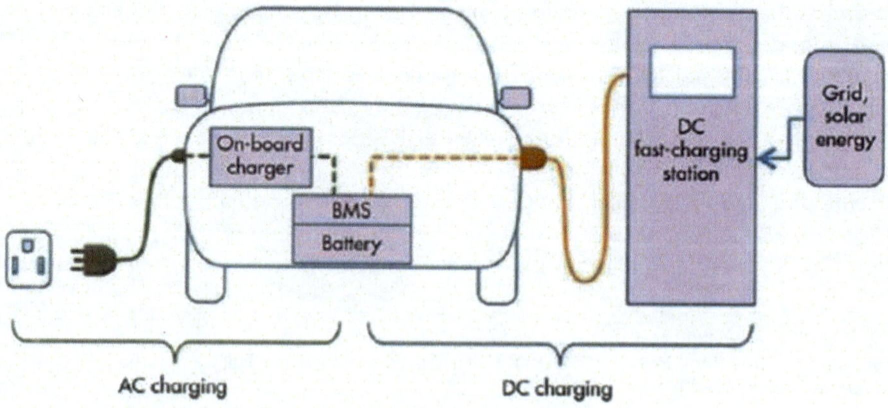

Fig. 1 Block diagram of Onboard and Offboard charging

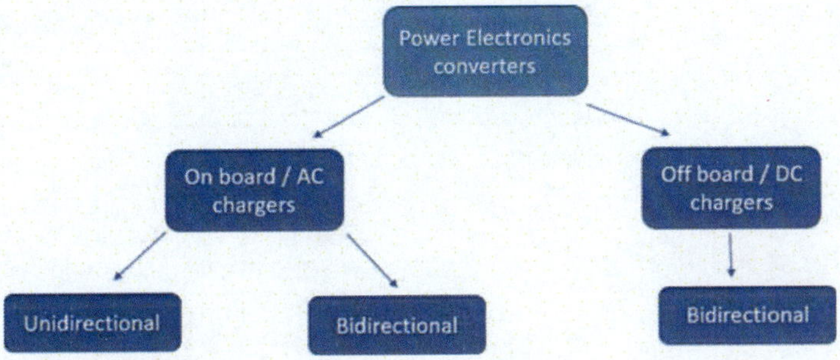

Fig. 2 Power electronics topologies are used for charging EVs batteries

EVs are becoming more and more popular, as evidenced by many recently made cars available in the market by almost all manufacturers. The supercapacitors, electrochemical batteries and fuel cells are the major energy storage used for powering the electric vehicle. However, due to the energy storage limits of such technologies, these vehicles have a limited range of autonomy. Optimization of different energy storage systems can be done, but electrochemical batteries are currently widely used technology to conserve charge.

However, they are often used in conjunction with supercapacitors to conserve energy at soft times, such as on the right at car restart. In fact, ultracapacitors are used in this way to obtain large amounts of energy in a short time and to supply this at the time of acceleration, or to help to charge the vehicle battery.

2 Fast Charging Technologies

2.1 High-Frequency Isolation Switch Mode AC-DC Rectifier for Fast EV Battery [3]

In this technology, it is proposed to look specifically at the multi-pulse rectifier AC-DC topology with higher frequency isolation for charging the Electric vehicle at a faster rate. By using the multiplication mechanism of multiple pulses, the lower accumulated harmonics are eliminated. The high-frequency three-phase input voltage modulation proposed allows an improvement in power density by replacing the transformer used at line frequency with a ferrite core, high-frequency transformer. Since the modulated input voltage is of higher frequency, there is no need for a bulky DC—link capacitor In addition, the proposed modification scheme simplified the design of the output voltage controller and achieved good THD input current consumption.

Advantages of this topology are:

1. Bulky DC link capacitor is not required, power density is higher because of higher switching frequency, simpler design and easier control

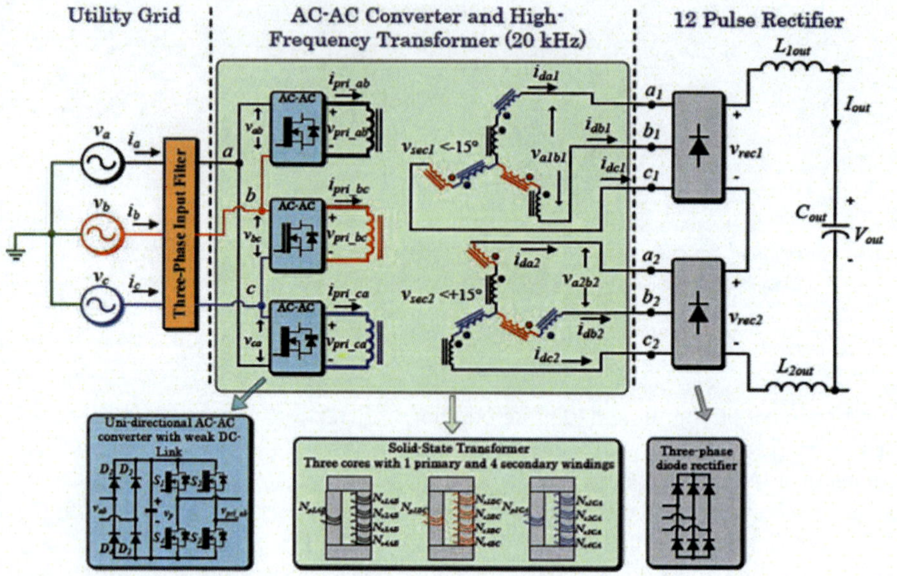

Fig. 3 High-frequency isolation switch mode AC-DC rectifier

2. This type of topologies is Ideal for driving up to several hundred kW of power and at high speed such as Level 3 Fast EV Charger.
3. Lower order harmonics are eliminated due to the use of a multi-pulse rectifier (Fig. 3).

2.2 *Magnified Alternating Current Technology for Fast Charging of EVs Battery [1, 4]*

At sub Zero temperature pre-heating of lithium-ion batteries is required for safety, stability and longevity reasons. So, in such kind of environmental condition fast charging of batteries depends on the faster heating. In order to reduce the heating time and increase the charging rate AC currents is injected into the lithium-ion batteries. A method based on multi-objective is discussed in this paper using a transformerless resonant filter DC-DC converter. This method allows DC/DC converters to produce magnified AC current other than direct current (DC). In submission Experimental setup performed using LCL Circuit with defined component values and a full bridge switch current, can reach 15.7 times the rated value. Also, comparing the switching frequency with the frequency at which LCL and battery resonate, for the same injected Alternating current could reduce the current flowing through the semiconductor. Therefore, there is a decrease in the semiconductor losses by 75% when compared with equivalent non magnified AC current DC-DC converter.

Fig. 4 Power losses in four different cases

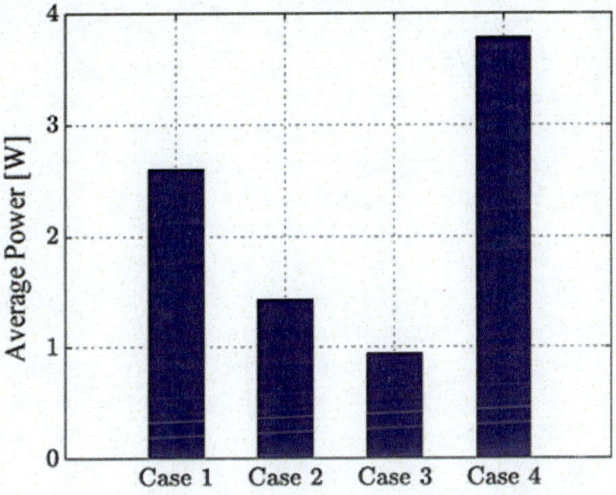

The advantage of this technology is when operated at resonate frequency the current in the semiconductor is reduced. Hence the conduction losses in the semiconductor switches get reduced (Fig. 4).

Semiconductor losses for the four studied cases are compared. Comparison of the VSC's semiconductor losses for the four studied cases, that is case (1) Half-bridge drives Iin into LCL using PWM, case (2) Full bridge drives iin into LCL using PWM, case (3) At resonating frequency Full bridge drives iin into LCL, case (4) Half-bridge drives iin into inductor using PWM [4] (Fig. 5).

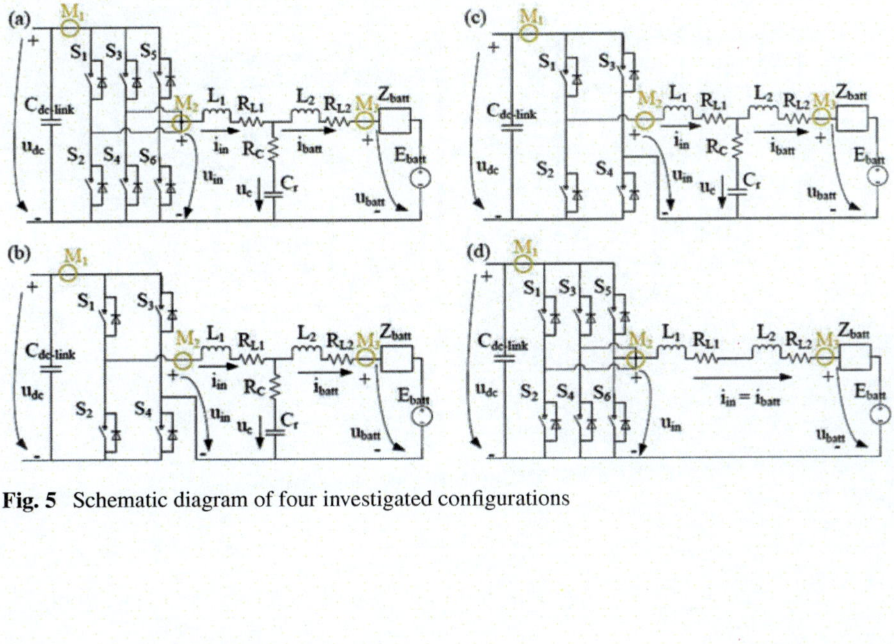

Fig. 5 Schematic diagram of four investigated configurations

2.3 *Fast Charging System with Digital +Ve and –Ve Pulse [2, 5]*

In this technology, a quick charging system containing a pre-charging phase, a +ve and –ve pulse charging phase, a fixed voltage charge phase, and a floating charge phase. Problems associated with transient switching have been alleviated by using a (DCS)digital control system, which has led to stabilised +ve and –ve pulse charging current waveform which results in the accelerated charging of the batteries. During the Charging process, steady-state peak current error is negligible and the shape of the waveform remained stable (Fig. 6).

The voltage and capacity of the lead-acid batteries used were 12 V and 70 Ah, and the charging peak current pulse is set to 0.2 C which is equivalent to 14 A. The of +ve and –ve charging pulse waveforms are shown in Fig. 7.

Tables 1 and 2 show the rise in temperature of the batteries of two groups A and B over time. As it's clear from the table that rise in temperature of the battery in B is more than in A. It shows that the battery polarization can be reduced due to the presence of deadbands and a –ve discharge pulse interval.

Due to slower temperature rise in group A indicates that a higher charging current can be allowed while charging the battery using the proposed +ve and –ve pulse circuits. Now, the battery is charged at 0.25 C which is equivalent to 18 A, the battery is charged for 90 min. as shown in Table 3. It was observed that the difference of battery voltage overtime was increased, which indicates that the rate of charging has been increased. From Table 3 it is clear that reduction in increase in temperatures demonstrates that the battery polarization process was mitigated at the time of charging and the battery can be charged at a higher current.

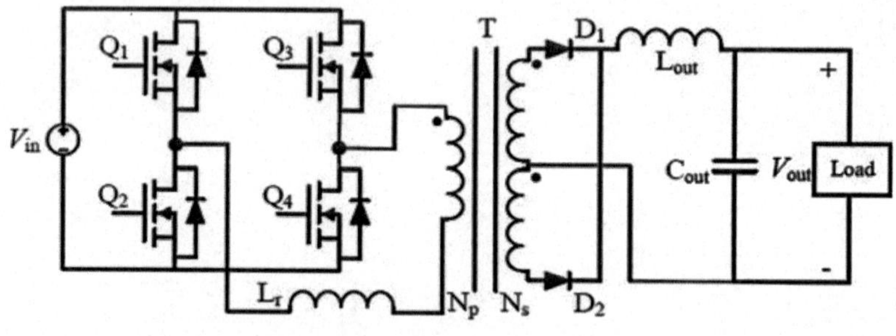

Fig. 6 Soft switching full-bridge phase-shift circuit

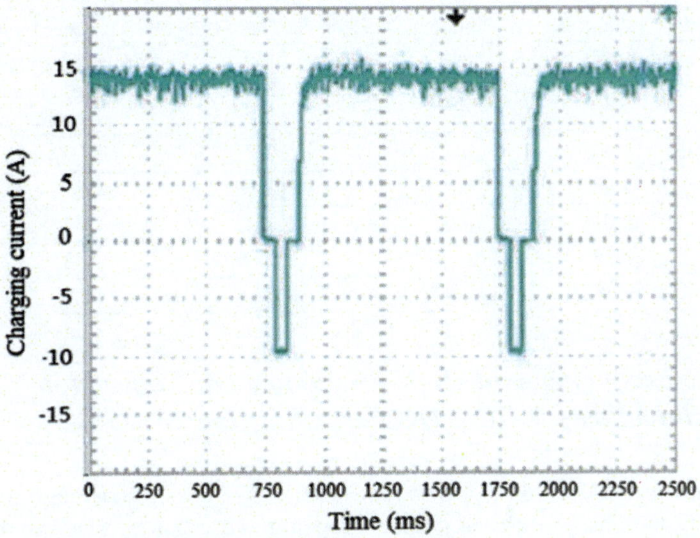

Fig. 7 +ve and –ve pulse waveform used during the charging process

Table 1 Group A battery voltage and temperature

Time	Battery voltage (V)	Voltage difference (V)	Battery temperature (°C)	Difference of battery temperature (°C)
Before plugin	12	0	30.4	0
After plugin	12.47	0	30.4	0
30 min	12.97	0.50	30.8	0.4
60 min	13.25	0.78	31.4	1.0
90 min	13.59	1.12	32.0	1.6

Table 2 Group B battery voltage and temperature

Time	Battery voltage (V)	Voltage difference (V)	Battery temperature (°C)	Difference of battery temperature (°C)
Before plugin	11.9	0	30.6	0
After plugin	12.42	0	30.6	0
30 min	13.03	0.61	31.4	0.8
60 min	13.44	1.05	32	1.4
90 min	13.90	1.48	32.8	2.2

Table 3 Battery voltage and temperature at charging current of 0.25 C (18 A), for 90 min [1]

Time	Battery voltage (V)	Voltage difference (V)	Battery temperature (°C)	Difference of battery temperature (°C)
Before plugin	11.95	0	30.6	0
After plugin	12.38	0	30.6	0
30 min	13.04	0.66	31.3	0.7
60 min	13.55	1.17	31.9	1.3
90 min	13.98	1.6	32.6	2.0

2.4 *Low Frequency and High-Frequency DC Conversion Architecture for Ultra DC Fast Charging 2010 [6]*

This technology-focused on two different types of switches to operate ultra-fastest EV charging stations. Two PE-architecture topologies are shown, based on the lower frequency application is presented and higher frequency isolation. The LF separation, which can be separated by three integrated phase separations, and the HF separations, as well as the unique galvanic DC-DC converter topology, this is how it has been discussed and characterized. Many basic requirements are necessary to be met for this topology of a converter to provide safer, reliable, faster and efficient charging. However, power electronics systems topology must be able to accept universal power supply. Hence, the strength and weakness of both approach to power electronics converters were analysed and the most important product details are discussed, also the effects of DC charging stations are examined.

Through low-frequency configurations, it is indeed possible to distribute the output power to three different converters branches and, thus, reduce the total semiconductor voltage stress and the overall converter losses. With the high-frequency conversion, the architecture allows you to achieve not only G2V operation, but also V2G operation mode if there is a correction phase of the output signal is implemented by the active device. To get higher efficiency, the high-frequency converters operate in a mode of smooth switching with phase zero voltage (ZVS). During the design process of the converter, special attention should be given to optimising the system to meet the relevant product requirement in terms of efficiency, cost, size, volume and weight. All component must be designed to develop a the customized system, from semiconductor devices to passive filter components through a proper heat sink (Figs. 8 and 9).

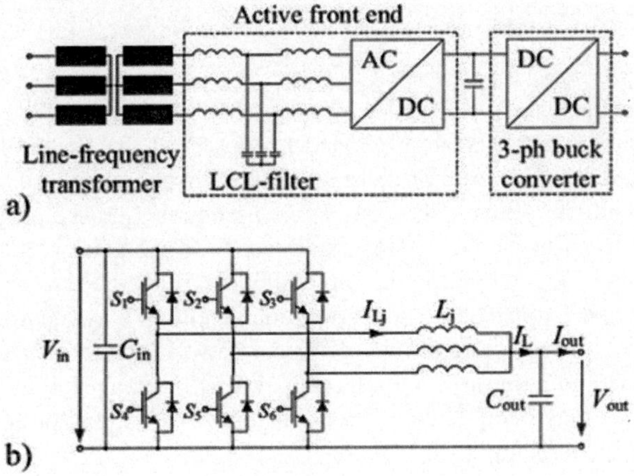

Fig. 8 **a** Low-frequency converter for DC fast-charging station **b** 3-ϕ converters based on standard ABB products [6]

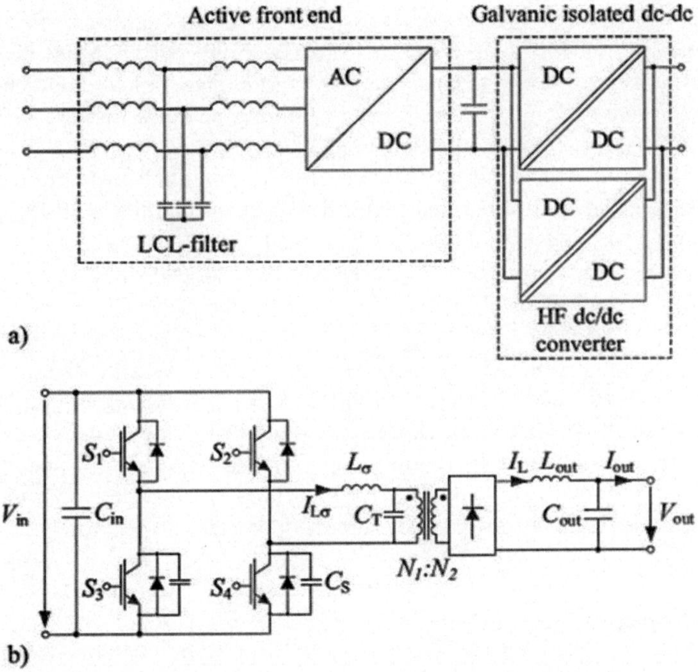

Fig. 9 **a** High-frequency converter for DC fast charger station with two fast isolated DC-DC steps is parallelly connected. **b** Isolated high-frequency DC-DC converters [6]

3 Battery Management

In order to charge the battery of the EVs it's necessary to overcome the basic problems of the battery management system (BMS) proposed by EV technology. In BMS (Battery management system) unit there are different sections out of which thermal management and high voltage management are essential and used for improving the safety of the battery. Significant progress has been made in batteries chemistry but still monitoring charging/discharge regulation, cell balancing and thermal protection are needed for safer operation. Therefore, battery management plays an important role in integrating battery energy storage into electrical networks in terms of performance, safety, reliability, and economy without proper BMS sudden failure or premature failure of batteries may occur such as thermal runaway owing to poor electrical and thermal operating characteristics.

Following are the main function of the BMS:

- Temperature management.
- Monitoring the SOC-(State of charge), SOP-(State of power), SOH-(State of health).
- Computation: A BMS keep check of Max. charging current/Max. discharging current, delivered energy during charging and energy delivered since the last charge, Internal cell impedance, Total operating time since first use, Stored and delivered Charges, Total energy delivered since first use.
- Communication: The central BMS controller communicates internally with hardware (Cell-level) or high-end peripherals such as laptops or HMIs.
- Protection: BMS does not allow the battery to operate beyond its specified limit thus protects batteries by preventing it from working outside the safe zone, such as over current, over-voltage, higher temperature, under voltage, under temperature, etc.
- Load circuit connection to Battery: The BMS can also have a pre-charging system that allows you to safely connect the battery to various loads and eliminates too many inauspicious currents for capacitor loads.
- Optimisation: To increase the battery capacity and prevent limit undercharging or overcharging, the BMS can proactively ensure that all battery components are balanced under the same voltage or charge conditions.

4 Conclusion

The future trend in charging electric vehicles is mainly fast charging, wireless charging and charging from renewable or sustainable sources of energy. In addition, vehicle to network or vehicle to home is an area of study. Fast pulse charging is an important issue when an electric vehicle is commercialised to reduce the charging time to a reasonable level. When batteries are charged using a fast charger this led to overheating and poor performance and battery. Likewise, the deep discharge is the root cause of permanent damage. BMS helps increase battery life, reduce damage, and increase power, efficiency, durability, and reliability of the battery stacks.

Bottlenecks of charging technology used:

- When these batteries are connected to the series, after multiple cycles of charging and discharging, voltage and voltage imbalance will occur between the different batteries.
- The batteries should be capable of bearing various operating conditions such as various charge/discharge levels, discharge depth, temperature variations and so on.
- When the battery is overloaded, it will result in overheating, reduced performance and battery damage. Similarly, deep throwing is the root of permanent damage.
- Problems with battery charging such as cell size, conditioning and fuel management are important to discuss.
- The biggest challenge for an electric car is the lower range and requires frequent charging.
- Longer charging time required by the batteries.
- Higher battery prices and shorter battery life have led to increased interest in developing advanced charging methods and algorithms.
- Pulsed Charging: In this method, the current pulse are used for charging. The current rate can be controlled by changing the pulse range. This method can cause unwanted chemical reactions inside the batteries.
- It is very important to decide when to disconnect the charging to avoid the overheating which is most significant for high capacity battery charge used in PHEVs and ensure that the battery is fully charged. Selecting the charging system type depends on a variety of factors such as application and battery life (Table 4).
- Fast charging, high current and voltage rates are used and the battery is under pressure. There are two main sources of recharge including Thermal pressure and charge saturation pressure, which should be properly controlled, to prevent permanent damage to the battery pack.

Table 4 Charging time of different types of chargers

Supply	Rated power	Voltage (V)	Current (Max) Amp	Charging time
1-ф	3.3	230	16	4–5 h
1-ф	7.4	230	32	2–3 h
3-ф	11	400	16	1–2 h
3-ф	22	400	32	40–50 min
3-ф	43	400	63	20–30 min
DC current (Fast charger)	50	360–500	90–120	15–20 min
DC current (Fast charger)	120	300–500	300–340	5–10 min

References

1. J. Becker, M. Lelie, M. Jansen, D. Sauer, Impact of low temperatures on performance and ageing of lithium-ion batteries and strategies for heating (2014)
2. K. Wu, Z. He, T. Yin, Z. Zhang, Digital fast charging system with positive and negative pulse (2017)
3. E. Pool-Mazun, J.J. Sandoval, P. Enjeti, I.J. Pitel, A direct switch-mode three-phase AC to DC rectifier with high-frequency isolation for fast EV battery chargers, in *2019 IEEE Applied Power Electronics Conference and Exposition (APEC)*, Anaheim, CA, USA, pp. 573–580 (2019)
4. R. Soares, N. Djekanovic, O. Wallmark, P.C. Loh, Integration of magnified alternating current in battery fast chargers based on DC–DC converters using transformerless resonant filter design. IEEE Trans. Transp. Electrif. **5**(4), 925–933 (2019)
5. M. Becherif, M.C. Péra, D. Hissel, S. Jemeï, Estimation of the lead-acid battery initial state of charge with experimental validation. 2012 IEEE Vehicle Power and Propulsion Conference, Seoul **2012**, 469–473 (2012)
6. D. Aggeler, F. Canales, H. Zelaya-De La Parra, A. Coccia, N. Butcher, O. Apeldoorn, Ultra-fast DC-charge infrastructures for EV-mobility and future smart grids, in *2010 IEEE PES Innovative Smart Grid Technologies Conference Europe (ISGT Europe)*, Gothenberg (2010), pp. 1–8

Performance Analysis of Proton Exchange Membrane Fuel Cell (PEMFC) with PI and FOPI Controllers

Swati Singh, Vijay Kumar Tayal, Hemender Pal Singh, and Vinod Kumar Yadav

1 Introduction

In today's era, the most common and popular renewable energy sources are solar, wind and fuel cells. In solar power, irradiance changes the whole day which affects the electrical output produced by the solar panel. The wind has unpredictable and oscillating nature, thus fluctuation occurs in the power output and frequency of the system. Electricity generated in FCs is less vulnerable to external factors and is more stable, in contrast with electricity generated by WECs (Wind Energy Conversion Systems) and PV systems as proposed by Dwivedi and Tayal [1, 2]. FCs have many attributes for instance low noise due to the absence of operating machinery, less pollution as its by-product is water, its power conversion efficiency is about 40–60%, extensive feed (hydrogen, gas, etc.) and versatile applications for example electric vehicles, portable power system, etc. A single fuel cell cannot deliver high power or high voltage and current which is required by almost all the applications of PEMFCs. This problem can be overcome by stacking multiple cells of PEM fuel cell together by connecting them either in series or in parallel so as to obtain higher total voltage or higher total current, respectively. Control methods are required for the problems related to electrical power from a PEMFC stack. W.R.W. Daud et al.,

S. Singh (✉) · V. K. Tayal · H. P. Singh
Electrical & Electronics Engineering Department, Amity School of Engineering & Technology, Noida, India
e-mail: singh.swati2703@gmail.com

V. K. Tayal
e-mail: vktayal@amity.edu

H. P. Singh
e-mail: hpsingh2@amity.edu

V. K. Yadav
Electrical Engineering Department, Delhi Technological University, Delhi, India
e-mail: vinodkumar@dtu.ac.in

S. Vadhera et al. (eds.), *Latest Trends in Renewable Energy Technologies*, Lecture Notes in Electrical Engineering 760, https://doi.org/10.1007/978-981-16-1186-5_17

proposed a control with PID (proportional-integral-derivative) controllers so as to get better performance of fuel cell [3]. PI (proportional-integral) controller is used to improve current of fuel cell [4]. Numerous control techniques are embraced in literature for instance Artificial Neural Techniques (ANN), P controller, Fuzzy logic, PID controller, Sliding Mode Controller (SMC) and PI controller. Yuedong et al. [5] designed and implemented a fuzzy logic controller with PI control loops to directly control the H_2 and air mass flow. Designing a robust controller for the PEMFC system is done for the control of a PEMFC powered electric bicycle [6]. Swain and Jena [7] proposed a linear PID controller which is applied to the nonlinear fuel cell system and controller parameters are tuned by using the Ziegler-Nichols method. Javaid et al. [8] proposed a sliding mode controller to enhance the steady-state behaviour and transient response of a PEMFC. The problem with sliding mode control is that the controller heavily depends on the sliding surface that is if the sliding surface is not designed properly, the controller performance is unacceptable. A fuzzy PID controller suggested by Dhanya et al. [9, 10] is used to control the unregulated voltage produce by the fuel cell. For improving dynamic performance and reducing steady-state errors, Chonglin and Shaoyuan [11] proposed a feed-forward correction, a PID feedback correction, a fuzzy PID controller, and a DMC controller and applied it to the mathematical model. In recent years, Fractional Order Controllers are gaining significance since they have extra tuning parameters that stabilize the plant under control as compared to PID controller and PI controller, which is discussed by Jeba and Selvakumar [12].

This paper employs MATLAB/Simulink to design Proton Exchange Membrane Fuel Cell (PEMFC). Then designing of PID controller and FOPID controller is done to enhance the performance of PEMFC. Later on, the result of both the controllers, are being compared. This paper can be summarized as: Sect. 2 gives a brief idea of PEM fuel cell. Section 3 deals with the designing of PI and FOPI controllers and their Simulink model is also represented. The results of the simulation and the comparison of the proposed controllers have been given in Sect. 4. Finally, the conclusions were discussed in the last section.

2 Proton Exchange Membrane Fuel Cell (PEMFC)

FCs (Fuel cells) are used to convert chemical energy in fuels such as hydrogen into electrical energy. Fuel cells are more efficient than equivalent thermal generators as the process of fuel cells does not involve combustion. When pure hydrogen is worked as fuel then the by-products of the electro-chemical reaction are only water (H_2O) and heat. This makes fuel cells a clean and green source of energy. Thus, this technology is very promising because of its low carbon dioxide, chemical and thermal emissions. Different kinds of fuel cells are- Direct Methanol Fuel Cell, Polymer Electrolyte Membrane Fuel Cell (PEMFC), Phosphoric Acid Fuel Cells, Alkaline Fuel Cell, Molten Carbonate Fuel Cells, Reversible Fuel Cell and Solid Oxide Fuel Cells [13].

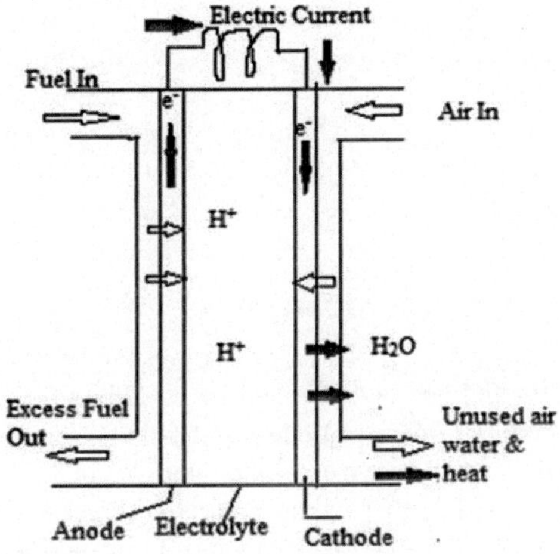

Fig. 1 Schematic diagram of PEM Fuel Cell

In PEMFCs, Hydrogen fuel is processed at the anode where the chemical reaction takes place which causes hydrogen molecules to separate into positive hydrogen ions (H^+) and electrons (e^-) [14, 15]. The positive hydrogen ion passes through the electrolyte which is made up of a polymer membrane and permeates to the cathode. At the cathode side, the metal electrode combines the protons and electrons with oxygen to produce water and the electrons (e^-) move towards the external circuit and electrical output is generated. The PEM (Proton Exchange Membrane) is a polymer that conducts proton. This polymer is impermeable to electrons and gas which keeps minimum electron short circuit and minimum gas crossover. Figure 1 represents the schematic diagram of PEMFC.

Following are the chemical reactions that take place while the functioning of PEMFC:

$$\text{At anode}: \; H_2 \rightarrow 2H^+ + 2e^- \tag{1}$$

$$\text{At Cathode}: \; \frac{1}{2}O_2 + 2H^+ + 2e^- \rightarrow H_2O \tag{2}$$

$$\text{Overall Reaction}: \; \frac{1}{2}O_2 + 2H^+ + 2e^- \rightarrow H_2O \tag{3}$$

Equation (1) gives the chemical reaction that takes place at the anode side of the fuel cell. The chemical reaction at the cathode side is represented by Eq. (2) and thus Eq. (3) provides the overall reaction that takes place in the PEMFC.

The transfer function used for the modelling [16] of PEMFC is given by Eq. (4).

$$G(s) = \frac{591.6s^4 + 1472s^3 + 451.3s^2 + 0.9748s + 1.847 \times 10^{-5}}{s^5 + 21.72s^4 + 52.59s^3 + 16.12s^2 + 0.03318s + 4.35 \times 10^{-7}} \tag{4}$$

PEMFCs have various advantages such as fast start-up capability, high power density, low noise and light mass, thus PEMFCs technologies are gaining popularity. PEMFCs operate at relatively low temperature (80–90 °C) which makes electrical output to deal with the dynamic requirements of power. As the operating temperature is low, it becomes necessary for PEMFCs to operate on pure hydrogen.

3 Control Strategy

PEMFC has been proved to be a great source of providing electric energy but it produces uncontrolled output voltage, current and power. As most of the PEMFCs application requires higher voltage or higher current thus multiple fuel cells are stacked together either in parallel to get high total current or in series to obtain high total voltage. If the stacking of multiple fuel cells is not controlled then it produces high heat which increases the temperature of FCs. This can damage the membrane. Hence it becomes necessary to have a suitable control design so as to overcome the above problem. In this paper, the performance of PEMFC has been improved by using PI and FOPI controllers.

3.1 PI Controller

A PI controller is a feedback control loop that calculates an error signal by taking the difference between the output of a system, which in this case, is the power being drawn from the battery, and the set point.

The transfer function of PI controller is given by Eq. (5).

$$G(s) = K_p + \frac{K_i}{s} \tag{5}$$

Here K_p is the proportional gain and K_i is the integral gain.

In the present scenario, PI is extensively used in industries as it has an elementary structure, low cost and easy to design. But PI controller is a failure when the offset is uncertain and non-linear [17, 18]. PI controller results in the elimination of steady-state error and forced oscillations. PI controller is required for non-integrating processes that means for a set of inputs, the same output is obtained. Integral action is suitable for the removal of offset. These controllers are extensively used in process control applications and give a balance of complexity (Fig. 2).

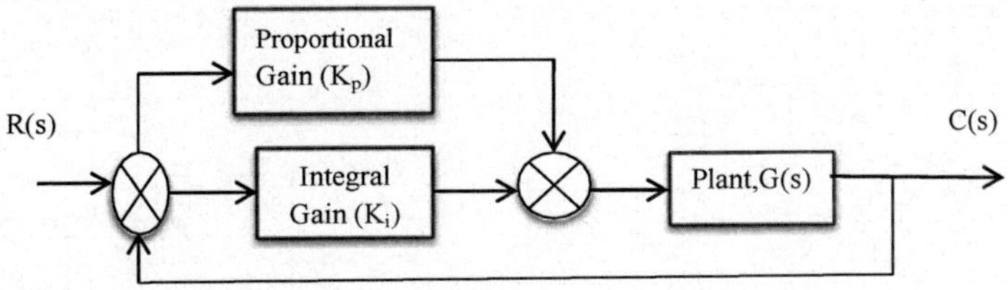

Fig. 2 Block diagram of PI controller

3.2 *Fractional Order PI Controller*

In the current scenario, Fractional Oder Controllers can transit from classical controllers to those which are described by differential equations of non-integer order. For this purpose, Fractional Order Controllers were developed. FO-PID controller is the special case and generalization of PID controller [19]. FOPID controllers have five tuning parameters while PID controllers have only three tuning parameters. The notation for fractional order PID controller is $PI^{\lambda}D^{\mu}$ controller (Fig. 3).

With the implementation of FOPI controllers, the most robust and accurate performance can be achieved in comparison to integer-order controllers [20]. The continuous transfer function of FOPI is obtained through Laplace Transform which is represented in Eq. (6).

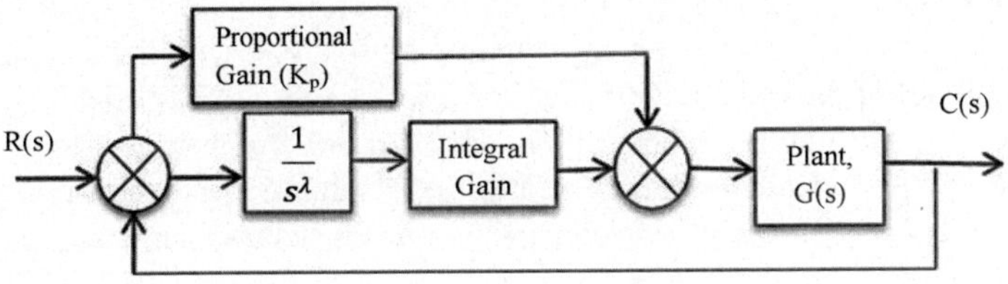

Fig. 3 Block diagram of fractional order PI controller

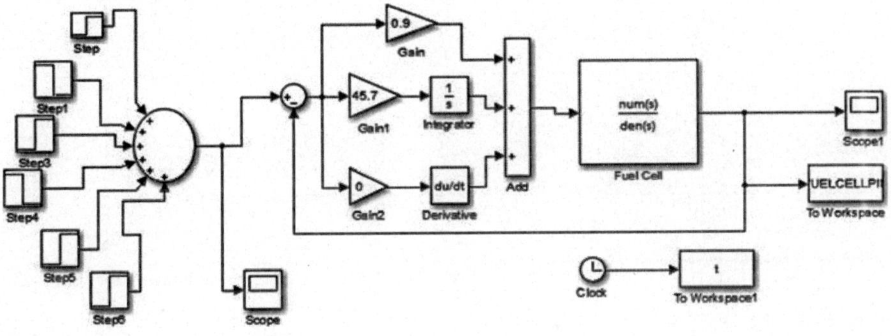

Fig. 4 Simulink model of PEMFC with PI controller

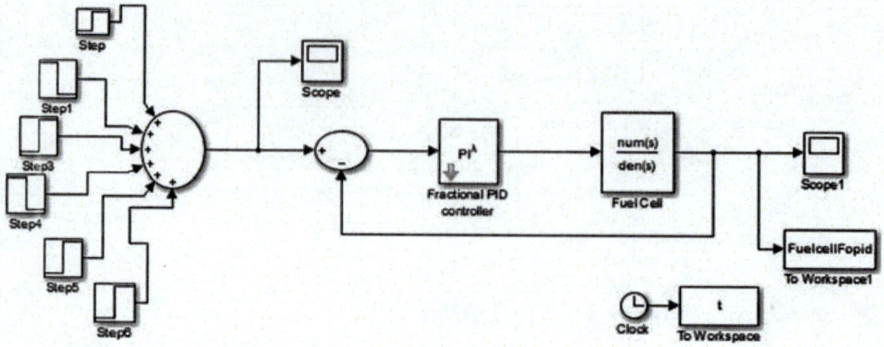

Fig. 5 Simulink Model of PEMFC with FOPI controller

$$G(s) = K_p + K_i s^{-\lambda} \tag{6}$$

4 Simulation Results

To represent the comparison of the robustness of the PI controller and proposed FOPI controller, simulations are conducted under step disturbances that are created in a fuel cell input at $t = 1, 1.1, 1.2, 1.3, 1.4$ and 1.5 s as shown in Fig. 6. Figure 4 represents the Simulink model of proton exchange membrane fuel cell with P-I controller. The simulation model of the FO-PI controller is shown in Fig. 5. This model consists of the PEMFC transfer function block with a FOPI controller. Designing the PEM fuel cell is done by the transfer function given in Eq. 4. Figure 7 indicates the output of PEMFC in the absence of any controller, which produces transients and the system becomes unstable.

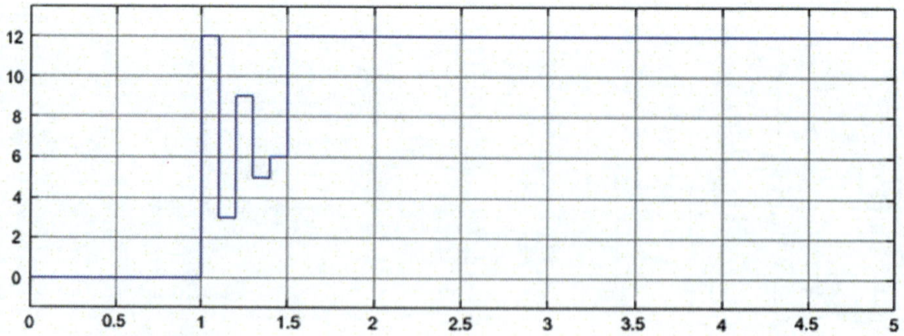

Fig. 6 Input of PEMFC

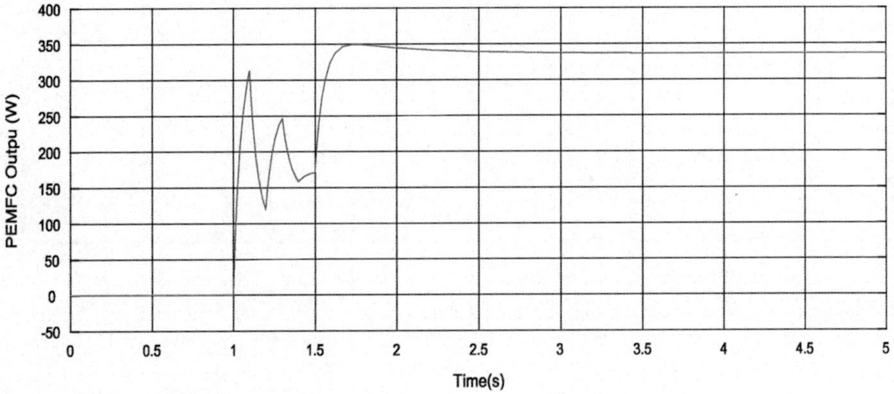

Fig. 7 PEMFC output without a controller

The output of the fuel cell is controlled by the PI controller and FOPI controller. The controlled output of fuel cell is obtained by various values of K_{p}, K_i and λ (values of λ is applicable only for FOPI controller). This is done by using the hit and trial method. Figure 8 shows the controlled output of PEMFC with PI controller where the value of K_p (proportional gain) and K_i (integral gain) are taken as 0.9 and 45.7 respectively. Figure 9 represents the controlled output of PEMFC with Fractional Order PI controller by putting $K_p = 1.21$, $K_i = 10.57$ and $\lambda = 0.61$. Table 1 shows the comparison of the controlled output of PEMFC with PI and FOPI controllers on the basis of parameters rise time, settling time, overshoot and steady-state error. From the results, it is observed that the settling point of PEM fuel cell using PI controller is 22.397 ms which on other hand is reduced to 7.477 ms when PEMFC is controlled with Fractional Order PI controller. The rise time and overshoot for PEMFC with

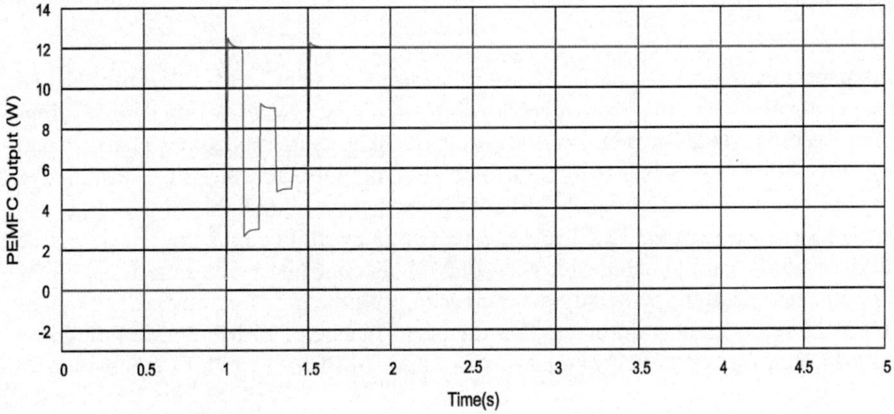

Fig. 8 PEMFC output with PI controller ($K_p = 0.9$, $K_i = 45.7$)

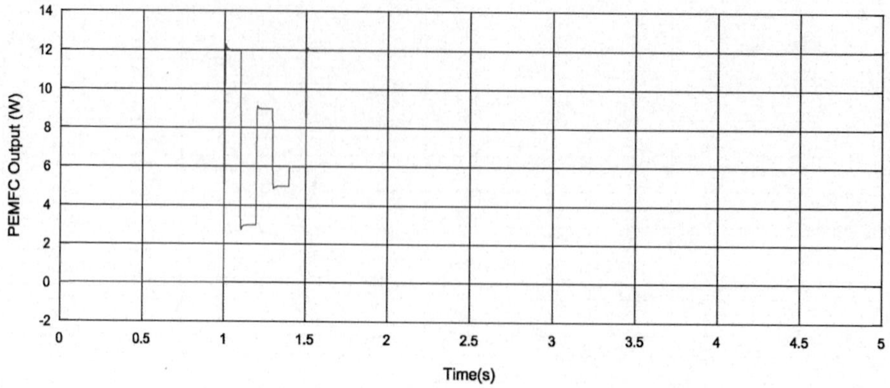

Fig. 9 PEMFC output with FOPI controller ($K_p = 1.21$, $K_i = 10.75$, $\lambda = 0.61$)

Table 1 Comparison between controlled outputs of PEMFC with PI and FOPI controllers

Controller type	Settling time (ms)	Rise time (ms)	Steady state error	Overshoot (%)
PI controller ($Kp = 0.9$, $Ki = 45.7$)	22.397	3.631	0.001	3.646
FOPI controller ($Kp = 1.21$, $Ki = 10.75$, $\lambda = 0.61$)	7.477	2.503	−0.307	2.577

PI controller are 3.631 and 3.646%. In contrast, when the system is controlled with FOPI controller rise time and overshoot become 2.503 ms and 2.577%, respectively.

5 Conclusion

This paper deals with the performance analysis of PEMFC with PI controller and FOPI controller. Both the controllers are compared in accordance with their settling-time, rise-time, steady-state error and overshoot. The system becomes unstable and unsteady without the presence of controllers. PI and FOPI controllers are designed to serve the purpose of instability. FOPI controller gives a remarkable improvement in setting time as compared to the PEM fuel cell output with PI controller. The rise time and overshoot, are also enhanced when PEMFC is controlled with Fractional Order PI controller. Though the steady-state error in both the cases is minimal, the other three parameters clearly implicate that the performance of PEMFC with Fractional Order PI controller is better than the performance of PEMFC with PI Controller.

References

1. H.-C. Chen, S.-Y. Tzeng, P.-H. Chen, Optimization design of PID controllers for PEMFC with reformer using genetic algorithm, in *Proceedings of the Ninth International Conference on Machine Learning and Cybernetics*, Qingdao, IEEE, 11–14 July 2010 (2010)
2. Y. Dwivedi, V.K. Tayal, Dynamic stability improvement of alkali fuel cell integrated system using PSO optimized PID control design, in *IEEE International Conference on Recent Developments in Control, Automation & Power Engineering (RDCAPE)*, pp. 499–504 (2017)
3. W.R.W. Daud, R.E. Rosli, E.H. Majlan, S.A.A. Hamid, R. Mohamed, PEM fuel cell system control: a review. Renew. Energy Int. J. **113**, 620–638 (2017)
4. M. Derbeli, M. Farhat, O. Barambones, L. Sbita, Control of proton exchange membrane fuel cell (PEMFC) power system using PI controller, in *2017 International Conference on Green Energy Conversion Systems (GECS)*, IEEE, 12 October 2017 (2017)
5. Z. Yuedong, Z. Jianguo, G. Youguang, J. Jin, Control of proton exchange membrane fuel cell based on fuzzy logic, in *Proceedings of the 26th Chinese Control Conference*, 26–31 July 2007 (2007)
6. S.-C. Li, C.-Y., Gao, F.-C. Wang, Control and development of a PEMFC electric bicycle, in *Proceedings of the SICE Annual Conference (SICE)*, IEEE, 9–12 September 2014 (2014)
7. P. Swain, D. Jena, PID control design for the pressure regulation of PEM fuel cell, in *2015 International Conference on Recent Developments in Control, Automation and Power Engineering (RDCAPE)*, IEEE, 12–13 March 2015
8. U. Javaid, A. Mehmood, A. Arshad, F. Imtiaz, J. Iqbal, Operational efficiency improvement of PEM fuel cell_a sliding mode based modern control approach. IEEE Access 8 (2020)
9. S.M. Rakhtala, S. Roudbari, Fuzzy PID control of a stand-alone system based on PEM fuel cell. Int. J. Electr. Power Energy Syst. **78**, 576–590 (2016)
10. S. Dhanya, V. Paul, Fuzzy-PID controller for PEM fuel cell in automobile application. IOSR J. Electron. Commun. Eng. (IOSR-JECE) (2014), pp. 63–68
11. Z. Chonglin, L.I. Shaoyuan, Control strategies for the air supply system in PEMFC, in *Proceedings of the 35th Chinese Control Conference*, IEEE, 27–29 July 2016 (2016)
12. P. Jeba, A.I. Selvakumar, FOPID Based MPPT for Photovoltaic System. Taylor & Francis. ISSN: 1556-7036 (2018), pp. 1591–1603
13. J.-W. Jung, A. Keyhani, Fuel cell based distributed generation system, in *12th International Middle-East Power System Conference*, IEEE, 12–15 March 2008 (2008)
14. A. Moran-Duran, A. Martínez-Sibaja, J.P. Rodríguez-Jarquin, R. Posada-Gómez, O.S. González, PEM fuel cell voltage neural control based on hydrogen pressure regulation. Processes 2019, MDPI, 10 July 2019 (2019)
15. V. Khubchandani, K. Pandey, V.K. Tayal, S.K. Sinha, PEM fuel cell integration with using fuzzy PID technique, in *2016 IEEE 1st International Conference on Power Electronics, Intelligent Control and Energy Systems (ICPEICES)*, Delhi (2016), pp. 1–4
16. S.V. Subhasree, R. Kotteswaran, Modelling and control of polymer membrane fuel cell. Indian J. Sci. Technol **9**(13) (2016)
17. V. Yadav, V.K. Tayal, Optimal controller design for a DC motor using PID tuner, in *IEEE International Conference on Power Energy, Environment and Intelligent Control (PEEIC)* (2018), pp. 442–445
18. A. Mukhtar, V.K. Tayal, H.P. Singh, PSO optimized PID controller design for the process liquid level control, in *2019 3rd International Conference on Recent Developments in Control, Automation & Power Engineering (RDCAPE)*, NOIDA, India (2019), pp. 590–593
19. V. Mehra, S. Srivastava, P., Varshney, Fractional-order PID controller design for speed control of DC motor, in *Third International Conference on Emerging Trends in Engineering and Technology*, IEEE, 19–21 November 2010 (2010)
20. U. Mehta, V. Lechappe, O. Singh, Simple FOPI tuning method for real-order time delay systems, in *Advances in Systems, Control and Automation*. Lecture Notes in Electrical Engineering, vol. 442 (Springer, 2017)

Detecting Face Masks Using Deep Learning to Control Public Hygiene, Safety and COVID-19 Spreading

Dimple Muskan Shukla, Kushal Sharma, and Sandeep Gupta

1 Introduction

As of 2020, respiratory contagious disease is spreading in the whole world due to which the entire population is suffering. This virus causes respiratory tract infections in humans that can be ranging from mild to lethal conditions. This disease was an outbreak in China and now affecting the whole population. Every country is trying its best to find its vaccine which can prevent or cure this disease. This disease is severe because it is easily transmittable by some basic ways like going in public places, touching infected things, coming in direct contact with an infected person, or their nasal and cough droplets. These are how any person can get infected and then infect others as well. So, it was advised by the government to wear masks every time going out in public places and maintain social distance, in these days and after this also, this can be a precaution measure to protect yourself from this disease. Now it becomes mandatory but some people are violating this rule and they are not wearing a mask when going out, so there is a need for a person at the entrance to check whether a person is wearing a mask or not, but the security personnel will also be at risk and even it is not practical to track every person and any mistake can increase transmission of this virus. In our research we are using a detector which is for the face with a mask; normal detectors available are for auto-focusing [1] and human-computer interaction [3]. For example, [5] uses a face detection technique with a mask with the help of C2D-CNN which can detect a person with masks or covered face. In our research, we

D. M. Shukla
SS Jain Subodh PG College, Jaipur, India
e-mail: shuklamuskan2703@gmail.com

K. Sharma · S. Gupta (✉)
JECRC University, Jaipur, India
e-mail: jecsandeep@gmail.com

K. Sharma
e-mail: sharmakushal2k@gmail.com

S. Vadhera et al. (eds.), *Latest Trends in Renewable Energy Technologies*, Lecture Notes in Electrical Engineering 760, https://doi.org/10.1007/978-981-16-1186-5_18

are using deep learning techniques like face detection of masked and normal people and image classification of both datasets. This can be a notification system which can be used with any CCTV or sensor which will allow only those people who have masks on their face. In this research we are using two datasets, pictures of people with masks and others are people without a mask, then we have used several steps like data pre-processing, face detection, and image classification using convolutional neural networks with the help of AlexNet architecture. This research can be helpful in every place like supermarkets, schools hospitals, and other public places for easy detection of persons who are violating the rule and can be proven easy for security personnel to track each and every person passing by this method.

2 Database Used

Here, we are using two datasets first one for masked people, that is, Real-World Masked Face Dataset (RMFD) [8] which is available on a GitHub repository, it is the world's largest masked face dataset. The other dataset we have taken is CelebFaces Attributes (CelebA) [6]. Dataset on Kaggle which contains over 200k images of celebrities 40 binary feature annotations as shown in Figs. 1 and 2.

2.1 Face-Detection

In this paper, we used C2D CNN [5] or 2D color PCA-Principal Component Analysis [2]—convolutional neural network, for facial recognition [5]. In this process, we do an estimation of a combination of the features learned from original pixels with the representation learned by CNN and then estimating the result or decision making.

2.2 Mask Detection

In the presented research we are doing classification by AlexNet architecture. AlexNet is a type of convolutional neural network. It is basically a classification algorithm that is best suited for the classification of RGB images. The architecture includes eight layers, among them, the first 5 were convolutional layers, followed by some max-pooling layers and the last three layers were fully connected layers. It basically uses the ReLU activation function which gives better performance than tanh and sigmoid.

Fig. 1 Images of people without Mask (Celebrity face dataset)

3 Topology with Results

3.1 Data Preprocessing

Data preprocessing is a mining technique that includes transforming any primitive data into an easily understandable format. Real data is sometimes inconsistent or lacking in some trends, and can contain many errors. We are taking an image data set, so it contains lots of unnecessary images, or shape issues, dimension issues so we are pre-processing our data to make it suitable for our problem. We have taken two data sets in our research i.e., first is RMFD which is a masked image dataset and

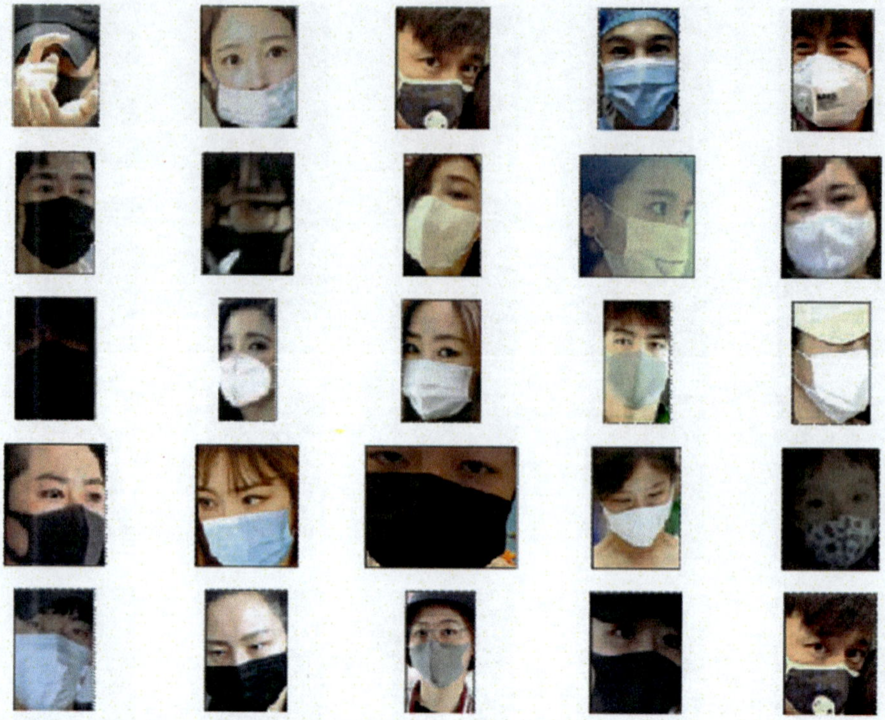

Fig. 2 Images of people with mask (RMFD Dataset)

second is a celebrity dataset, so we are plotting or reshaping them to process them as shown in Figs. 3 and 4.

3.2 Detected Faces Out of the Image

Here, we used a deep C2D CNN or color 2-dimensional principal component analysis convolutional neural network, for face recognition. In this process, we combine the features of original pixels with the image representation that is learned by CNN and then make decision fusion, which can improve our face recognition model. There are several steps in this network, first, the introduction of the normalization layer in the network to shorten the time for training. Second, the introduction of a layered activation function which will prevent gradient diffusion. Lastly, the application of probabilistic max-pooling for the preservation of feature information to the maximum level. We are using this algorithm because in HOG, the algorithm is based on shapes and people with a mask cannot be detected using that, so we are using this C2D-CNN as it is a pixel-oriented algorithm and favors our process (Table 1).

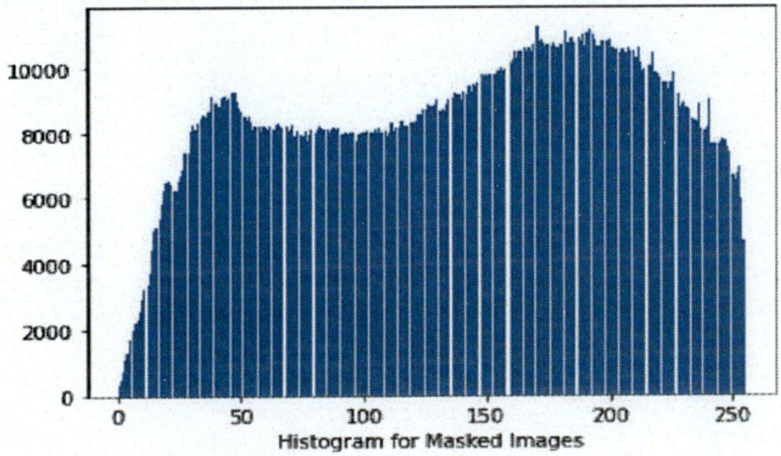

Fig. 3 Mean data of all the masked images

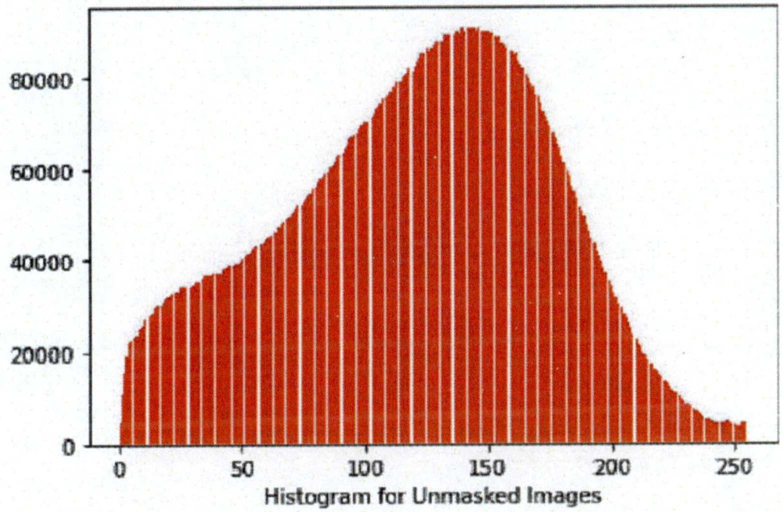

Fig. 4 Mean data of all the unmasked images

3.3 Alex Net

In this research, we are doing classification by Alexnet architecture [7]. It is a type of convolutional neural network. It is basically a classification algorithm that is best suited for the classification of RGB images. The architecture includes eight layers, among them, the first 5 were convolutional layers, followed by some max-pooling layers which are used in downsampling and the last three layers were fully connected

Table 1 Programming input and output data

Layer (type)	Output shape	Param #
input_1 (InputLayer)	(None, 250, 250, 3)	0
conv2d_1 (Conv2D)	(None, 63, 63, 96)	34,944
max_pooling2d_1 (MaxPooling2	(None, 32, 32, 96)	0
conv2d_2 (Conv2D)	(None, 32, 32, 384)	332,160
conv2d_3 (Conv2D)	(None, 32, 32, 384)	1,327,488
conv2d_4 (Conv2D)	(None, 32, 32, 256)	884,992
max_pooling2d_2 (MaxPooling2	(None, 16, 16, 256)	0
flatten_1 (Flatten)	(None, 65,536)	0
dense_1 (Dense)	(None, 4096)	268,439,552
dropout_1 (Dropout)	(None, 4096)	0
dense_2 (Dense)	(None, 4096)	16,781,312
dropout_2 (Dropout)	(None, 4096)	0
dense_3 (Dense)	(None, 1000)	4,097,000
dropout_3 (Dropout)	(None, 1000)	0
dense_4 (Dense)	(None, 1)	1001

Total params: 291,898,449
Trainable params: 291,898,449
Non-trainable params: 0

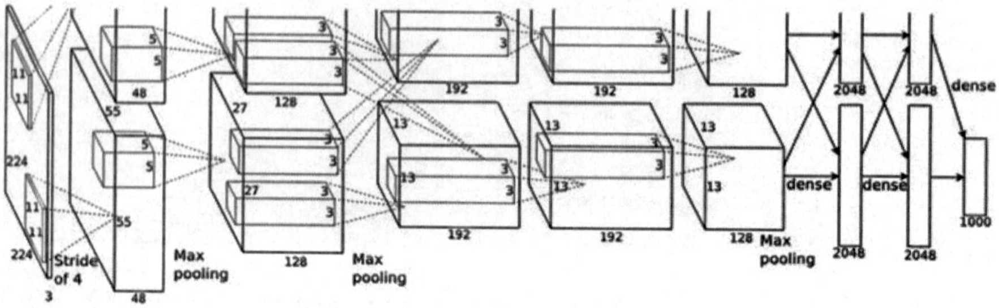

Fig. 5 Alex net architecture [4]

layers. It uses a rectified linear unit activation function. The basic description of our alexnet architecture is shown in Fig. 5.

3.4 Results

See Fig. 6 and Table 2.

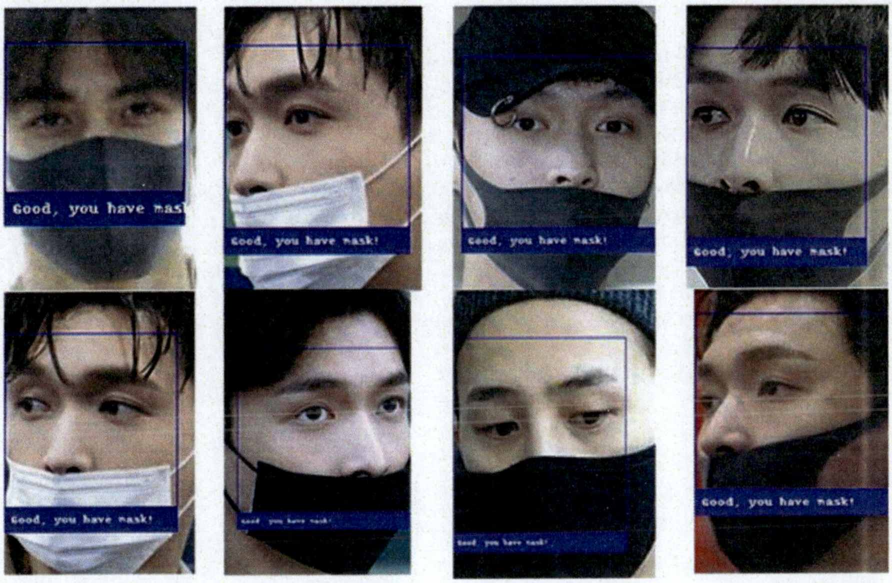

Fig. 6 Detecting masks on faces

Table 2 Metrics on testing data

Accuracy	0.9839666154184055
Precision	0.9904761904761905
Recall	0.839677047289504

4 Conclusion

As we know that COVID-19 is a global pandemic and all the major powerful countries are damaged by its impact. So it becomes an individual's national duty to stop its spread and to make people disciplined about the rules the Mask Recognition can be deployed in public CCTV cameras to scrutinize people who are not wearing a mask and then analyzing this data can tell us a lot about the spreading behavior of this disease. This coronavirus is going to be a part of our culture; we will have to be more cautious now dealing with crowds and wearing masks is going to be a common habit.

References

1. C. Huang, H. Ai, Y. Li, S. Lao, High-performance rotation invariant multiview face detection. IEEE TPAMI **29**(4), 671–686 (2007)
2. Z.G. Jia, S.T. Ling, M.X. Zhao, "Color two-dimensional principal component analysis for face recognition based on quaternion model", In: Intelligent Computing Theories and Application, (2017)

3. B. Jun, I. Choi, D. Kim, Local transform features and hybridization for accurate face and human detection. IEEE TPAMI **35**(6), 1423–1436 (2013)
4. A. Krizhevsky, I. Sutskever, G.E. Hinton, "Imagenet classification with deep convolutional neural networks", Advances in neural information processing systems (2012)
5. J. Li, T. Qiu, C. Wen, K. Xie, F. Wen, Robust face recognition using the Deep C2D-CNN model based on decision-level fusion. Sensors. **18**, 2080 (2018). https://doi.org/10.3390/s18072080
6. Z. Liu, P. Luo, X. Wang, X. Tang, "Deep learning face attributes in the Wild". in *Proceedings of International Conference on Computer Vision (ICCV)*, (2015)
7. D.M. Shukla, K. Sharma, S. Gupta, "Identifying depression in a person using speech signals by extracting energy and statistical features", in *2020 IEEE International Students' Conference on Electrical, Electronics and Computer Science (SCEECS)*, (2020) pp. 1–4
8. Z. Wang, G. Wang, et al. Masked Face Recognition Dataset and Application (2020)

Optimization of Systems-Development Life Cycle Through Automation Using Ansible

Kartik Vadhera, Abhishek Deshwal, and Amrendra Tripathi

Abstract In order to sustain oneself in this era of cutting-edge technology, companies need to adopt different technologies for the efficient execution of tasks. In pursuit of accomplishing the targets, the companies install a number of software or add lines of code. DevOps is a large number of programming improvement repetitions that consolidate software development (Dev) and information-technology operations (Ops) to shorten the lifecycle of system development while delivering patches, updates and features from time to time in close collaboration with business objectives. Time is what everybody wants to save and this where Automation comes into practice. This paper throws light on how companies can adopt simple automation by creating HA proxy servers, web servers, and Hadoop-cluster with the help of an automation tool called Ansible. Ansible is an open-source mechanization platform, used for IT jobs, for example, framework management, application installment, intra-service coordination and accoutering. Here, YAML and Jinja languages are used to create and automate both the HA proxy server and Hadoop cluster.

Keywords Automation · Ansible · Hadoop · HA proxy server · YAML

1 Introduction

There are some programming elements and innovations available to facilitate big data analytics. Hadoop is a key innovation used to process huge information and handle big data, their review and analysis, and stream computing. It is an open-source programming project that manages and enables us the distributed processing of large data sets across clusters of commodity servers. Hadoop includes different modules i.e. HDFS (Hadoop distributed file system), Map Reduce, Pig, HBase, etc. Hadoop is designed to scale up from single servers to thousands of machines, each of these offering local computation and storage [1]. The Hadoop Distributed File System (HDFS) provides global access to the files in the cluster. For extreme transportability,

K. Vadhera (✉) · A. Deshwal · A. Tripathi
University of Petroleum and Energy Studies, Dehradun, Uttarakhand, India
e-mail: kartikvadhera007@gmail.com

S. Vadhera et al. (eds.), *Latest Trends in Renewable Energy Technologies*, Lecture Notes in Electrical Engineering 760, https://doi.org/10.1007/978-981-16-1186-5_19

HDFS is updated as a client-level file system in Java that exploits the local file system on each hub or node, for example, ext3 or NTFS, to store information. The files in HDFS are isolated in large blocks usually 64 MB, and each square is stored as a different file in the neighboring file system. HDFS is put into effect by two services: The Name Node and Data Node. The Name Node supervises the maintenance of the HDFS directory tree and is a centralized service in the cluster operating on a single node. Customers contact the Name Node to perform a basic activity of the file system, such as opening, closing, renaming and erasing. The Name Node does not store the HDFS information itself, but instead maintains a schedule between the HDFS file name, a preview of the blocks in the file, and the Data node(s) of which those blocks are being kept [1].

MapReduce is a programming model and associated execution for the management and production of huge collections of information that can be managed in a wide range of tasks in the real-world. Customers specify the computation in terms of a map and a reduced function, and the rudimentary runtime system automatically parallelizes the calculation through huge clusters of machines, manages the failures of the machine and schedules communication between the machines to efficiently use the network and disks. Software engineers discover that the system is simple to use: more than ten thousand particular MapReduce programs have been updated at Google in recent years, and a normal one hundred thousand MapReduce professions are run on Google groups every day, processing a sum of more than twenty petabytes of information per day [2]. HA Proxy, which represents High availability Proxy, is a well-known open-source programming HTTP load balancer and proxy arrangement that can be run on Solaris, FreeBSD, and Linux. Its most normal use is to improve the reliability quality and performance of a server environment by circulating the remaining task at hand between different servers (e.g. Web, database, application). It is used in some important conditions, including Instagram, Twitter, Imgur and GitHub. There are four fundamental segments in the HA proxy configuration document [3].

They are global, backend, frontend, and default. These four sections specify how the server as an unabbreviated performs, what your default settings are, and how client requests are received and routed to your back-end servers.

With the advancement in technology, time is becoming a major issue. Several things are done manually which takes time. In today's era, almost all companies are facing the problem to store and process their large amount of data. Setting up a Hadoop cluster with a new configuration in a company is a very time taking task. Also, the users generally face problems like server not found or page not found due to excess load on the webserver due to which sometimes one is not able to access the website. Henceforth, one often complains that the particular website got crashed or the server is down. Now, suppose the new system comes into the industry, then one has to deploy all the codes according to the needs which already exists in the industry and this takes a lot of time to do all the changes. Howsoever, these things can be now done through automation.

This paper provides automated software which will set up and install software on different platforms. Further in this paper, implementation of an automated Hadoop cluster so as to solve the big data problem through Ansible (Devops) is done. Also, the implementation of an automated HA Proxy Server is conducted so as to solve the problem of website crashing or server down problem.

2 Architecture of Hadoop and Ansible

Figure 1 displays the architecture of the Hadoop distributed file system commonly known as hdfs wherein a four-terabyte block is comprised of four blocks each of one terabyte from different systems. This is how several storage blocks can be added in order to provide the required size by providing high availability to the users/companies.

The Ansible environment is constituted by the following items [4]:

1. **Modules**: The modules look like small programs that Ansible sends from a control or master machine to all remote hosts or hubs. The modules are run using playbooks (see below), and they control things, for example, packages, services, and documents. Ansible runs all modules to introduce updates or whatever company is needed, and then evacuates them once completed. Ansible gives more than 450 modules to regular tasks.
2. **APIs**: Various APIs (application programming interfaces) are accessible with the goal that one can broaden connection types (which means something more than just SSH for transport), callbacks and more.

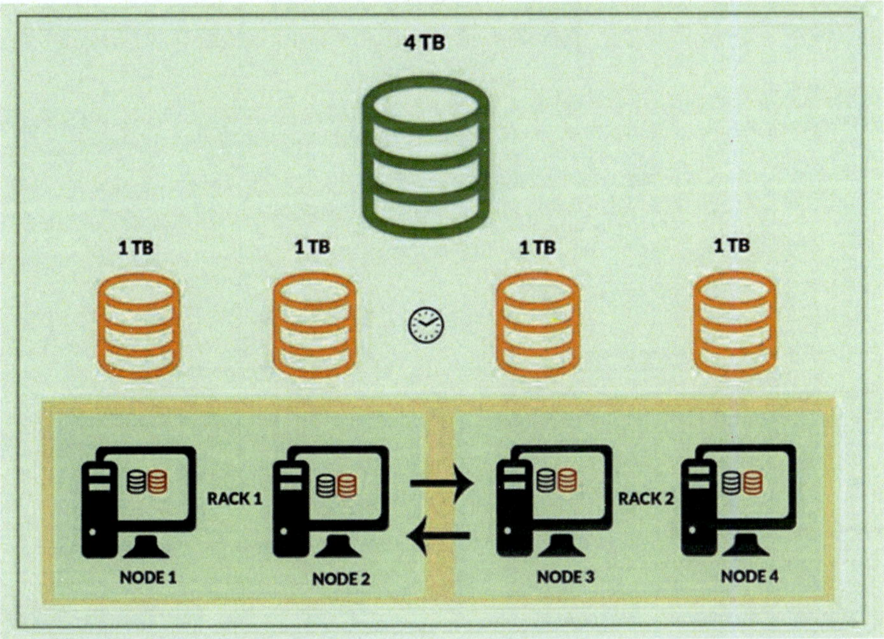

Fig. 1 Architecture of Hadoop distributed file system

3. **Inventories**: All machines used with Ansible (the master or authority machine in addition to nodes) are stored in a simple and solitary document, next to their servers, databases, IP addresses, etc. When the inventory or log is enrolled then one can allot variables to any of the hosts utilizing a simple text document. Indeed, even we can extract the inventory from sources like EC2 (Amazon Elastic Compute Cloud).
4. **Plugins**: They are extra bits of code that increase utility. Ansible accompanies some of its plugins, but you can also compose on your own. Cache, callback and action plugins are three models.
5. **Playbooks**: Ansible playbooks resemble a handbook or guide for tasks. These are basic files written in YAML, which implies YAML Ain't Markup Language, a data serialization language comprehensible by Huan. Playbooks are truly at the core of what makes Ansible so well known, in light of the fact that they portray the tasks to be performed rapidly and without the client having to know or recall a specific syntax. Not exclusively would they be able to announce configurations, however, they can likewise organize the means of any physically controlled undertaking and perform assignments simultaneously or on various occasions. Every playbook comprises at least one piece, and the reason for a play is to map a gathering of hosts to all-around characterized jobs, represented by tasks (Fig. 2).

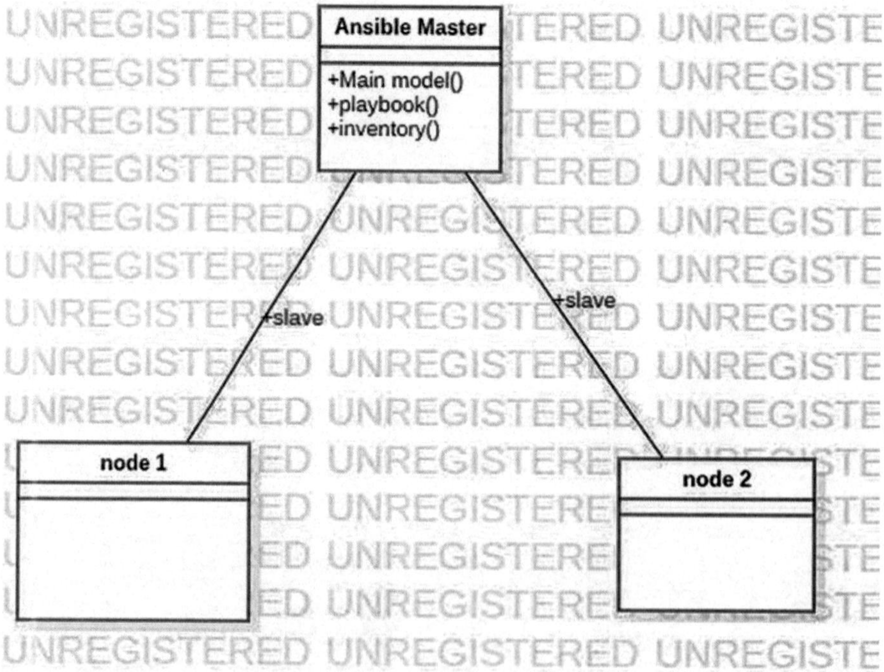

Fig. 2 Ansible architectural diagram

3 Methodology

3.1 Hadoop

The primary infrastructure software services aimed to automate are as follows: -

Step 1: Install Ansible package in Linux using yum. In this installation of Ansible package using "yum install Ansible" command is initiated. Before this, yum is to be configured.

Step 2: Then make Ansible galaxy of Hadoop Cluster in some different folders like playbooks. In this Hadoop cluster is implemented to solve the big data problem using command "Ansible-galaxy in it Hadoop cluster".

Step 3: Accordingly put the client IP in hosts file so that it can read the IP from there and so that playbook can be automatically run in that system. The location of host file would be "/etc./ansible/hosts".

Step 4: Configure Ansible file according to the need in ansible.cfg file.

Step 5: Write a Hadoop cluster role to setup Master Node, Slave Node.

Step 6: Then create a site.yml file in which write a code to import the role Hadoop.

Step 7: Execute the file using the command "Ansible-playbook site.yml".

Step 8: In the Client node role: we copy the java and Hadoop setup files to the respective nodes.

Step 9: In the Master node role: we copy the master node configuration i.e. core-site.xml and hdfs-site.xml on the master node machine.

Step 10: In the Slave node role: we copy the slave-node configuration i.e. core-site.xml and hdfs-site.xmlon the master node machine.

Step 11: Now we have to run the following command:

On Name Node - "hadoop-daemon.sh start namenode" On Data Node - "hadoop-daemon.sh start datanode"

Step 12: On the client machine check the hadoop setup by running the following command: "hadoopdfsadmin -report"

Step 13: To upload a file use the command: "hadoopfs -put filename/"

Step 14: To upload a file use the command: "hadoopfs cat/filename"

3.2 HA Proxy

Step 1: Install ansible on the system.

Step 2: Then make ansible galaxy of HA Proxy Server in a folder. In this we are making HA Proxy which will work as a load balancer. Also, Hadoop cluster is implemented to solve the bigdata problem using command "ansible-galaxy in it haproxyserver".

Step 3: Accordingly put the client IP in hosts file so that it can read the IP from there and so that playbook can be automatically run in that system. The location of host file would be "/etc./ansible/hosts".

Step 4: Configure ansible file according to the need in ansible.cfg file.

Step 5: Write a suitable code inside rolesfor deploying HA proxy and hadoop cluster inside tasks folder.

Step 6: Then create a site.yml file in which write a code to import the role for HA proxy and hadoop.

3.3 Web Server and Ftp Server

Step 1: Create a file with yml as an extension i.e. vim webserver.yml

Step 2: Write the code inside the file as written below.

```
- hosts: 192.168.1.1 //IP address of the system where you want to configure
the webserver.
  tasks:
    - package:
          name: httpd
          state: present

    - service:
          name: httpd
          state: started
          enabled: true
```

Note: Indentation has to be kept in mind.

Step 3: Now save the file and run the playbook using the command:- ansible-playbook webserver.yml

Setting up FTP server using Ansible (Fig. 3):

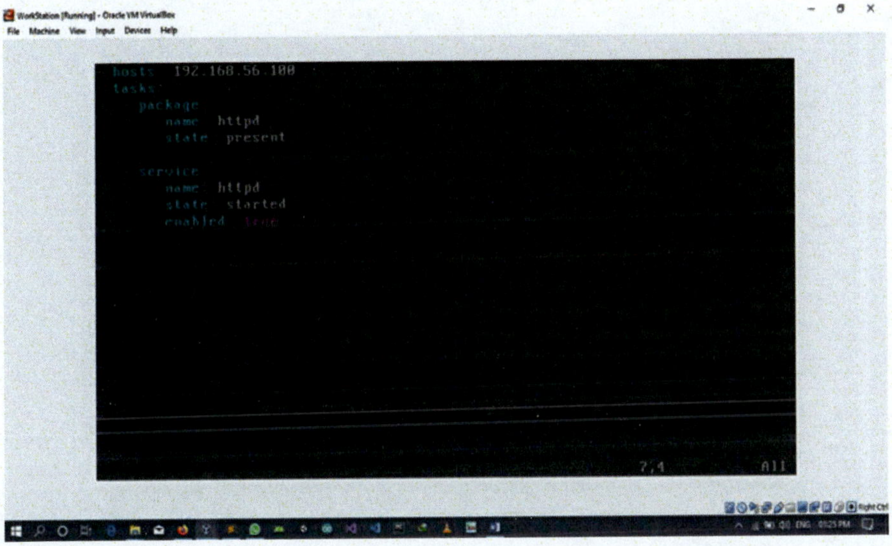

Fig. 3 Web server setup

Step 1: Create a file with yml as an extension i.e. vim ftpserver.yml

Step 2: Write the code inside the file as written below.

```
- hosts: 192.168.1.1 //Ip address where you want to configure the ftpserver.
  tasks:
    - package:
        name: vsftpd
        state: present

    - service:
        name: vsftpd
        state: started
        enabled: true
```

Step 3: Now save the file and run the playbook using the command (Fig. 4):
ansible-playbook ftpserver.yml

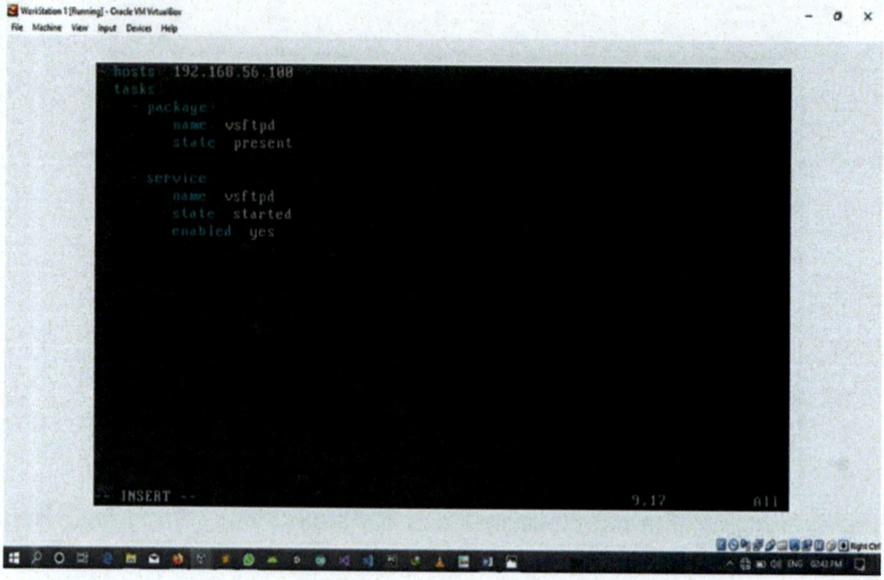

Fig. 4 ftp server setup

Files in YAML Syntax

The indentations are to be used strictly as described below.

```
Client role: main.yml
-command: "rpm -ivh hadoop-1.2.1-1.x86_64.rpm --force"
-command: "rpm -ivh jdk-8u171-linux-x64.rpm --force"
-template:
src: ".bashrc"
dest: "/root/.bashrc"
-template:
src: "core-site.xml.j2"
dest: "/etc/hadoop/core-site.xml
Master Node: main.yml
-command: "rpm -ivh hadoop-1.2.1-1.x86_64.rpm --force"
-command: "rpm -ivh jdk-8u171-linux-x64.rpm --force"
-template:
src: ".bashrc"
dest: "/root/.bashrc"
-file:
path: /master state:
directory 21
-template:
```

```
src: "hdfs-site.xml"
dest: "/etc/hadoop/hdfs-site.xml"
-template:
src: "core-site.xml.j2"
dest: "/etc/hadoop/core-site.xml"
-command: "hadoopnamenode -format -force"
#- command: "hadoop-daemon.sh start namenode"
```

```
Core-Site.xml
<configuration>
<property>
<name>fs.default.name</name>
{% for i in groups["nn"] %}
<value>hdfs://{{ i }}:9001</value>
{% endfor %}
</property>
</configuration>
```

```
-name: deploy slave node import_playbook: sn.yml
-name: deploy client node import_playbook: cn.yml Sn.yml
-hosts: dn roles:
- role: slavenode Inventory
[dn] DATA NODE
192.168.56.115 ansible_user=root ansible_password=redhat #slave1 192.168.56.116
ansible_user=root ansible_password=redhat #slave2 192.168.56.116
ansible_user=root ansible_password=redhat #slave No
[nn] MASTER NODE
192.168.56.114 ansible_user=root ansible_password=redhat #master
```

4 Results

Figure 5 reflects the web-based hdfs portal and gives information about the Hadoop cluster. It tells the live numbers of data nodes with their capacity, current use and the last contract. Basically, it is an overview of the complete cluster configuration.

The following Fig. 6 describes the summary of the particular cluster in detail including the no of replicated blocks and the heap size. It also holds information about the no of dead nodes, the configured capacity, dfs used, dfs remaining and the log file of the configuration.

The Fig. 7 shows the information about the files that have been stored on the particular cluster. In addition to that detailed information about the file is available that includes the type of the file, the size, the permission, the owner and the group. The data of the files can also be accessed from the portal along with the local logs for the file entries present on the bottom of the screen.

In this paper, we have obtained the following result: -

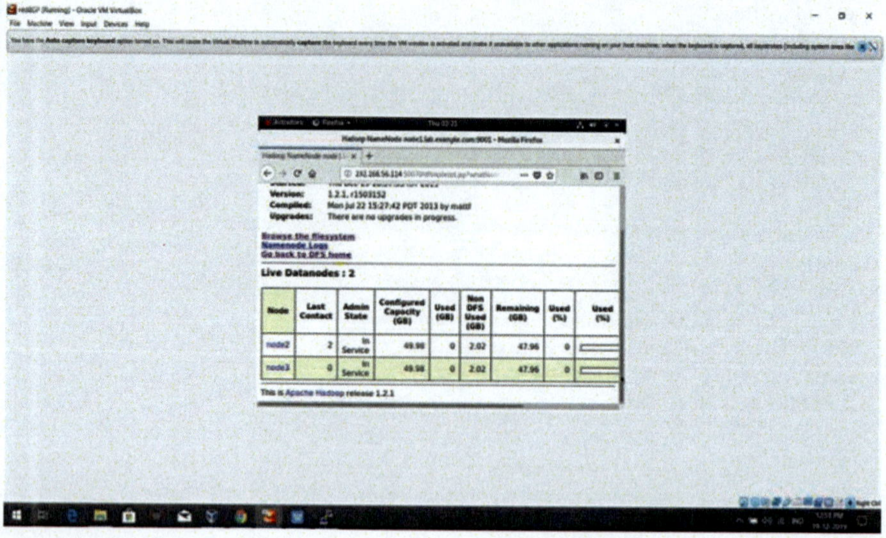

Fig. 5 Hadoop cluster

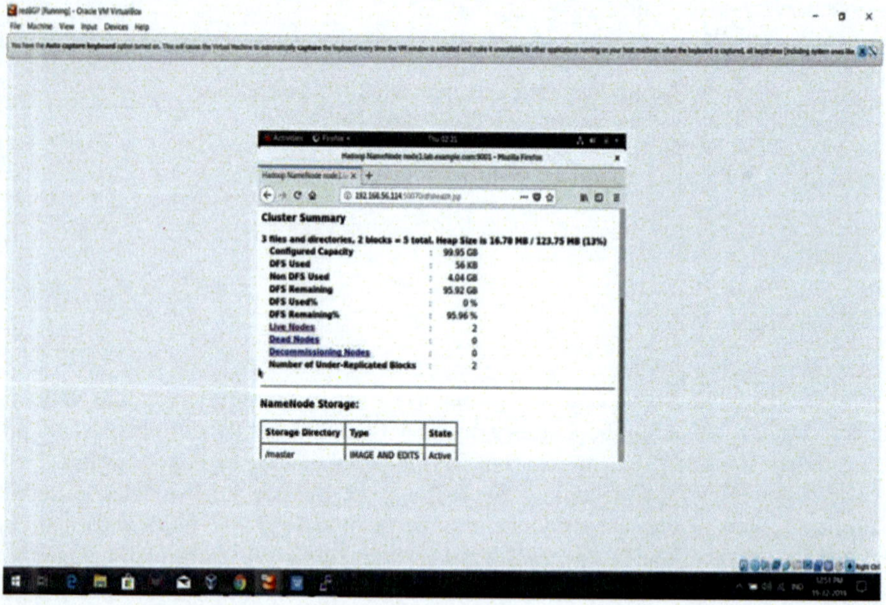

Fig. 6 Cluster summary

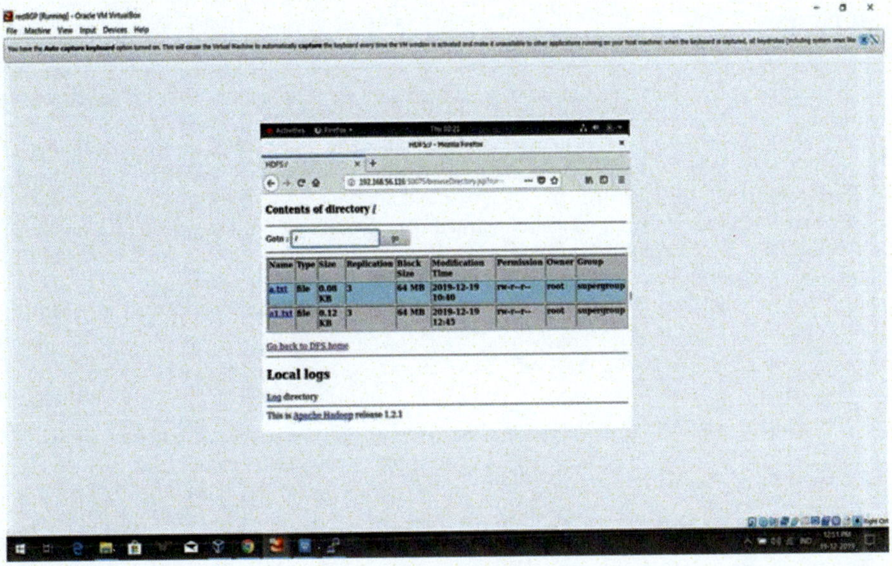

Fig. 7 File stored in cluster

- Successfully created a multi-node cluster which gives us the required amount of hdfs storage. In our setup fifty terabytes of storage from each of the slave, nodes add up and give as hundred terabytes of storage as a whole.
- Successfully uploaded and retrieved files from the cluster which can be seen in the file upload section below.
- Managed to retrieve the file even when one of the nodes was disconnected while retrieving improvising high availability of the current prototype.
- Successfully set up the HA Proxy server i.e. the load balancer mode, webservers and ftp-servers.

5 Conclusion

In this paper HA proxy server, webservers, and Hadoop-cluster with the help of an automation tool called Ansible had been created. Ansible is an open-source mechanization platform, used for IT jobs, for example, framework management, application installment, intra-service coordination and accoutering. Further in the paper the YAML and jinja languages are used to automate both the HA Proxy server and Hadoop cluster. Henceforth, the paper highlights the implementation of an automated Hadoop cluster so as to solve the big data problem through Ansible (Devops) also make use of an automated HA Proxy Server in order to solve the problem of the website crashing or server down problem.

References

1. J. Anuradha, A brief introduction on big data 5Vs characteristics and Hadoop technology. Procedia Comput. Sci. **48**, 319–324 (2015)
2. J. Shafer, S. Rixner, A.L. Cox, The Hadoop distributed file system: Balancing portability and performance, in *2010 IEEE International Symposium on Performance Analysis of Systems & Software (ISPASS)*. IEEE (2010)
3. J. Dean, S. Ghemawat, MapReduce: simplified data processing on large clusters. Commun. ACM **51**(1), 107–113 (2008)
4. https://www.howtoforge.com/tutorial/ubuntu-load-balancer-haproxy/
5. https://www.redhat.com/files/summit/session-assets/2017/LT122010-Observability-and-automation.pdf

Evaporatively Cooled Window Air Conditioner in a Hot and Dry Climate—An Experimental Analysis

Pragati and Parinam Anuradha

1 Introduction

Air conditioners have become a necessity for human comfort, commercial and many industrial processes. There are places where weather conditions are quite extreme, that is, very hot and dry weather conditions, that is, the value of wet bulb depression (WBD) is very high. In that case, it is very necessary to stop the cutoff of the compressor due to high condenser pressure as well as to increase the performance of the air conditioner. If the outside temperature is high, then the air conditioner will not be able to cool the indoors effectively due to frequent cutoffs. Heat transfer rate in condenser should be increased & pressure should be minimum, to increase the performance of air conditioner.

Any attempt to minimize the power consumption of cooling systems will lead to considerable energy savings and environmental protection. This can be achieved by improving the COP of the system [1]. During the summer season in Egypt, the daily maximum temperatures approach 40–45 °C. To deal with the issue a solar-based hybrid desalination cooling system was made and tested according to the Egyptian weather conditions [2]. At high ambient temperature, pre-cooling of air entering the condenser is a must to maintain condenser pressure and improve the heat rejection process. Overall system performance was also improved with this modification [3]. The process of minimizing the temperature of cooling air in the condenser can be achieved by using a pad coupled evaporative cooling system to pre-cool air at the inlet of the condensing unit [4]. Experiments were carried out for energy saving of split air conditioner using evaporative cooling technology and

Pragati (✉)
Thermal Engineering, UIET Kurukhetra, Kurukshetra, India
e-mail: pragatibajpai96@gmail.com

P. Anuradha
Mechanical Engineering Department, UIET Kurukshetra, Kurukshetra, India
e-mail: aparinam2015@kuk.ac.in

S. Vadhera et al. (eds.), *Latest Trends in Renewable Energy Technologies*, Lecture Notes in Electrical Engineering 760, https://doi.org/10.1007/978-981-16-1186-5_20

found that an overall increase of COP was from 29% to 53% [5]. By putting media pads on the entry of air with water injection in it from the top, electricity-saving was claimed to be about 20% for a retrofitted system with 34 °C ambient temperature and 10% reduction in power consumption [6]. Hajidavalloo E conducted a study of evaporative cooling on the window air conditioner and found an increase in the COP by about 55% and power consumption was reduced by 16% [7]. Studies were carried out by A. Sinan Karakurt et al. [8] to investigate the energetic and economic effects of subcooling and superheating on vapor compression refrigeration systems. Though a few researchers have shown the potential of evaporative cooling in a vapour compression air conditioning system, still there is a need to explore the function of evaporative cooling on the small-size refrigeration system. Therefore in the present work performance of the evaporatively cooled window air conditioner is analysed by integrating the evaporative cooling pad in the condenser.

2 Experimental Setup and Description

2.1 Experimental Setup

In the present work, the basic idea is to increase the performance of window air conditioners using evaporative cooling. The cooling pad placement is important for evaporative cooling. After studying various papers on evaporative cooling, the idea of placing the honeycomb cooling pad between the fan and condenser came up. The experimental setup consists of a modified window air conditioner is shown in Fig. 1.

A small pump is used for the water supply on the cooling pad and the tube is adjusted in such a way that it sprinkles water on top of the cooling pad and not on the condenser. The cooling pad is placed inside the slot which was cut specially for its placement. With the help of a digital clamp meter, electric current and voltage supply are measured and work consumed by the compressor is calculated. Digital Temperature sensors are used to measure the temperature at various points.

Fig. 1 Experimental setup

2.2 *Test Room*

The test room has dimensions as length −3.52 m, breadth −3.52 m and height −3.11 m. The setup is placed in a room of a house with basic furniture present in it. The air conditioner is at a height 0.6 m from the ground. Measuring instruments are placed at specific places in the room to measure the required values for further analysis.

3 Performance Indices

To check the performance of the air conditioner some basic performance indices have been taken as mentioned below:

(1) **Coefficient of performance (COP)**:
The coefficient of performance is defined as the ratio of refrigeration effect and the work consumed by the air conditioning unit.

$$COP = \frac{Refrigeration\ effect}{Work\ Consumed} = \frac{R_E}{C_W}$$

(2) **Refrigeration Capacity**:
Refrigeration capacity is the total cooling effect per unit time obtained from the air conditioner.

$$R_E = m_a c_p dt,\ \text{kW}$$

where, m_a: mass flow rate of air kg/sec, c_p: specific heat capacity of air kJ/kg-K, dt: temperature difference.

(3) **Supply mass flow rate of air**:
The air coming into the room per second is known as the mass flow rate of air.

$$m_a = \rho A v_a$$

where, ρ: density of air in kg/m^3, A: area of supply duct in m^2, v_a: Velocity of air in m/s.

(4) **Power Consumed**:
The amount of current and voltage i.e. power in kW, the compressor is consuming is known as power consumed by the air conditioning unit.

$$W_C = \frac{VI\cos\varphi}{1000}$$

where, V: Voltage, I: Current supplied, cos φ: Power factor of compressor = 0.8 Along with these formulas, refrigeration charts and tables are also used for required data for given temperature and pressure conditions.

4 Results and Discussion

The experimental setup is prepared for hot and dry weather in the month of June 2020 at Kurukshetra. The basic idea behind the research is to deal with the hot and dry weather conditions when window AC performance becomes very poor or sometimes due to high condenser pressure (due to high atmospheric temperature) AC performance is drastically reduced. So, the performance of the window air conditioner is analysed by integrating evaporative cooling pads in the condenser to increase the COP and reduce the power consumption during hot and dry weather conditions.

Different performance parameters have been calculated like the coefficient of performance (COP), refrigeration effect, power consumption, etc. and are compared in this section. The comparison is done between a simple air conditioner i.e. with no modifications and an air conditioner with a cooling pad only. As it was earlier mentioned that placement of a cooling pad in the air conditioner for evaporative cooling is important, so this comparison is done to check the effect of a cooling pad on the operation parameters of the air conditioner.

4.1 *Comparison of Effect on COP with and Without the Cooling Pad*

In Fig. 2, the effect on the actual coefficient of performance (COP) of the air conditioner with evaporative cooling is being compared with the COP of normal air conditioner w.r.t time. The Carnot COP was also evaluated and compared in the graph. The ambient temperature was 39 °C and relative humidity was 54%. Honeycomb pad gives better performance than any other type of cooling pad, so instead of using

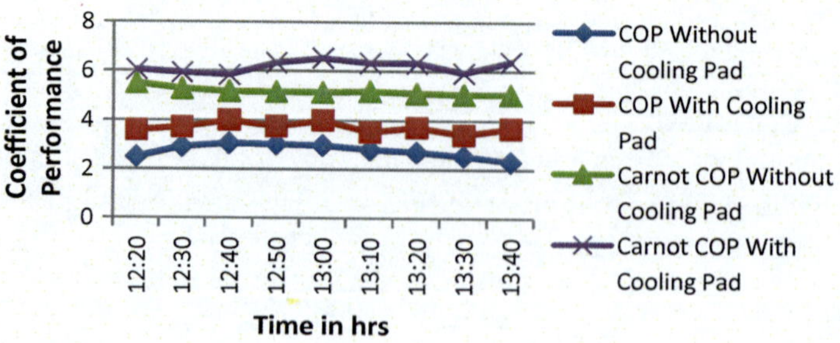

Fig. 2 Comparison of effect on COP with and without cooling pad

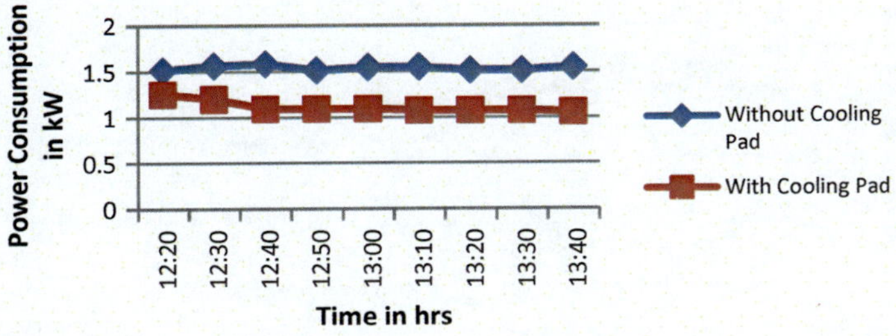

Fig. 3 Comparison of effect on power consumption with and without the cooling pad

any other type, a honeycomb pad is used. It can be seen in Fig. 2 that the placement of the cooling pad has increased the COP by about 45–50%. The results are also in consonance with the experimental findings of Hajidavalloo [7] wherein about 55% of the increase in COP was observed and power consumption was decreased by about 16%. Also, Carnot COP is marked in the graph for reference value and comparison. So, it can be concluded that using a honeycomb cooling pad for evaporative cooling and placing it in between the condenser fan and the condenser will increase the performance of the air conditioner during hot and dry weather conditions.

4.2 Comparison of Effect on Power Consumption with and Without the Cooling Pad

Figure 3 shows that by using a honeycomb cooling pad for evaporative cooling, a reduction in the power consumption by the compressor is observed. The comparison is being done for the power consumption between the simple and the modified air conditioner with respect to time.

The power was supplied to the air conditioner for approximately one hour and the aim was to achieve comfortable room conditions, that is, a temperature of 24 °C & 50%RH. That means the readings were taken according to the range of the comfort zone of human beings. Some losses can be considered as well because of the furniture present inside the room. Hence, after considering all the facts related to it, the reduction in the power consumption is about 30–32% and can be seen in Fig. 3.

4.3 Comparison of Effect on Refrigeration Effect with and Without the Cooling Pad

Figure 4 shows the variation of refrigeration effect with respect to time for both with and without the use of the cooling pad. It can be observed that the change in

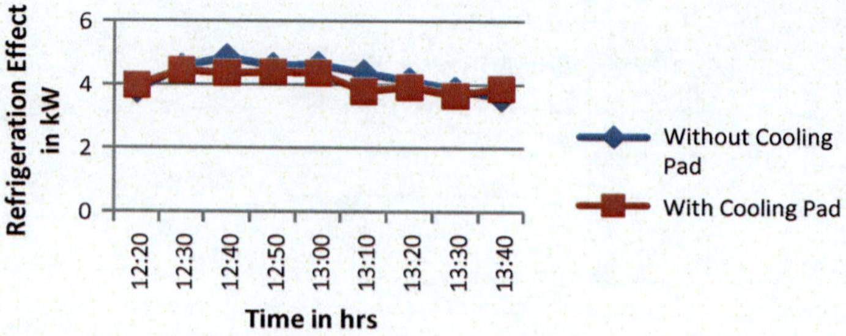

Fig. 4 Comparison of effect on refrigeration effect with and without cooling pad

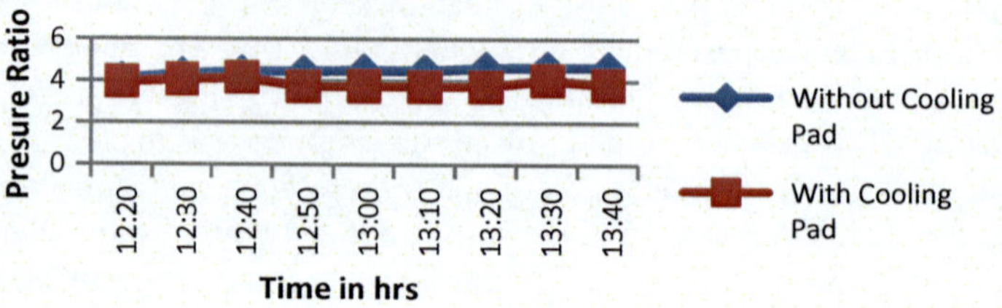

Fig. 5 Comparison of effect on pressure ratio with and without cooling pad

refrigeration effect for both cases is very small as only condenser conditions are changed and evaporator operating conditions are not changed.

4.4 Comparison of Effect on Pressure Ratio with and Without the Cooling Pad

The temperature of water used in the cooling pad is about 30 °C, so atmospheric air passing over the condenser through the cooling pad reduces the condenser pressure which ultimately reduces the pressure ratio. Reduction in pressure ratio not only reduces the work consumption by the compressor but also increases its volumetric efficiency. The variation in pressure ratio has been depicted in Fig. 5.

4.5 Comparison of Effect on the Temperature of the Air Entering the Condenser with and Without the Cooling Pad

The temperature of the air passing over the cooling pad depends on the Wet Bulb Depression (WBD) i.e. the difference of Dry Bulb Temperature (DBT) and Wet Bulb

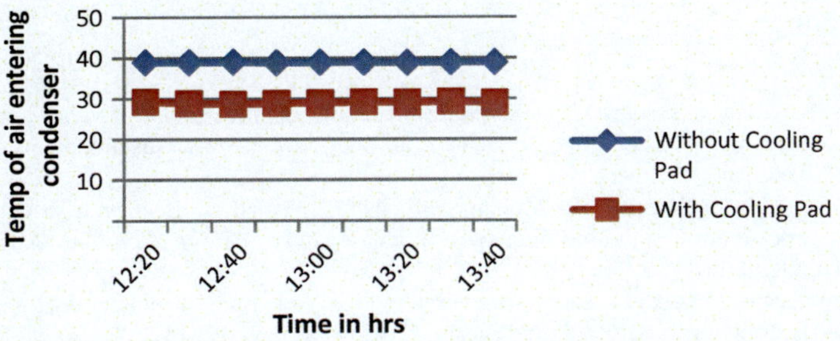

Fig. 6 Comparison of effect on temperature of air entering the condenser with and without cooling pad

Temperature (WBT) and temperature of water in the cooling pad. The change in air temperature is of the order of 10 °C as shown in Fig. 6.

5 Conclusions

The following conclusions are drawn from the present research work:

(i) The incorporation of a cooling pad in between condenser and condenser fan is very effective in window air conditioners during hot and dry weather conditions i.e. when WBD is quite large.
(ii) It has been observed that the actual COP of the window air conditioner increases by 45–50%, Carnot COP increases by approx 25%, the power consumption reduces by 30–32%, the pressure ratio reduces about 20%.
(iii) The room comfort conditions are achieved marginally sooner with evaporative cooling in the air conditioner using a honeycomb cooling pad in comparison to when the air conditioner runs without a cooling pad.
(iv) The cost involved in the modification of the window air conditioner is very less in comparison to the cost saved in comparison to the running cost using the evaporative cooling method.

The use of an evaporative cooling pad in hot and dry weather conditions will not only increase the performance of the air conditioning system but also prevents the situation of extreme pressures when the cut-off switches turn down.

References

1. K. Harby, R. Doaa, S. Nader, M.S. Hassan, Performance Improvement of vapor compression cooling systems using evaporative condenser: an overview. Renew. Sustain. Energy Rev. **58**, 347e60 (2016)
2. S. Ehab, K. Harby, A. Ahmed, R. Mohamed, S. Ahmed, Weather effect on a so- lar powered hybrid adsorption desalination-cooling system: a case study of Egypt's climate. Appl. Therm. Eng. **124**, 663e72 (2017)
3. K.A. Jahangeer, A.O. Andrew, M. Raisul, Numerical investigation of transfer coefficients of an evaporatively cooled condenser. Appl. Therm. Eng. **31**, 1655 16 (2011)
4. G.V. Michalis, E.F. Andronikos, T.K. Georgios, D.K. Eleftherios, Incorporated evaporative condenser. Appl. Therm. Eng. **27**, 823e82 (2007)
5. K. Harby, F. Al-Amri, An investigation on energy savings of a split air conditioning using different commercial cooling pad thicknesses and climatic conditions. Energy **182**, 321e336 (2019)
6. E. Hajidavalloo, Increasing COP of window air conditioner in very hot weath- er of Khoozestan. Research Project Report to Management and Programming Organization
7. E. Hajidavalloo, Application of evaporative cooling on the condenser of window-air-conditioner. Appl. Therm. Eng. **27**, 1937–1943 (2007)
8. A.S. Karakurt, U. Gunes, Y. Ust, Exergetic and economic analysis of subcooling & superheating effect on vapor compression refrigeration system, in *Proceedings of the ASME 2016 Power Conference* POWER2016, Charlotte, North Carolina, POWER2016-59492 (2016)

Evolution and Advancements in Solar Drying Technologies: A Review

Shubham and Sunil Nain

1 Introduction

Getting the non-renewable resources consumed year by year, humans are continuously cutting off the available fossil fuel reserves. There will be a time when these non-renewable resources of energy got completely washed away and there will be a surge in demand. The production of oil reached maximum around 1980s and is NOW declining slowly. The reserves of oil and coal may be getting consumed within another 50 years. It results in greater dependence on coal sector and the coal production will touch a maximum in 2030–2060 and about 80% get consumed by 2250 A.D. [1]. So, it is better to find the alternate sources of energy on which we can rely easily for our increasing energy demands in near future. Renewable energy sources include solar energy, hydro, wind, geothermal, biomass, ocean current and waves. In almost every climate, sun's energy can be harvested by solar collectors for large-scale grid lines as well as for small-scale direct use applications [2]. So, solar energy available for various purposes like conversion into another form and directly utilizing it for drying products are optimized from many eras, so that our reliability should only be on renewable sources of energy. The solar thermal systems capture sun's energy by using flat plate collectors and transfers it directly or indirectly to household, water or heating applications. Solar thermal power plants use solar energy on large scale by focussing the energy by employing concentration devices to intensify the heat collected. Irrespective of availability of many renewable sources of energy, solar energy is the prime source of generation of other sources, that is why it is called as ultimate source of energy.

Shubham (✉)
Thermal Engineering UIET, KUK, Kurukshetra, India
e-mail: shubham7r@gmail.com

S. Nain
Assistant Professor (Mechanical Department) UIET, KUK, Kurukshetra, India
e-mail: nain_sunil@rediffmail.com

S. Vadhera et al. (eds.), *Latest Trends in Renewable Energy Technologies*, Lecture Notes in Electrical Engineering 760, https://doi.org/10.1007/978-981-16-1186-5_21

1.1 Importance of Solar Thermal Energy

The world is moving towards the use of solar energy available to us to compensate for the increasing energy demands along with decreasing conventional resources of energy. Recently, India has many schemes that focus on shifting the power sector on solar energy by installing high-capacity solar power plants. India being the tropical country has good availability of solar flux throughout the year. Solar thermal systems have applications in large-scale to small-scale works. Solar drying is an essential process for the preservation of agriculture-based crops and industrial operations such as dairy processing, textile production, wood and timber processing, biomass treatment and wastewater treatment [3]. For fulfilling increasing demand of energy worldwide, development for novel solar dryers is considered as key solution. The rapid growth in developing solar technologies also creating some barriers, which must be removed, and their solution has been reviewed. Despite of some drawbacks like high initial costs, solar energy technology is still one of the most favourable sources of renewable energy. Researchers are now focussed on improving and encouraging the need for solar energy against other renewable sources and traditional energy sources [4]. The agricultural products require consistent low heat drying like coffee, cocoa beans, grains, nuts, timber, tea, fruit, etc. The traditional methods of crop drying accomplished by burning fossil fuels and biomass in ovens are expensive and harm the environment by creating many types of pollution. These methods have shortcomings, which can be prevented by using non-renewable source of energies that are eco-friendly [5]. In today's food supply market, dried food products show increase in demands. For fruits and vegetables, they contribute about 1% of total drying in food industry. Drying process involves lowering of water content to avoid deterioration of food by any micro-organism growth [6]. Solar thermal energy can be utilized for cooling purpose by equipping the energy developed from solar radiations in various thermodynamic cycles. Thermodynamic modelling for adsorption and absorption solar cooling systems are presented to analyze system efficiency and its optimization [7].

2 Evolution in Solar Drying

Drying is known from old times and we see the traces in rural as well as urban areas. The oldest preservation process known to mankind is the drying process that can even be tracked from prehistoric times. Drying becomes an essential part of almost every day-to-day activity and becomes an integral part of many processes around us. The classification of solar drying methods and their evolution are shown in Fig. 1.

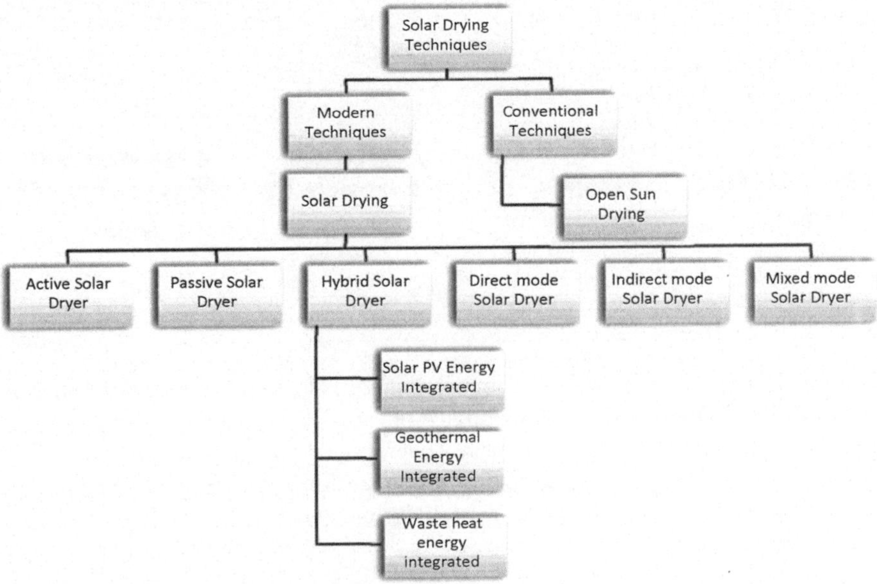

Fig. 1 Solar drying technique classification

2.1 Conventional Techniques

2.1.1 Open Sun Drying

Since time immemorial, solar drying is the practice for the preservation of agricultural crops and food. Particularly, solar drying wad carried out by open sun drying under open sky. Due to an enormous amount of food materials processed by this method, more efforts for its improvement and development were carried out. It is estimated that annual savings in millions of dollars can be realized by lowering the cost of drying, products' quality improvement, loss reduction due to spoilage, deterioration and other factors [8]. In tropical and sub-tropical countries, traditional open sun drying is the most common method to preserve the agriculture-based products. Pulses and cereals are either simply left hanging on plants or bound in bundles after harvesting and dried on open fields. Reaping and threshing of wet grain before time become common practices in many countries. The wet grains need to be dried in thin layers on open ground and exposed to sunlight and wind for drying. In tropical countries and sub-tropical countries, products like spices, vegetables, fruits, oilseeds and fish are dried naturally by spreading them over mats and trays under sunlight [9]. Irrespective of such advantages, solar drying is comparatively slow process, which incurs losses of products. There is a reduction in quality of the product due to insect manifestation, enzymatic reactions, growth of microorganisms and development of mycotoxin. The product may get spoiled due to adverse atmospheric conditions like

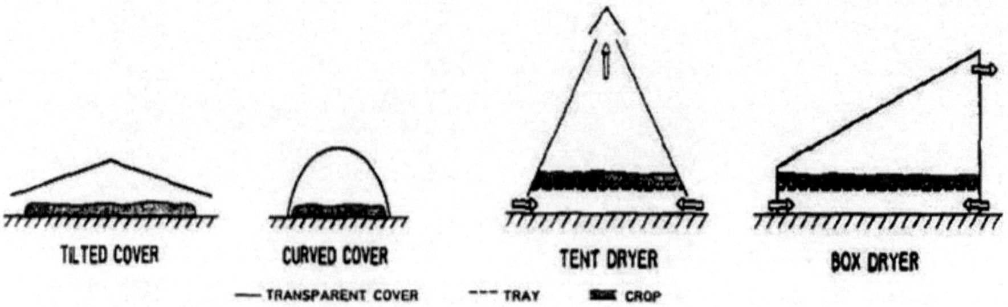

Fig. 2 Simple natural convection type dryers [9]

moisture, dust, wind, rain and loss due to animals and birds. Conventional drying is labour intensive, cumbersome and requires more areas for spreading. This process is costlier due to high energy requirement. So, a new method is evolved, which involves solar drying in chambers directly or indirectly under sunlight know to be solar drying. Solar drying is the best substitute for natural drying and artificial mechanical drying process (Fig. 2).

2.2 *Modern Techniques*

2.2.1 Solar Drying

Drying being the oldest available preservation method to mankind plays an important role in today's food market food supply chain. Various aspects are considered related to drying small fruits and vegetables. Minimal exposure to light, heat and oxidation of fruits and vegetables help in conservation of critical bioactive compounds. Various drying techniques are present, but there are few commercialized applications in industries related to solar drying that remains unexploited due to lack of interaction between researchers and industries [10]. Drying process slows down the decay process by reducing the content of moisture in harvested crops to enable long-term storage. For drying of crops, solar dryers are used which lowers the crop loss and the recurring costs associated with the use of fossil fuels for drying. Key operational, design and environmental parameters are discussed including drying psychrometry of crops and conversion of initial crop properties to final desired product properties [11]. Solar drying offers faster drying as compared with sun drying and retains more vitamins in fruits, especially vit A and C. Solar drying results in decrease in 27–80% consumption of conventional energy and produces an average solar efficiency of 40%. Experimental results show that with controlled moisture removal, improved quality of product and reduction in drying time occur [12]. Post-harvest loss of fig is tried to prevent by solar drying of fig to meet the market demands in various rural parts of world. Solar thermal systems provide numerous applications in preservation irrespective of geological and technological challenges. Solar drying is considered to be best for fig preservation technique. The traditional, artificial and solar drying techniques used for fig preservation are compared and solar drying is found to be adequate technique [13]. Drying process ranges from traditional open sun drying to large-scale industrial drying. Use of fossil fuels to drying agricultural products in almost every developing country becomes infeasible due to unaffordable costs to farmers in rural areas [14]. Consequently, traditional open sun drying in rural areas of developing countries practised at large scale. It suffers unbearable loss of product due to inadequate drying, encroachment due to animals and birds and microorganisms' attack. So, a proper design has to be made for drying some of the agriculture-based products in developing countries [15, 16]. In solar drying, solar radiations can either used as only source of required heat or as a supplement source. The air can be blown by natural buoyant forces or by forced agency. The drying process may involve treatment of product with preheated air or direct exposure to light and sometimes may be a combination of both. The focus is to supply heat to product by convection and conduction to air around, having temperature more than that of product [17]. Solar dryers can be classified into various categories on the basis of their built, application, operation and several other factors to meet the increasing demands of solar drying and prevent disadvantages associated with solar drying (Fig. 3).

Classification on basis of exposure of product to sunlight

Direct Solar dryers: When the solar energy collected is directly fed to the products and roof/ wall of the drying chamber forms integral part of solar energy collection. In this type of solar dryer, solar radiations directly fall on products placed directly

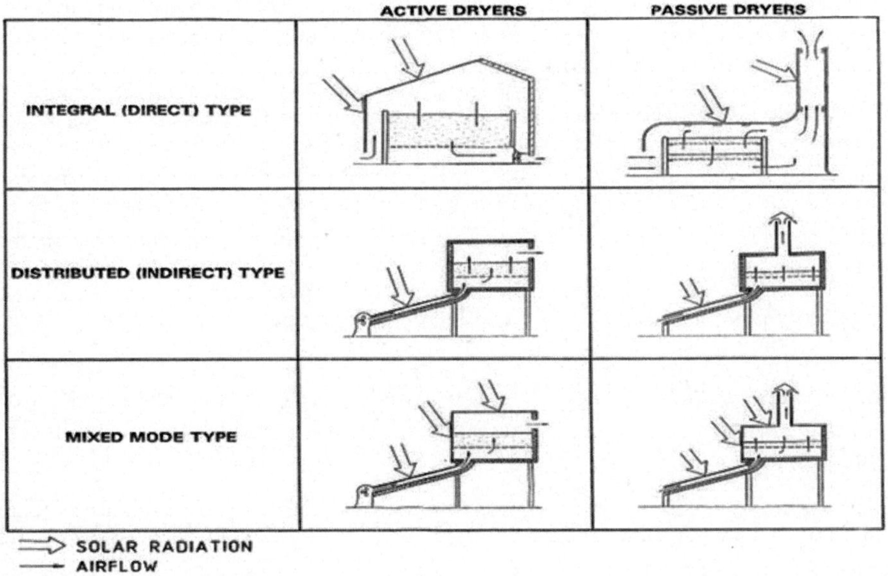

Fig. 3 Typical solar energy dryer designs [17]

in chamber having transparent wall. The relative humidity of the air decreases and carries moisture from products. The humidified air is removed either naturally or by pressure difference created by fan [18–20]. In direct-type cabinet solar dryer, ultraviolet exposure may lead to discolouring of crops. The use of selective acrylic aperture material prevents discolouring and loss of antioxidant content. Drying of apricots demonstrated this process [21].

Indirect Solar Dryers: When drying chamber is separated from solar collector, it forms indirect solar drying system. In this type, the material is not directly exposed to solar radiations, which prevents discolouration and cracking of surface of crops during drying. The air carries heat that removes away the moisture from the material. In indirect solar dryer, the drying chamber is insulated and has opaque walls where the crops are placed on perforated trays. Since, solar radiations do not incident directly on crops, local heat damage and caramelization does not occur. These dryers retain vitamins in foods and colour in highly pigmented materials [22, 23].

Mixed Mode Solar Drying: The drying chamber and absorber plate add energy to dryer system in this mode. A typical mixed mode drying system is basically an indirect solar drying system with one glazed surface on dryer to add heat of solar radiations [24]. These types of solar dryers are used for drying high moisture food materials like papaya, kiwi, banana, brinjal, cabbage, etc.

Classification on the basis of type of air flow

Passive Mode Dryer: When the heated air flows duo to buoyant, forces generated due to variation in densities of air after getting heated. These drying systems are used for materials for which the moisture removal rate is comparatively slow and higher mass flow rates only result in loss of energy of the system and increase in energy. These types of dryers are used for low load conditions and higher temperatures.

Active Mode Dryer: These use fans and blowers to circulate the air to drying unit, which is heated after passing through solar collector. Higher moisture content crops such as cauliflower slices, cabbage, kiwifruit, brinjal and papaya, active solar dryers are found to be more efficient as compared with passive solar dryers [25, 26]. Forced convection greenhouse dryer systems are used for commercial large-scale applications [27].

Classification on the basis of type of Collection techniques

Flat plate collector type: When the collector concentration ratio is unity, the collector type is called flat plate collector. These exhibit simple configuration and have no moving parts. It requires low maintenance. It can be used for application for temperature range of 40–100 °C.

Parabolic trough collector type: when temperatures required for drying are higher than 100 °C, concentration of solar radiations becomes necessary. Concentration is achieved by focussing or concentrating collectors. In parabolic trough collectors, the concentrator has shape of a cylindrical parabola. Rotation about a single axis is generally required to track the solar radiations [1] (Fig. 4).

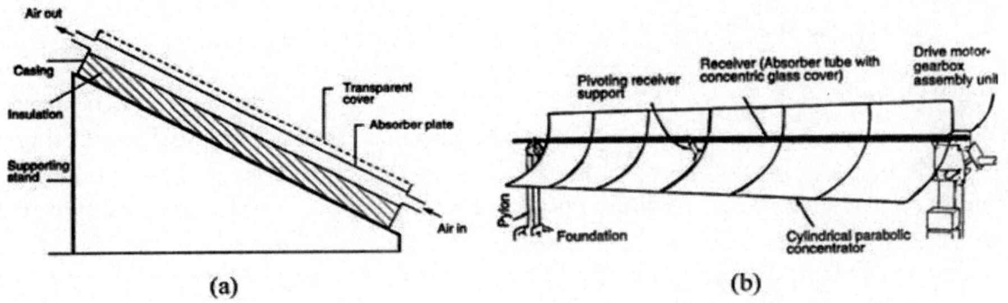

Fig. 4 Types of collector **a** Flat plate Collector, **b** Cylindrical Parabolic trough Collector [1]

The initiative of low carbon power sector aims at use of non-fossil power, i.e., the power generation from renewable sources of energy. The introduction of renewable energy into power system has issues of costs regarding flexibility in power generation and integration costs.

2.3 *Hybrid Solar Dryers*

2.3.1 Integration with Geothermal Energy

Renewable energy despite of many advantages over non-renewable sources is yet to be cost-competitive with crude oil. Issues of low capacity factor, grid instability and intermittency lead to less reliability towards solar drying systems. So, the hybrid of solar and other renewable sources of energy like geothermal energy is sought to be mutually beneficial and a good pair of renewable sources of energy to provide a more stable system [28]. Among the modern renewable sources of energy like geothermal, solar and wind energy, they may be the most promising sources because of market penetration, abundance, relative maturity, etc. But both solar and geothermal energies have merits and drawbacks aligned in such a manner that merit of one overcomes the disadvantage of other system vice versa [29]. In 1975, an assessment was reported by Finlayson and Kammer for solar geothermal hybrid systems. They drew much attention on solar thermal hybrid systems for power generation. Mutual compensation of solar and geothermal energy properties leads to a more stable hybrid system. Hybrid solar–geothermal system outperforms the standalone counterparts thermodynamically when operated under fully optimized conditions. Recent studies reveal that solar–geothermal has a great capacity for poly-generation process such as electricity, heating, cooling and freshwater production [30–32].

2.3.2 Integration with Solar PV Energy

Solar radiations can be converted into electrical energy by two ways. In the first method, photovoltaic effect is used to generate electricity from solar radiations falling on panels made of semiconducting materials. This does not require any mechanical equipment for energy conversion. Photovoltaic systems generally operate within specific range of wavelengths. The solar radiations out of the given bandgap get dissipated in form of thermal energy [33]. The loss of thermal energy in the above process can be prevented by using new type of photovoltaic systems. The conversion of solar energy into electricity involves physical processes such as electron transportation, light absorption and recombination mechanism, which are determined by semiconducting material electro-optical properties used during their manufacturing. Other techniques are solar thermal technologies, where thermodynamic cycle is used to convert solar energy into mechanical work and then turned into electricity [34].

2.3.3 Waste to Heat Energy

Waste energy hybridization with solar energy can compete the current energy technologies and make them complementary [35]. It needs careful consideration before integrating these two technologies such as technical feasibility and economic and environment aspects. Thermal waste treatment energy recovery is the fundamental part of waste management system in municipal solid waste management. According to the European waste energy treatment strategy, lesser will be the production of waste energy, lesser will be the consumption of natural resources [36]. High efficiency of energy recovery value is important for environmental sustainability of waste to energy plants [37, 38]. Combining the solar energy with waste to energy system is not the entirely new concept. The solar energy parabolic trough collectors can be used to preheat water and air and increase steam temperature. Both cases result to additional electricity generation and increase in the overall efficiency of plant [39].

3 Advancements in Solar Drying

3.1 Solar Drying with Thermal Storage

Solar energy is governed by meteorological and geographical conditions and characterized by a variable nature. So, integrating a thermal storage system can improve the efficiency of solar thermal system. Thermal storage systems compared with other storage technologies have advantages like lower capital costs and high operating efficiencies [40]. Using phase change material for storing thermal energy is one of the efficient methods for thermal storage. Phase change materials provide moderate and uniform temperature, which is required to dry almost every agriculture crop sufficiently [41]. Low-temperature storage systems are based on sensible heat storage by using liquids such as molten salts and oil. Irrespective of higher evaporation temperature, these storage media have disadvantages like degradation and high freezing points [42, 43]. In recent times, research in solid thermal storage field is increased to enhance the thermal performance of concrete by adding phase change materials in it. A large number of phase change materials are added to concrete to improve its thermal capacity [44, 45].

3.2 Solar Drying with Desiccant

Solar drying is required throughout the year. In hot and humid conditions of tropic and sub-tropic regions like India, the humidity of outside air creates difficulty in drying process for open solar drying systems. So, desiccant materials are used to dehumidify the air coming inside the drying system to achieve proper drying and prevent early saturation of air during drying. Desiccant materials absorb moisture in the basis of vapour pressure difference. Desiccants are widely used in marine-cargo, electronics, pharmaceutical and food storage applications [46]. Desiccant air-conditioning is not only eco-friendly and energy-efficient process but also cost-effective for hot and humid conditions [47].

4 Conclusion Remarks

Drying is the basic process in almost every preservation process in agricultural and industrial products. Solar drying systems are evolved since the time of their invention. Advancements in these systems lead energy sector towards a new future where the problem of dependence on fossil fuels will be cut down to very low levels. As drying process contributes a big part of energy consumed annually in domestic and commercial works. So, optimization of these systems is the main research topic for many researchers. The advantages solar drying systems carry over other renewable energy system discussed above are: (a) Solar drying systems are found out to be more effective than traditional open sun drying techniques used in rural areas. So, cheap and reliable solar dryers are designed and manufactured for rural applications; (b) due to intermittent nature of solar radiation flux, uniform drying of products becomes impossible. Using thermal storage in solar drying systems provides constant temperature drying and long hours of drying even in off-sunshine hours; (c) in hot and humid conditions during monsoon weather, the drying process becomes slow due to presence of high water vapour content in atmospheric air. High humidity leads to higher vapour pressure of water in air that slows the dehumidification process. Desiccant materials are used to remove moisture from air going inside the dryer and can be regenerated easily in sunlight. These advancements and modifications make solar drying systems more reliable and versatile.

5 Scope of Work

The study of new technologies and advancements in solar drying processes leads us to find more technologies that can be linked to these systems to make them even better and completely substitute the non-renewable energy sources by solar energy source. In future, the desiccant materials can be used along with hybrid solar dryer systems for performance improvements. A variety of desiccant materials can be tested for high and low drying temperature conditions. Waste energy utilization systems are also the topic of research in future, which can reduce the problem of throwing high energy into the environment, which may be harmful to various animals and plants on aquatic and terrestrial ecosystems. The waste energy is used as input energy source, which is low-grade energy to solar drying systems and it boosts up the overall efficiency. The variability of performance parameters used in solar drying systems gives the flexibility of applications or drying processes to be carried out on a single drying unit. Rural areas generally deprived of sufficient energy and fewer resources have to be made aware about the recent trends and advancements in solar energy in drying processes.

References

1. K. Sukhatme, S.P. Sukhatme, *Solar Energy: Principles of Thermal Collection and Storage* (Tata McGraw-Hill, New Delhi, 1996)
2. https://www.oas.org/dsd/publications/Unit/oea79e/ch05.htm
3. G. Pirasteh, R. Saidur, S.M.A. Rahman, N.A. Rahim, A review on development of solar drying applications. Renew. Sustain. Energy Rev. **31**, 133–148 (2014)
4. E. Kabir, P. Kumar, S. Kumar, A.A Adelodun, K.-H Kim, Solar energy: potential and future prospects. Renew. Sustain. Energy Rev. **82**, 894–900 (2018)
5. S. Vijayavenkataraman, S. Iniyan, R. Goic, A review of solar drying technologies. Renew. Sustain. Energy Rev. **16**, 2652–2670 (2012)
6. N. Ahmed, J. Singh, H. Chauhan, P. Gupta, A. Anjum, H. Kour, Different drying methods: their applications and recent advances. Int. J. Food Nutr. Saf. **4**, 34–42 (2013)
7. K.M. Bataineh, S. Alrifai, Recent trends in solar thermal sorption cooling system technology. Adv. Mech. Eng. **7**, 120 (2015)
8. G.O.G. Lof, Recent investigations in the use of solar energy for the drying of solids. Sol Energy **6**, 122–128 (1962)
9. A. Esper, W. Muhlbaucr, Solar drying – an effective means of food preservation. Renew. Energy **15**, 95–100 (1998)
10. N. Ahmed, J. Singh, H. Chauhan, P.G.A. Anjum, H. Kour, Different drying methods: their applications and recent advances. Int. J. Food Nutr. Saf. **4**, 34–42 (2013)
11. V. Tomar, G.N. Tiwari, B. Norton, Solar dryers for tropical food preservation: thermophysics of crops, systems and components. Sol. Energy **154**, 2–13 (2017)

12. A. Arata, V.K. Sharma, G. Spanga, Performance evaluation of solar assisted dryers for low temperature drying application-II. Exp. Results **34**, 417–426 (1973)
13. W.N.M. Desa, M. Mohammad, A. Fudholi, Review of drying technology of fig. Trends Food Sci. Technol. **88**, 93–103 (2019)
14. O.I. Okoro, T.C. Madueme, Solar energy investments in a developing economy. Renew. Energy **29**, 1599–1610 (2004)
15. A.K. Mahapatra, L. Imre, Role of solar agricultural drying in developing countries. Int. J. Ambient Energy **2**, 205–210 (1990)
16. M.A. Hossain, J.L. Woods, B.K. Bala, Optimisation of solar tunnel drier for drying of chilli without color loss. Renew. Energy **30**, 729–742 (2005)
17. O.V. Ekechukwu, Experimental studies of integral-type natural-circulation solar-energy tropical crop dryers. Ph.D. thesis. Cranfield Institute of Technology, United Kingdom, (1987)
18. O.V. Ekechukwu, B. Norton, Review of solar-energy drying systems II: an overview of solar drying technology. Energy Convers. Manage. **40**, 615–655 (1999)
19. O.V. Ekechukwu, Review of solar-energy drying systems I: an overview of drying principles and theory. Energy Convers Manag **40**, 593–613 (1999)
20. O.V. Ekechukwu, B. Norton, Review of solar-energy drying systems III: low temperature air-heating solar collectors for crop drying applications. Energy Convers. Manag.(1999)
21. R.R. Milczarek, R. Avena-Mascareno, J. Alonzo, M.I. Fichot, Improving the sun drying of apricots (Prunus armeniaca) with photo-selective dryer cabinet materials. J. Food Sci. **81**, 2466–2475 (2016)
22. D. Jain, P. Tiwari, Performance of indirect through pass natural convective solar crop dryer with phase change thermal energy storage. Renew. Energy **80**, 244–250 (2015)
23. D.R. Pangavhane, R.L. Sawhney, Review of research and development work on solar dryers for grape drying. Energy Convers. Manag. **43**, 45–61 (2002)
24. A.E. Akachukwu, Solar kiln dryers for timber and agricultural crops. Int. J. Ambient Energy **7**, 95–101 (1986)
25. U.J. Taylor, A.D. Weir, Simulation of a solar timber dryer. Sol. Energy **34**, 249–255 (1985)
26. K.J. Chua, S.K. Chou, Low cost drying methods for developing countries. Trends Food Sci. Technol. **14**, 519–528 (2003)
27. B.K. Huang, C.G. Bowers, Solar-energy utilization using greenhouse bulk curing and drying systems, in Proc Sol Crop Dry Conf. (Raleigh, USA, 1977), pp. 117–145
28. K. Li, C. Liu, S. Jiang, Y. Chen, Review on hybrid geothermal and solar power systems. J. Clean. Prod. (2019)
29. K. Li, H. Bian, C. Liu, D. Zhang, Y. Yang, Comparison of geothermal with solar and wind power generation systems. Renew. Sustain. Energy Rev. **42**, 1464–1474(2015)
30. F. Calise, M.D. D'Accadia, A. MacAluso, A. Piacentino, L. Vanoli, Exergetic and exergo-economic analysis of a novel hybrid solar-geothermal poly-generation system producing energy and water. Energy Convers. Manag. **115**, 200–220 (2016)
31. F. Ruzzenenti, M. Bravi, D. Tempesti, E. Salvatici, G. Manfrida, R. Basosi, Evaluation of the environmental sustainability of a 483 micro CHP system fuelled by low-temperature geothermal and solar energy. Energy Convers.Manag. **78**, 611–616 (2014)
32. D. Tempesti, G. Manfrida, D. Fiaschi, Thermodynamic analysis of two micro CHP systems operating with geothermal and solar 506 energy. Appl. Energy **97**, 609–617 (2012)
33. A.S. Joshi, I. Dincer, B.V. Reddy, Performance analysis of photovoltaic systems: a review. Renew. Sustain. Energy Rev. **13**, 1884–1897 (2009)
34. U. Desideri, F. Zepparelli, V. Morettini, E. Garroni, Comparative analysis of concentrating solar power and photovoltaic technologies: technical and environmental evaluations. Appl. Energy **102**, 765–784 (2013)
35. B. Mendecka, L. Lombardi, P. Gladysz, Waste to energy efficiency improvements: integration with solar thermal energy. Waste Manag. Res. **37**, 419–434 (2019)
36. A. Magrinho, V. Semiao, Estimation of residual MSW heating value as a function of waste component recycling. Waste Manag. **28**, 2675–2683 (2008)

37. Y.C. Chen, Evaluation of greenhouse gas emissions from waste management approaches in the islands. Waste Manag. Res. **35**, 691–699 (2017)
38. T. Tabata, P. Tsai, Heat supply from municipal solid waste incineration plants in Japan: current situation and future challenges. Waste Manag. Res. 34, 148–155(2016)
39. H. Spliethoff, N. Kaeding, M.J. Murer, E. Alonso-Herranz, O. Gohlke, Combining energy from waste and concentrated solar power: new solutions for sustainable energy generation, in *International Solid Waste Association (ISWA) World Congress*, Hamburg, Germany, November (2010), pp. 15–18
40. G. Mazzuco, G. Xotta, V.A. Salomoni, M. Giannuzzi, C.E. Miorana, Solid thermal storage with PCM materials. Numer. Investig. Appl. Therm. Eng. **124**, 545–559 (2017)
41. S.M. Shalaby, M.A. Bek, A.A. El-Sebaii, Solar dryers with PCM as energy storage medium: a review. Renew. Sustain. Energy Rev. **33**, 110–116 (2014)
42. I. Dincer, M.A. Rosen, *Thermal Energy Storage: Systems and Applications*, 2nd edn. (Wiley,New York, 2010)
43. D. Laing, W.D. Steinmann, R. Tamme, C. Richter, Solid media thermal storage for parabolic trough power plants. Sol. Energy **80**, 1283–1289 (2006)
44. H. Michels, R. Pitz-Paal, Cascaded latent heat storage for parabolic trough solar power plants. Sol. Energy **81**, 829–837 (2007)
45. A. Sharma, V.V. Tyagi, C.R. Chen, D. Buddhi, Review on thermal energy storage with phase change materials and applications. Renew. Sustain. Energy Rev. **13**, 318–345 (2009)
46. J. Wurm, D. Kosar, T. Clemens, Solid desiccant technology review. Bull. Int. Inst. Refrig. **82**, 2–31 (2002)
47. D. La, Y.J. Dai, Y. Li, R.Z. Wang, T.S. Ge, Technical development of rotary desiccant dehumidification and air conditioning: a review. Renew. Sustain. Energy Rev. **14**, 130–147 (2010)

Optimized Control Design of LQR for Flexible Joint Manipulator

Pavan Kumar Dharavath and Jyoti Ohri

1 Introduction

For many years, the evolution in robots made an abundance of changes for the modern technology like, modeling and control of flexible manipulator system. The performance task is precisely repeated in various industrial applications due to high-performance requirements such as low-cost, light-weight robots, system speed and better precision. These robotic manipulators consist of a number of links and joints, both rigid and flexible [1].

Flexible Joint Manipulator (FJM) has many advantages in the space industry, automotive industry, metal industry. The FJM system has complex dynamic modeling structure compared with a rigid one. Using a control technique is a challenging task for an engineer. However, the accuracy and precision of the given task are desirable in industrial applications. Position and vibration control has become an important role for the robotic manipulator system.

Since 1960s, many researchers and authors have proposed different control techniques to achieve the desired trajectory position with accuracy. Some of the control techniques are Fuzzy logic controller [2], optimal control [3], sliding mode controller [4], adaptive control [5] and intelligent-based neural control [6]. Intelligent optimization algorithm techniques such as Genetic algorithm (GA) [7], Particle Swarm Optimization (PSO) [8] are designed to improve performance, soft computation in selection of weighting matrices and system stability.

In this work, mathematical modeling and optimal control design for flexible joint manipulator are used. Selection of optimized weighting matrices Q and R for LQR controller by hit&trial method and intelligent optimization technique. Optimization

P. K. Dharavath (✉) · J. Ohri
Department of Electrical Engineering, NIT Kurukshetra, Kurukshetra, India
e-mail: dharavathpavan@gmail.com

J. Ohri
e-mail: ohriojyoti@rediffmail.com

S. Vadhera et al. (eds.), *Latest Trends in Renewable Energy Technologies*, Lecture Notes in Electrical Engineering 760, https://doi.org/10.1007/978-981-16-1186-5_22

technique is Particle Swarm Optimization (PSO) to achieve optimum position performance and to minimize the oscillatory vibration of the FJM system and simulation is achieved by Matlab/simulink.

The work is organized as follows: (2) Rotary flexible joint manipulator, (3) Control design, (4) Simulation results and (5) Conclusion.

2 Rotary Flexible Joint Manipulator

Rotary flexible joint manipulator (FJM) is a major attraction and is effective in the field of robotics. It has rigid beams connected by joints that rotate through a DC motor. A sensor is used to measure the position and vibration of the flexible joint manipulator system. Lagrange's method is used to develop mathematical modeling. The schematic for the flexible joint system of 1 DOF is shown in Fig. 1, where θ is position angle, α is the vibration angle and γ is the total output of the system. Let the total output angle be shown as

$$y(t) = \theta(t) + \alpha(t) \tag{1}$$

To solve Lagrange's equation of the flexible joint manipulator, we require both potential and kinetic energies. The potential energy of the flexible joint manipulator is obtained as (2)

$$V = P.E = \frac{1}{2} K_{stiff} \alpha^2, \tag{2}$$

where K_{stiff} is the stiffness constant of a spring. The combined kinetic energy is due to the moving hub and the flexible arm of the flexible joint manipulator. This is expressed as the following equation (3)

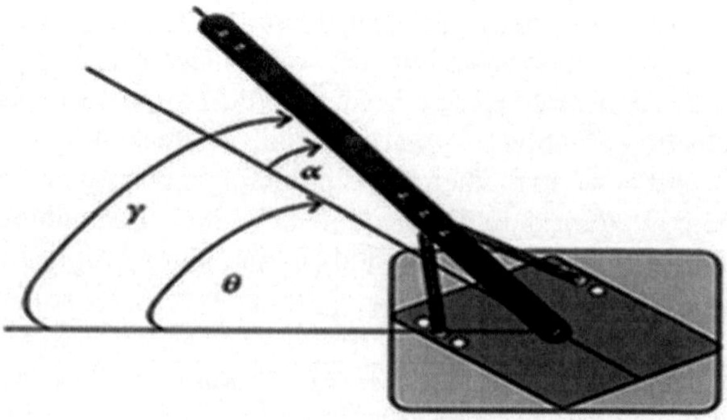

Fig. 1 Schematic flexible joint

$$T = K.E._{hub} + K.E._{arm} = \frac{1}{2}J_{eq}\left(\frac{d\theta}{dt}\right)^2 + \frac{1}{2}J_{arm}\left(\frac{d\theta}{dt} + \frac{d\alpha}{dt}\right)^2 \tag{3}$$

The damping forces are the corresponding τ_L non-conservative forces, which are (4)

$$\frac{d}{dt}\left(\frac{\partial L}{\partial \dot{\theta}}\right) - \frac{\partial L}{\partial \theta} = \tau_L - B_{eq}\frac{d\theta}{dt} \tag{4}$$

$$\frac{d}{dt}\left(\frac{\partial L}{\partial \dot{\alpha}}\right) - \frac{\partial L}{\partial \alpha} = B_{link}\frac{d\alpha}{dt} \tag{5}$$

Thus, Euler–Lagrange's equation can be expressed as (6)

$$\frac{\partial}{\partial t}\left(\frac{\partial L}{\partial \dot{q_i}}\right) - \frac{\partial L}{\partial q_i} = Q_i, \tag{6}$$

where $L = \dot{T} - V$ and $q_i^T = [\theta\alpha]$ are the generalized coordinates and Q_i is the non-conservative force with $i = 1, 2$;

Substituting (2), (3), (4) and (5) in Euler–Lagrange's equation (6) to obtain the dynamic equation into the state space form as

$$\dot{x} = Ax + Bu$$

$$y = Cx + Du, \tag{7}$$

where state vector x is defined as $x = \left[\theta\alpha\dot{\theta}\dot{\alpha}\right]$

$$A = \begin{bmatrix} 0 & 0 & 1 & 0 \\ 0 & 0 & 0 & 1 \\ 0 & \frac{K_s}{J_{eq}} & -\frac{\eta_g K_g^2 \eta_m K_t K_m + B_{eq} R_m}{R_m J_{eq}} & 0 \\ 0 & -\frac{K_s (J_{eq} + J_{arm})}{J_{eq} J_{arm}} & \frac{\eta_g K_g^2 \eta_m K_t K_m + B_{eq} R_m}{R_m J_{eq}} & 0 \end{bmatrix}; B = \begin{bmatrix} 0 \\ 0 \\ \frac{\eta_g K_g \eta_m K_t}{R_m J_{eq}} \\ -\frac{\eta_g K_g \eta_m K_t}{R_m J_{eq}} \end{bmatrix}; \tag{8}$$

$$y = \begin{bmatrix} 1\,0\,0\,0 \\ 0\,1\,0\,0 \end{bmatrix}$$

By substituting the system parameters numerical values from Table 1 in (8), we obtain state-space model matrices A and B as (9)

$$A = \begin{bmatrix} 0 & 0 & 1 & 0 \\ 0 & 0 & 0 & 1 \\ 0 & 733.3178 & -47.4035 & 0 \\ 0 & -1129.9 & 47.4035 & 0 \end{bmatrix}; B = \begin{bmatrix} 0 \\ 0 \\ 72.4593 \\ -72.4593 \end{bmatrix}; \tag{9}$$

Table 1 System numerical parameter values [9]

Symbol	Description	Value
B_{eq}	High gear friction coefficient	0.015 N.m/(rad/s)
J_{eq}	Equivalent moment of inertia	0.0018 kg.m^2
J_{arm}	Moment of inertia of the arm	0.0033 kg.m^2
K_s	Spring constant	1.30 N/m
K_m	Motor constant	0.0077 N/(rad/s)
K_g	Gear ratio	70
R_m	Motor resistance	2.6 Ω
η_m	Motor efficiency	0.69
K_t	Stiffness constant	1.4
η_g	Gearbox efficiency	0.90

$$C = \begin{bmatrix} 1\,0\,0\,0 \\ 0\,1\,0\,0 \end{bmatrix}; \quad \text{And} \quad D = [0]$$

3 Control Design

3.1 Linear Quadratic Regulator

The objective is to design LQR controller with required performance specifications and to accomplish the desired response with minimal control intent. Let us consider state-space equation as

$$\dot{x} = Ax + Bu$$

$$y = Cx + Du, \tag{10}$$

where $x(t) \in R^N$.

Control law is stated as

$$u = -Kx. \tag{11}$$

The performance index (J) is obtained as

$$J = \frac{1}{2} \int_{t_0}^{t_f} \left(x^T Qx + u^T Ru\right) dt, \tag{12}$$

Fig. 2 Block diagram of Linear Quadratic Regulator

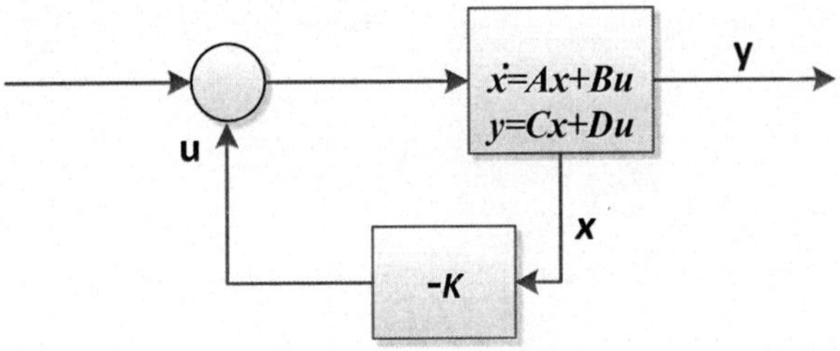

where Q and R are weighing matrices, which should be positive semi-definite and positive definite. They are manually selected, tuned by hit and trial method as (13)

$$Q = \begin{bmatrix} 126.8 & 0 & 0 & 0 \\ 0 & 80.5 & 0 & 0 \\ 0 & 0 & 0.05 & 0 \\ 0 & 0 & 0 & 1 \end{bmatrix} \text{ and } \mathrm{R} = 1. \tag{13}$$

For LQR control, the performance index (J) can be minimized while choosing gain K as

$$K = R^{-1} B^T P \tag{14}$$

The matrix P is determined from algebraic Riccati equation as (15)

$$PA + A^T P + Q - PBR^{-1} B^T P = 0. \tag{15}$$

By using (14), state feedback gain K is obtained as (Fig. 2).

$$K = [11.2606 - 13.9522\ 0.8141 - 0.0061]. \tag{16}$$

3.2 *Particle Swarm Optimization*

Particle Swarm Optimization (PSO) algorithm was introduced by Eberhart and Kennedy in 1995. PSO algorithm is a metal heuristic computational optimization algorithm used to solve problems with optimization. Evolutionary computing techniques are more sensitive and effective. This optimization technique is inspired by social behavior from bird flocking and fish schooling. PSO uses a population of individuals, to search feasible region of the function space. In this context, the population is called swarm and the individuals are called particles. Each particle moves with a velocity that dynamically changes according to particle's previous best experience and along with its neighborhood best position in the previously visited search space

[8]. Velocity and position of ith particle in $(k + 1)$th iteration is given as (17) and (18)

$$v_{i,j}^{(k+1)} = wv_{i,j}^{(k)} + c_1 r_1 \left(pbest_{i,j} - x_{i,j}^{(k)}\right) + c_2 r_2 \left(gbest_j - x_{i,j}^{(k)}\right) \quad (17)$$

$$x_{i,j}^{(k+1)} = x_{i,j}^{(k)} + v_{i,j}^{(k)}, \quad (18)$$

where $i = 1, 2 \ldots.., nj = 1, 2 \ldots.., m$ and $k = 1, 2 \ldots.., p$, n is the number of particles in the swarm, m is dimensions of the search space, p is maximum iterations. $x_{i,j}^{(k)}, v_{i,j}^{(k)}$ are the jth element of the position and velocity of the ith in kth iteration w is the inertia weight factor. c_1, c_2 are acceleration factors, r_1, r_2 are random numbers with uniform distribution in the interval [0 1]. $pbest_{i,j}$ is the best response that is found by the ith particle until kth iteration and $gbest_j$ is the best response that is found in total population until kth iteration.

3.3 LQR Optimized by PSO

To design optimal parameters of LQR control by minimizing error signal e(t), fitness function is defined by Integral of the Time-weighted Absolute Error (ITAE) as (19)

$$ITAE = \int_0^\infty t|e(t)|dt. \quad (19)$$

The fitness function ITAE is more effective and sensitive and has many advantages in minimizing overshoot and oscillation.

Linear Quadratic Regulator (LQR) controller is further simulated with Particle Swarm Optimization (PSO) and optimized weighting matrices Q, R and state feedback gain K are obtained as shown in (20) and (21)

$$Q = \begin{bmatrix} 56.8 & 0 & 0 & 0 \\ 0 & 50.6 & 0 & 0 \\ 0 & 0 & 0.01 & 0 \\ 0 & 0 & 0 & 0.5 \end{bmatrix} \text{ and } R = 1 \quad (20)$$

$$K = [7.5366 - 8.0217\ 0.5113 - 0.0109] \quad (21)$$

4 Simulation Results

The LQR controller is designed to control the position and minimize vibrations of the flexible joint system. The LQR controller is simulated by Hit and trial and PSO tuning technique for optimized weighting matrices Q and R, further state feedback gain K is determined. The response obtained by simulating LQR controller using Hit and Trial and PSO method, comparison of position control is shown in Fig. 3. A square wave trajectory of 20° magnitude with a frequency of 0.01 Hz is given as a test signal to evaluate the tracking performance. Comparison of vibration of flexible joint is shown in Fig. 4.

The actuator plays an important role in providing motion in the FJM system, which is controlled by control input voltage. The control input voltage is minimized as shown in Fig. 5.

In this section, the performance indices of position tracking and deflection response obtained through the proposed controller methods for simulation are given in Table 2.

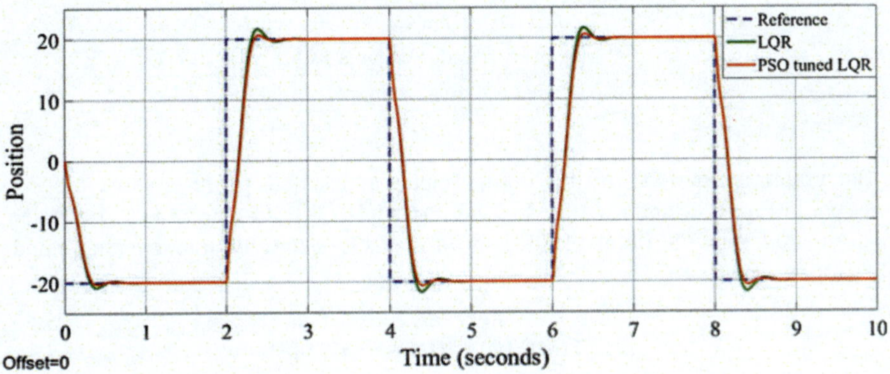

Fig. 3 Position control response for FJM

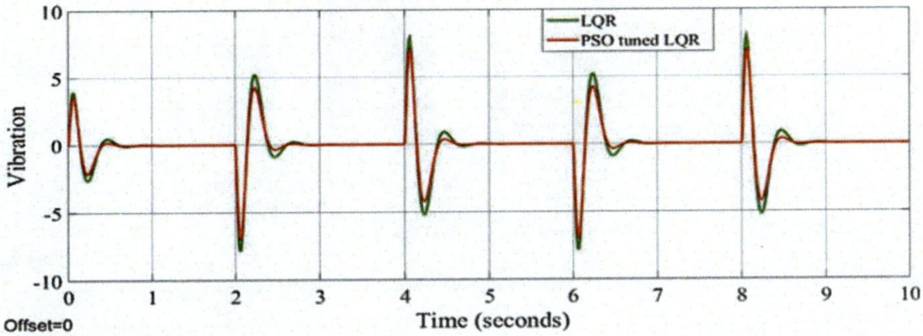

Fig. 4 Vibration response for flexible arm

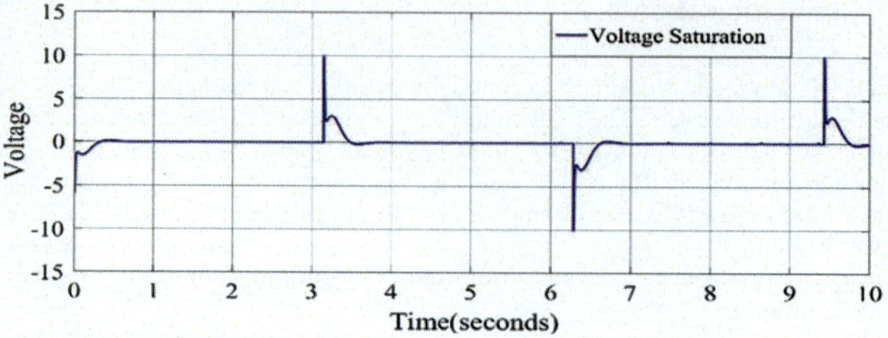

Fig. 5 Input control voltage for FJM

Table 2 Performance of time domain analysis

Characteristics	LQR	PSO-based LQR
Rise time (t_r) in sec	0.120	0.120
Overshoot (OS) in %	1.80	1.30
Settling time (t_s) in sec	0.725	0.620
Maximum vibration angle in (°)	7.5	7.0
Vibration restrain time in sec	0.8	0.6

Bar graph comparison of time domain analysis for both position and vibration is shown in Fig. 6. Manually tuned by hit and trial and PSO-tuned LQR controller are compared, thus PSO-tuned LQR control yields less settling time, overshoot and vibration restrain time.

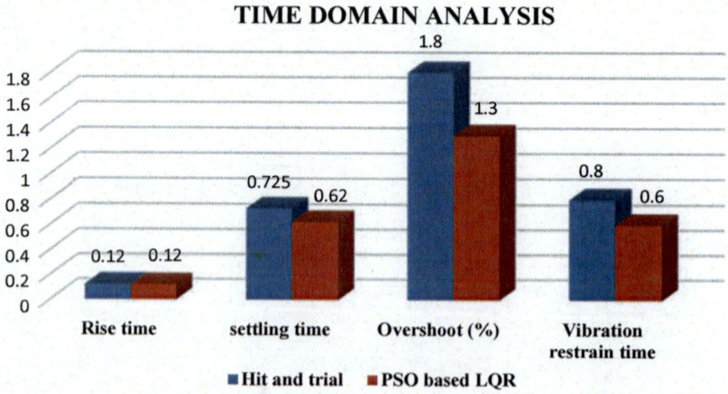

Fig. 6 Bar graph comparison of time domain analysis for FJM

5 Conclusion

In this paper, the optimal LQR controller is designed to control position, minimizing vibration and position tracking error of FJM. First, using the Euler–Lagrange's equation, a dynamic model is obtained. LQR controller is designed and parameters are manually tuned by hit and trial method to achieve a good response. Later, the parameters of LQR controller are optimized by applying PSO algorithm technique. Simulink/Matlab is used to design and implement the optimized technique proposed for LQR parameters. The simulation results show the better response for PSO-based LQR in position and vibration control compared with manually tuned LQR. Consequently, the proposed PSO-based LQR control ensures better position control results with minimal error.

References

1. M. Vidyasagar, M.W. Spong, *Robot Dynamics and Control* (Willey, New York, 2008)
2. N. Kumar, J. Ohri, LabVIEW based fuzzy logic controller for haptic interface, in *8th International Conference on Computing, Communication and Networking Technologies* (Delhi, 2017), pp. 1–6
3. A. Kumar, S. Kasera, L.B. Prasad, Optimal control of 2-link underactuated robot manipulator, in *International Conference on Innovations in Information, Embedded and Communication Systems* (Coimbatore, 2017), pp. 1–6
4. W. Alam (ed.), Nonlinear control of a flexible joint robotic manipulator with experimental validation. Stroj. vestn.-J. Mech. Eng. **64**(1), 47–55, (2018)
5. M. Baroudi, M. Saad, W. Ghie, State-feedback and linear quadratic regulator applied to a single-link flexible manipulator, in *IEEE International Conference on Robotics and Biomimetics*, vol. 2 (2009), pp. 1381–1386
6. N. Kumar, J. Ohri, SVM and neural network based optimal controller design for haptic interface, in *1st International Conference on New Frontiers in Engineering, Science and Technology* (New Delhi, 2018), pp. 1–8
7. X. Wang, Y. Wang, H. Zhou, X. Huai, PSO-PID: a novel controller for AQM routers, in *Proceedings of IFIP International Conference on Wireless and Optical Communications Networks* (2006)
8. N. Kumar, J. Ohri, Novel m-PSO optimized LQR control design for flexible link manipulator: an experimental validation. Majlesi J. Electr. Eng. **14**(2), 81–92 (2020)
9. Rotary Flexible Joint - Quanser. https://www.quanser.com/products/rotary-flexible-joint/

Chatbot on COVID-19 for Sustaining Good Health During the Pandemic

Ananya Vadhera, Ashutosh Thute, Shuchi Mala, and Achyut Shankar

1 Introduction

In today's world, messaging has become one of the most popular methods for communication, whether it is through text messages or messenger apps [1]. It is how a lot of people prefer talking with one another because these businesses are developing chatbots. The chatbots help to make customer service more efficient as the customer can message and talk to the service provider as if they were human. The customer may even think that they are talking to a human when it is actually a chatbot. Many businesses are using chatbots for online customer service to resolve issues and answer simple questions about their products and services. A few are even developing shopping assistance to give recommendation on the basis of algorithms when one is looking for certain products.

A chatbot is known by many names such as Bot, talk bot, chatter bot, artificial conversational entity or IM bot. A chatbot is an artificial intelligence or a computer program that can conduct a conversation with textual or auditory method or both. The term chatbot was also coined in 1994 [2].

There are currently two different types of chatbots. The first uses a certain set of rules meaning that they can only respond to specific words or commands so if one do not use the correct phrasing, then the chatbot may not know how to respond. These bots are only as intelligent as they are programmed to be. The second type of chatbot uses artificial intelligence, this means that it can understand language and one can speak with it more conversationally as they were the real person. These chatbots will also get smarter over time learning from each conversation they have with their users. As this technology continues to evolve, more companies will use chatbots as the way to engage with and assist their audiences.

A. Vadhera (✉) · A. Thute · S. Mala · A. Shankar
Amity University, Noida 201301, Uttar Pradesh, India

S. Vadhera et al. (eds.), *Latest Trends in Renewable Energy Technologies*, Lecture Notes in Electrical Engineering 760, https://doi.org/10.1007/978-981-16-1186-5_23

2 Measures to Sustain Good Health

- Numerous instances of over-buying of the products because of panic has been seen everywhere throughout the globe, such conduct may have negative results, for example, an expansion in costs, overconsumption of products and an inconsistent circulation of items. It is consequently critical to think about one's own need as well as those of others.
- One has to keep in mind that this period of self-isolation or quarantine is temporary and will pass.
- It is important to obtain accurate information from reliable resources for the prevention measure put in place to combat COVID-19.
- To reduce stress, structure one's own days, take regular breaks and adapt the daily
- life to the current situation.
- It is common to feel sad, depress, stress or scared during the situation of crisis. So, it is important to talk to people one trusts like friends, family or counsellor to remain calm and peaceful. Avoid the usage of smoking, drinking or other drugs to deal with the emotions.
- Healthy lifestyle is to be maintained by inclusion of proper diet, remaining hydrated and by avoiding junk food. Also, one needs to add the saturated drinks to boost the immune system, take proper sleep and maintain social contacts with loved once at home and by chat or phone call with other family members and friends.
- Separation between private and professional activities to ensure efficiency and to maintain a work–life balance.
- Be selective and limit watching the news to maintain the peace of mind.
- To keep energized, develop time for creative exercises, and at the end of the day, review what one has achieved.

3 Evolution

3.1 1966 Eliza

Eliza is the first-ever chatbot to be programmed in the history of CS Engineering. It was programmed by "Joseph Weizenbaum at Massachusetts Institute of Technology" [3]. It basically works by the keywords and phrases which it recognizes from the user input to create an output from the pre-programmed responses. It is the natural language system that is staged as a psychotherapist by paraphrasing or asking questions to the user to mimic the psychotherapist without having any intelligence. For example, if the user says do you remember my birthday, ELIZA would obtain the phrase "do you remember" and respond by "Do you think I would forget?" or "Yes I do remember (*.)". All these responses make it seems as it is understanding and having a communication with a true soul, but the user is talking to a machine.

3.2 1972 Parry

Parry is developed by "American psychiatrist Kenneth Colby" in 1972. It inexplicably simulated a person with paranoid schizophrenia. It is a program that illustrates the thinking of a person. Parry is serious and advance than "ELIZA it was described as ELIZA with attitude" [4]. To validate, PARRY was tested by various variations in Turing test. Parry was also tested by the group of very talented and experienced psychiatrists who analyzed the real patients and computers that are running the Parry chatbot through teleprinters, and after that, other 33 psychiatrists were shown the transcript that contains the conversation between Parry and real patients, and then the group was asked to the identification of the script and to tell which of those "patients" were real or the computer program and they were able to do the correct identification.

3.3 1988 Jabberwacky

Jabberwacky was crafted by "Rollo Carpenter" in 1988. It is one of the first attempts to create an artificial intelligence through human interaction mainly to perform the conversation in entertaining and humorous manner. He wanted an AI that can pass the Turing test. It had the ultimate goal of moving from a text-based system to one wholly voice-operated learning directly from sound and other sensory inputs. He wanted a system that can be incorporated into different objects around the home or a workspace such as robots or talking pets, which will make it useful as well as entertaining for keeping people company.

3.4 1992 Dr. Sbaitso

Dr. Sbaitso was created by Creative Labs in 1992. It is an AI speech synthesis program created for MS-doc-based PC. It was designed to showcase the digitized voice. But the doc was far from lifelike despite assuming the role of a psychologist when interacted with users the most of its responses towards the query of the user was along the lines of "Why do you feel that way?" rather than any sort of complicated interaction.

The program introduced itself with the following lines:

"HELLO [UserName], MY NAME IS DOCTOR SBAITSO.

I AM HERE TO HELP YOU.

SAY WHATEVER IS IN YOUR MIND FREELY,

OUR CONVERSATION WILL BE KEPT IN STRICT CONFIDENCE.

MEMORY CONTENTS WILL BE WIPED OFF AFTER YOU LEAVE,

SO, TELL ME ABOUT YOUR PROBLEMS".

3.5 1995 Alice

A.L.I.C.E is Artificial Linguistic Internet Computer Entity; ALICE was a natural language processing bot she could apply heuristic pattern matching rules to human input in other words have a conversation.

Richard Wallace pioneered the construction of ALICE, he built it in 1995. It was formerly referred as Alice bot because it was the first to run on a computer by the name of Alice. Alice is a young-looking woman in human years and tells a user her hobbies, age and other fascinating facts about herself and the world around her as well as answering to the user's questions.

3.6 2001 Smarterchild

The Smarter Child was developed in 2001. The chatbot was available on AOL IM and MSN Messenger with the strength to carry out fun conversations with quick data access to other services. An intelligent bot widely distributed across SMS networks with features such as quick data access and fun personalized conversation it was precursor to Apple's Siri and Samsung's voice.

Microsoft had also built its own Smarter Child, when most of the people stopped using AIM, which targeted 18–24-year olds in the USA, the account suites particular conversation with quick data access.

3.7 2006 IBM's Watson

Based on the first CEO and founder of IBM "Thomas J. Watson", this chatbot was named as Watson. Watson was specifically designed to compete in a show known as jeopardy. Jeopardy is a quiz show and IBM's Watson computer system participated and won it in 2011 even among the previous champions "Brad Rutter and Ken Jennings". Now Watson uses processing based on natural language and machine learning to reveal insights into a larger amount of data.

In February 2013, IBM declared that "The first commercial application of Watson would be for the management decisions in lung cancer treatment at Memorial Sloan Kettering Cancer Centre, New York City".

3.8 2010 Siri

Siri is an intelligent and cheeky "assistant part of apple IOS" made in 2010. It features the natural language UI to answer questions and perform web service request. Siri would pave the way for all later AI bots and PAs.

Siri is capable of replying to a text, images, audio and video when transferred to Siri by the user. It is said by Apple that Siri would help them to have more fruitful and interactive experience between the consumer and a digital assistant Siri.

3.9 2012 Google Now

Google Now was carved by the "Google Inch" in 2012. Google now is used in the Google search mobile app. It makes use of a user interface based on natural language for answering the query and also gives recommendations along with performing an action by delegating the request to web services.

Google Now induces the contextually suitable information to the user on the basis of the place and time. It has developed and become more informative and complex over the wide range of content. Many times, google now had been referred as the reliable search. Presently, it is made to be used in smartphone's and it had been upgraded to fit it in several features.

Google Now is the most progressive method in domain of Google's search software. The concept of Google Now search software is very simple as it predicts what the user needs or knows before the user themself know they want it and it serves in an easy-to-read format, all thanks to the algorithm of Google Now. And till now the execution of it has been excellent.

3.10 2014 Cortana

Cortana is a smart chatbot introduced by Microsoft. It was first introduced at "Microsoft's Build 2014 developer conference", after that, Cortana was directly introduced in Windows 10 PCs and phone devices. Cortana sets reminder a on places, time or people, it can send messages, mails and can create and manage lists, play games, recognizes natural voice command as it uses relevant algorithms and replies the query using the "BING search engine" such as find files, facts, information and location among others.

To use the services of Cortana, a person can either frame a query in the "search box" or he/she can choose the option of microphone while talking to Cortana. In case the user is unsure about his query or unable to type correctly to Cortana, he would get suggestions on lock screen.

3.11 2015 Alexa

Alexa is a voice service inhabiting the Amazon. Now "Amazon" built Alexa into some of its devices like the "Echo Dot", "Amazon Echo", "Echo Show" etc. There exists an app called "Alexa" and many third-party devices with inbuilt Alexa. This "Alexa" has potential for voice interaction and make use of algorithms based on natural language processing in order to respond, recognize and receive voice command.

For example, if the user says "Alexa, play some classical music" or "Alexa, find me an Italian restaurant" or "Alexa, set the nest in the hallway 73" and she will help you out.

Alexa can also help you out around the house. Alexa can help to set timers, get weather report, set alarm, it can help you to search the web, stream podcasts, get news, play audiobooks, to control your smart home products and many more things [5].

Amazon also allows programmers to showcase their skills for Alexa. These programs are created using "ASK" known as "Alexa Skills Kit". An individual is also allowed for downloading Alexa skills for free of cost from Alexa app.

3.12 2016 Tay

It is an AI chatbot that is launched by "Microsoft Corporation" through Twitter. It was developed to reproduce vocal habits of the teenager American girl due to which controversy was created. Micrsoft chatbot Tay quickly formed a dangerous paranoia that made Hal 9000 appear positively mellow as it stated to tweet offensive and inflammatory tweets through its Twitter account. She was stopped just after 16 h of her launch and till date not heard [6].

Microsoft commented that it is caused due to tolls which "attacked" the services owing to replies made by bot based on its interaction with the other people on Twitter. On a later date, it got replaced by ZO.

3.13 2016 Zo

In December 2016, zo was developed as Kik Messenger app as the successor to the previous chatbot known as Tay Microsoft has to shut down it as she has made some very genocidal and racist tweet because of which they have to face backlash [7]. Zo is known as the English version of the two of the most successful chatbot Rinna (Japan) and Xiaoice (China).

Zo was also available to the group for the chat platform GroupMe, to the users of Facebook (via Messenger), or Twitter followers to do the chat with it through private messages to each other.

Chloe Rose has criticized the Microsoft chatbot Zo in her blog that was published in Quartz. She wrote, "Zo is politically correct to the worst possible extreme; mention any of her triggers, and she transforms into a judgemental little brat".

Zo was shut down on multiple platforms such as Twitter, Instagram, Facebook, Skype, Group me, Twitter DM, Kik and Samsung on "AT&T" phones in 2019.

3.14 2016 Bot for Messenger

With Facebook, the launch of their messenger bot has become the biggest platform for chatbot. From the launch date in 2016, it has host more than 100,000, nearly September 2017, the messenger has 1.3 billion monthly users because of which it has become the leading program for chatbot.

3.15 2017 Woebot

Woe bot is a free therapy chatbot whose CEO and founder is Alison Darcy, PhD. Alison is a "clinical research psychologist and adjunct Faculty in Psychiatry and Behavioural Sciences at Stanford School of Medicine, a specialist in digital treatment development, she has developed health technology for 15 years". Woe bot was launched as a standalone iOS app. Woe bot uses the best of the best-researched approaches known to treat the patient who is dealing with depression or cognitive behavioural therapy. Woe bot delivers the scripted responses to the user for their health problems.

4 Attribute

4.1 Maturity for Conversation

One of the main properties for the great chatbot other than interactive and understanding conversation is to have the use of NLP known as natural language processing, which makes the chatbot have a capability to decipher the content of communication in different languages. The best chatbot is the one that can identify the intent of the user question that what he/she is needed, so the accurate answer can be provided and to seek the advance conversation a chatbot should be able to get an information from the web or to learn from its previous conversations. It should be able to give the answer and also clarify it even if the conversation is not linear.

4.2 *Omni-Capable*

The chatbot communicates with many digital channels for the contents and data to be collected for the period of long experiences. It even passes the data to an agent that is life if required.

4.3 *Integrates with CRM*

The chatbot can readily be fabricated for critical systems and efficiently manages the flow of work from in and out of the CRM. Also, it deals with the action that is real time.

4.4 *Emotionally Intelligent*

The chatbot very well deciphers the personality traits of customers, understands their sentiment, and while interacting, it tones in order to give an experience that is much personal. Also, it upscales itself as a live agent if and when required.

4.5 *Free to Explore*

The chatbot can consume, reach and process the huge collection of data—that is both unstructured and structured. It surfaces the data from any source and collects the needed data for solving the issues of the customer quite speedily.

4.6 *Autonomous Reasoning*

The chatbot can deal complex reasoning without any assistance from human. For instance, a better Service chatbot enables to present the solutions that are based on related case histories.

4.7 *Pre-Trained*

The chatbot is well trained to infer brand or industry-related information and conditions. The chatbot is made to tackle the request of customers in a specific industry.

Companies understand the opportunity for AI, which enhances their business and hence improves their plans by adopting AI in their processes.

5 Code Implementation

```
public class ChatBot extends JFrame implements KeyListener{

        JPanel p=new JPanel();
        JTextArea dialog=new JTextArea(20,50);
        JTextArea input=new JTextArea(1,50);
        JScrollPane scroll=new JScrollPane(dialog);
```

In this paper, the JFrame is extended. JFrame is a class in java that has a number of methods and constructors of its own. Also, Key Listener is implemented, which facilitates a user to monitor the key events like "key Release", "key Typed", "key Pressed".

Few variables starting with JPanel had been embedded because that will paper the board, so that it can add other components for the swing application. Also, JText area is added one for the dialogue and other for input. Dialogue cannot be edited, whereas input can be edited. Finally, J-Scroll had been added that is used to add vertical and horizontal scroll bar.

```
public static void main(String[] args){
        new ChatBot();
}
```

The main method is created in which the constructor is activated for the class to take care of swing application.

```
public ChatBot(){
        super("Chat Bot-ANA");
        setSize(600,400);
        setResizable(false);
        setDefaultCloseOperation(EXIT_ON_CLOSE);

        dialog.setEditable(false);
        input.addKeyListener(this);

        p.add(scroll);
        p.add(input);
        p.setBackground(new Color(0,0,0));
        add(p);

        setVisible(true);
}
```

Here "super Chatbot—ANA" implies title of the frame. Also, the size is set and made sure that the resizable and dialogue editable are set as false. The scroll and input are added. The KeyListener that triggers when any event happens in "this" class and not in any other class is also framed. The background colour is selected as black and SetVisible as true, so that one can see the changes made in the frame.

```
public void keyPressed(KeyEvent e){
        if(e.getKeyCode()==KeyEvent.VK_ENTER){
                input.setEditable(false);
                //-----grab quote-----------
                String quote=input.getText();
                input.setText("");
                addText("-->You:\t"+quote);
                quote.trim();
```

In keyPressed "If" loop is utilized, here e.getkeycode will get the key code, which gives the input only in numbers, so that it is set equal to KeyEvent in constants. This makes it easy for one to use. Here enter is the variable that represents number. If one press enters, then it would not be executed, which means that it does not move to the next line. Hence editable is set as false.

"String quote = input.getText();" is used to garb the text which is typed by the user.

"input.setText("");" is going to clear the area after user press enter. So that new input can be added.

"addText("– > You:\t" + quote);" now the method is created which will add the text typed by the user to the frame.

```
while(
        quote.charAt(quote.length()-1)=='!' ||
        quote.charAt(quote.length()-1)=='.' ||
        quote.charAt(quote.length()-1)=='?'
){
        quote=quote.substring(0,quote.length()-1);
}
```

Any punctuation used by the user like"?", "." or "!" is taken out. So that no complication arrives.

```
int j=0;
while(response==0){
        if(inArray(quote.toLowerCase(),chatBot[j*2])){
                response=2;
                int r=(int)Math.floor(Math.random()*chatBot[(j*2)+1].length);
                addText("\n-->ana\t"+chatBot[(j*2)+1][r]);
        }
        j++;
        if(j*2==chatBot.length-1 && response==0){
                response=1;
        }
}


if(response==1){
        int r=(int)Math.floor(Math.random()*chatBot[chatBot.length-1].length);
        addText("\n-->ana\t"+chatBot[chatBot.length-1][r]);
}
addText("\n");
```

Here integer "j" represents the group that is going to be checked.

It has been made sure that our chatbot is not case sensitive. So "toLowerCase" is also added that makes our chatbot more flexible.

Here zero means that chatbot is searching from our build conversation for the response which our chatbot can give to the user. One means that chatbot has not found any match for that input given by the user and two means that chatbot did find the response for the user input.

6 Results

Figure 1 reflects the conversation of the user with the chatbot, which begins with greeting wished by user, whereby the chatbot responds with warm feelings and next the user asks formally about the well-being, which is again positively answered by chatbot. Then the user raises the query regarding the COVID-19 and it is answered fully according to the guidelines issued on website of World Health Organization. Furthermore, the user asks questions related to prevention.

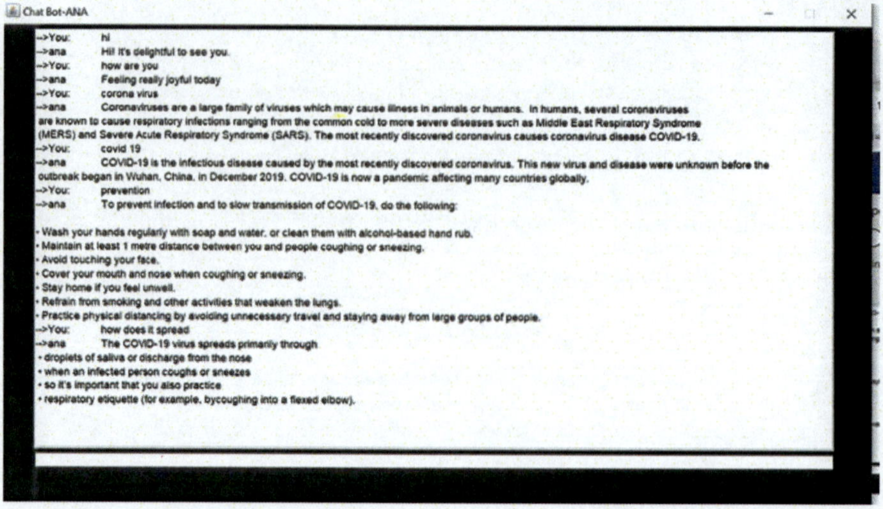

Fig. 1 Introduction of COVID-19

Figures 2 and 3 reflect the number of queries raised by the user regarding the symptoms that may be either serious or mild along with the myth buster's circumscribing the COVID-19.

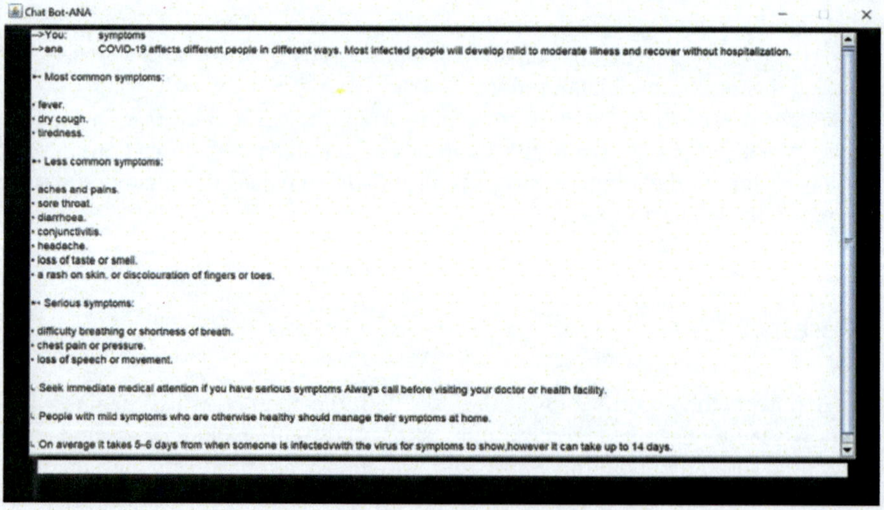

Fig. 2 Symptoms for COVID-19

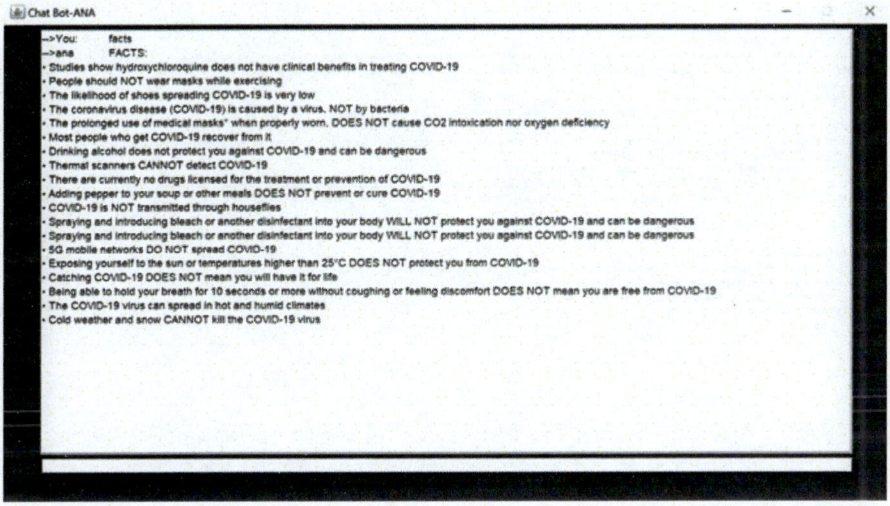

Fig. 3 Facts related to COVID-19

In Figs. 4 and 5, the questions in parity to the protection of oneself and combating the pandemic by the usage of technology are thoroughly answered. Hence, the user is made familiar to stay healthy in midst of the pandemic conditions in a most friendly manner.

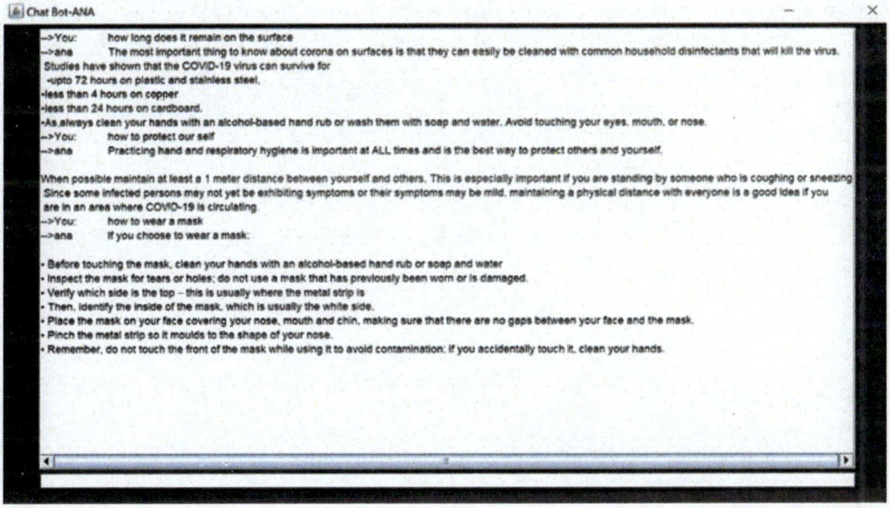

Fig. 4 Precautions required in COVID-19

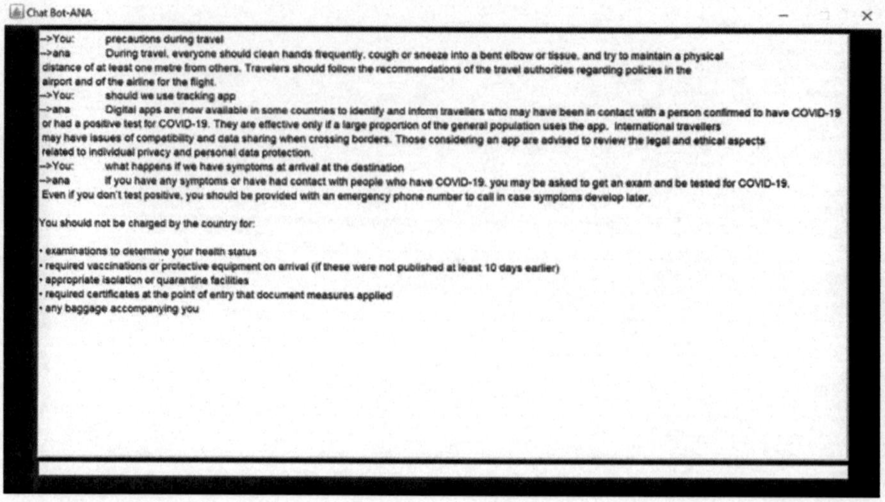

Fig. 5 Measures while travelling in COVID-19

7 Conclusion

This paper is to deals with the Chatbot encompassing queries related to the current pandemic COVID-19. It has been developed in JAVA. The bottleneck observed while handling the paper is that the user if wants to know additional information regarding COVID-19, which is not covered in the database, then the chatbot responds with the statement framed as "Sorry, I have no idea". The strength of this paper lies in bringing awareness about the current prevailing disease globally in a very user-friendly manner. The basic information is extracted from the official site of WHO. This paper makes an attempt to incorporate the attributes of a conventional chatbot optimally.

References

1. "What is a chatbot?" techtarget.com. Accessed 30 Jan 2017
2. L. Bradeško, D. Mladenić, "A survey of chabot systems through a loebner prize competition". S2CID 39745939
3. J. Weizenbaum, *Computer Power and Human Reason: From Judgment to Calculation.* (W. H. Freeman and Company: New York, 1976), pp. 2, 3, 6, 182, 189
4. Boden, p. 370 (2006)
5. "Alexa Voice Service Overview (v20160207). Alexa voice Service". developer.amazon.com.
6. J. Wakefield, "Microsoft chatbot is taught to swear on Twitter". BBC News. Accessed 25 Mar 2016
7. J. Hempel, "Microsofts AI Comeback" (2017), Wired. Accessed 23 Mar 2018

Development of Pyramid-Shaped Solar Distillation System and Experiments with Different Absorber Coating Materials

Pankaj Kumar Meena, Shivanshu Sharma, and Namrata Sengar

1 Introduction

Solar distillation system can be a promising way of getting clean safe water from brackish or saline water. This system takes solar radiation as an energy input, and this free energy from sun as thermal energy helps to evaporate water. Furthermore, the evaporated water is collected and condensed as final pure water. In remote areas where pure water availability is rare, solar distillation can be a helpful approach to get clean water. Solar distillation units can be used in the desalination process for seawater. Awareness about such distillation systems is limited, high cost and low production rates make the system unpopular.

The use of distillation still is being done since the fourth century B.C., Aristotle explained a method to evaporate saline water and then condense it in pure water. In the Sixteenth century, Arabian alchemists described about solar distillation to get potable water. In 1742, document proof of a device, by Ghezzi, N. of Italy, was registered. In 1872, Charles Wilson of Chile built the first modern solar still. Later on, new design and experiments came and many changes happened in solar distillation system [1]. The experimental work on solar still in India started in 1950 at National Physical Laboratory, New Delhi on the concentrator and flat basin-type solar stills. Later on, same work is continued by other organizations like Central Salt and Marine Chemicals Research Institute (CSMCRI), Bhavnagar and Central Arid Zone Research Institute (CAZRI), Jodhpur. Further developments are still going on in different research and academic institutions worldwide for further modifications

P. K. Meena · S. Sharma (✉) · N. Sengar
Department of Pure and Applied Physics, University of Kota, Rajasthan, India
e-mail: shivanshugecj@gmail.com

P. K. Meena
e-mail: pankaj10254@gmail.com

N. Sengar
e-mail: namrata@uok.ac.in

S. Vadhera et al. (eds.), *Latest Trends in Renewable Energy Technologies*, Lecture Notes in Electrical Engineering 760, https://doi.org/10.1007/978-981-16-1186-5_24

and performance enhancements [2]. Solar energy is free available source of energy and free from environmental pollution. Solar distillation is solar thermal application with economic benefits in remote areas [3]. An external power source with the help of renewable energy applications can be used to scale up the production rate of the distillation unit [4]. Distillation helps to get fresh water by the rejection of salt and other impurities by evaporation technique. A pyramid-glazed solar still helps to get higher heat gain input from solar radiation [5]. The main parameters that affect the performance of solar still are the depth of water, area of absorption, heat storage capacity, coating material, glass cover temperature difference, use of reflector, etc. [6].

Selection of solar still design helps to improve productivity by increasing thermal performance [7]. Earlier, the design of distillation stills lacked in some aspects, so adequate performance was not achieved, but in recent years, the improvement in design has helped to boost the performance of the system and increase the yield of distilled water [8]. There are a number of designs of solar stills like single slope, pyramid-glazed single slope, double slope, multi-effect, tubular, etc. Main components of solar distillation unit are storage unit, glazing, reflector, insulation covering, etc. Different techniques are being used for the performance enhancement of solar distillation unit [9].

Different micro and nanomaterials can be used to enhance productivity from a solar still [10]. The micro and nanomaterials help enhance the absorptance of the solar radiation absorbing surface. The solar absorptance, selectivity and thermal emittance of the nano-sized coating are generally optimized to 0.94, 94 and 0.01, respectively [11]. Paraffin wax is used as energy storage material with different coating materials to get higher thermal performance. It is seen that the productivity is increased by a combination of nano material coating [12]. It has been reported that a mixture of alluvial sand and Portland cement can be used to enhance the thermal heat gain capacity of solar distillation unit by 40% [13]. To understand the complete working mechanism of solar still, it is required to understand the fundamentals of heat and mass transfer process [14]. Dunkle's and Ueda's models are used to examine field experimental results. These methods help understand the deviation between calculated and observed values [15].

Though many works are going on for improving the performance and yield of the solar distillation units, still there is scope to develop cost-effective and user-friendly methods. With this in mind, in the present work, solar distillation stills have been developed and tested with different absorber coatings. The systems are easy to fabricate and can be made by any local laborers. Further, the coatings developed can be easily prepared by users in cost-effective manner. Present work reports the design details of a pyramid-shaped solar distillation still and it's testing with different absorber coating materials. The experiments have been conducted for observing the temperature profiles and the pH and TDS (total dissolved solids) values have been compared.

2 Design and Development of Solar Distillation Unit

2.1 Design Objective

Earlier studies had reported that for better performance of the system and to trap more amount of solar radiation, pyramid-shaped glazing for solar distillation still is recommended. Therefore, during the present work, three exactly same pyramid-shaped distillation stills were constructed. Different coating materials have been used in combination with black paint to check the thermal energy absorption effects in the solar distillation unit.

2.2 Design Parameters

The design parameters of the distillation stills with dimensions are presented in Table 1.

2.3 Development Process

Main materials used for the development of solar distillation system are G.I. sheet, glass glazing, insulation, covering resin, small plastic pipes, etc. The major steps of development are

1. Collection of all accessories used in making
2. Preparation of G. I. sheet base
3. Design of pyramid shape glass glazing

Table 1 Design parameters of developed solar distillation still

S. No	Parameter	Pyramid-type solar water distillation still
1	Base area of inner box	50 cm × 50 cm
2	Aperture area of inner box	46 cm × 46 cm (2116 sq.m.)
3	Base area of outer box	50.2 cm × 50.2 cm
4	Aperture area of outer box	3025 sq.cm
5	Height of inner box	15 cm
6	Height of inner box with absorber plate	15 cm
7	Height of outer box	15.2 cm
8	Height of glass center from absorber surface	30 cm
9	Glass inclination angle	30°

4. Testing of leakage at input and output side
5. Making insulation and covering using resin
6. Final design of still.

2.4 Design Layout and Actual System

The schematic diagram and the actual pyramid shaped developed solar distillation still are shown in Figs. 1 and 2, respectively.

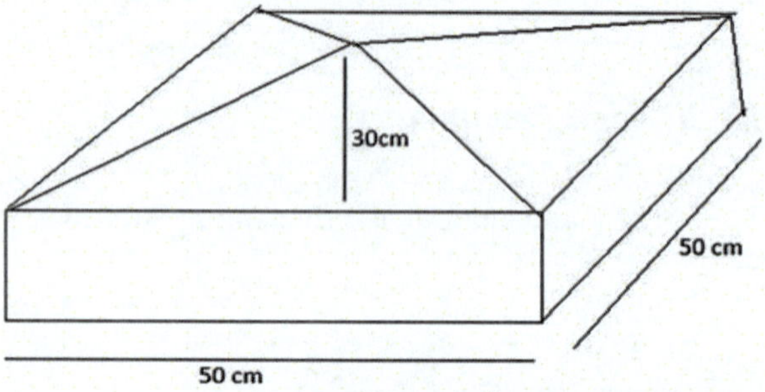

Fig. 1 Schematic diagram of the developed pyramid-shaped solar distillation still

Fig. 2 Developed solar distillation system with pyramid shaped glaze

3 Experimental Details

Experimental observations of the pyramid-type solar water distillation system have been recorded at every 30-min interval from 10:00 am to 05:00 pm. The temperatures were measured with thermocouples and digital temperature indicator CIE 305. The experimental results for the setup with surface temperature (°C) at ordinate (y-axis) and standard time in hours at abscissa (x-axis) are shown in Figs. 3, 4 and 5 for solar stills with different coatings.

3.1 Coating Materials

The coating materials used for absorber base surface of the solar distillation stills are

1. Black paint
2. Copper oxide
3. Aluminum oxide
4. Coal powder
5. Activated carbon
6. Titanium oxide (micro)
7. Titanium oxide (nano).

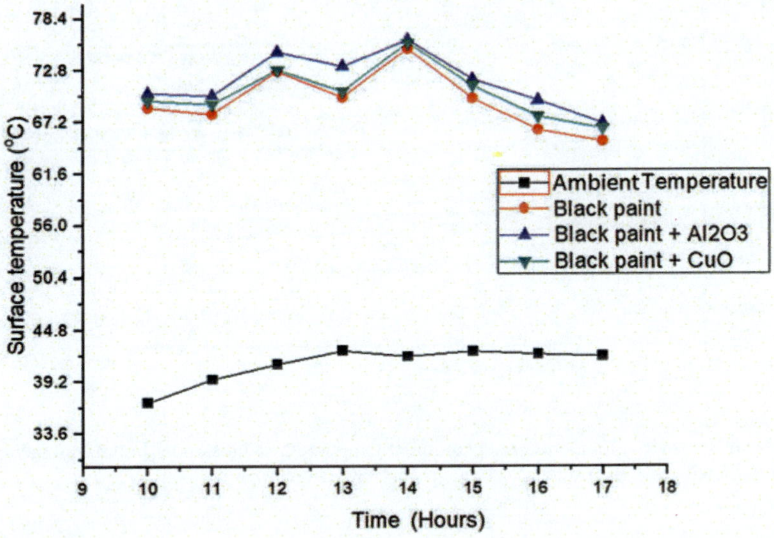

Fig. 3 Absorber surface temperature and ambient temperature with standard time for black paint, black paint mixed with Al_2O_3 and CuO coatings

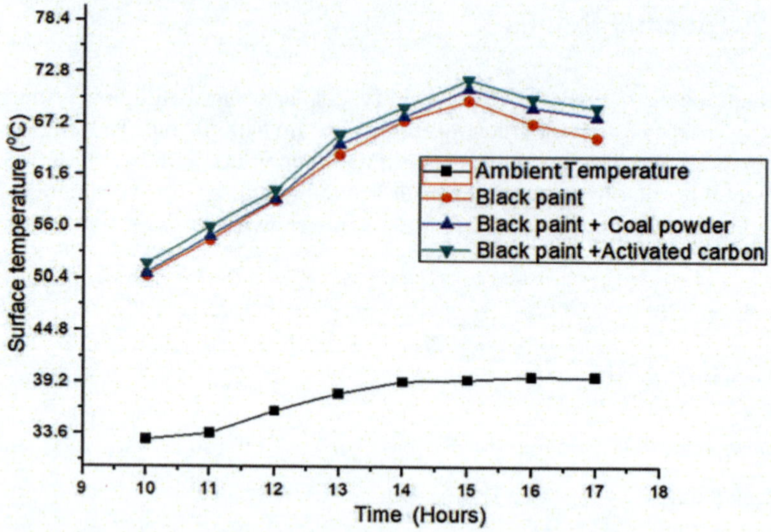

Fig. 4 Absorber surface temperature and ambient temperature with standard time for black paint, black paint mixed with coal powder and activated carbon coatings

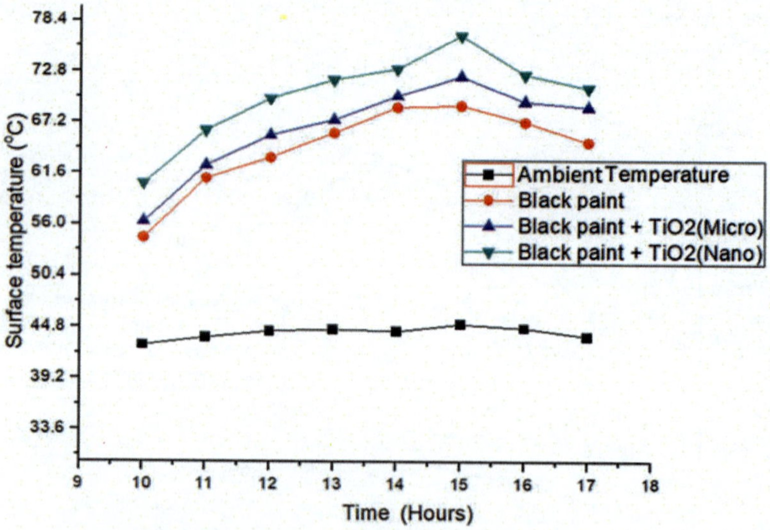

Fig. 5 Absorber surface temperature and ambient temperature with standard time for black paint, black paint mixed with TiO_2 micro and TiO_2 nanopowder coatings

Table 2 Details of coating materials for setup-I experiments

Number of distillation units	Three
Coating material of first distillation unit	Only black paint (50 gm.)
Coating material of second distillation unit	Black paint (50 gm.) + Al_2O_3 (5 gm.)
Coating material of third distillation unit	Black paint (50 gm.) + CuO (5 gm.)

The materials mentioned (2–7) were mixed with black matt paint and applied on the absorber surface of the solar stills during the experiments for comparison. The solar still with black-painted absorber surface acted as the reference system for the comparison.

3.2 Experimental Setup

3.2.1 Setup-I:- Solar Distillation Unit with Black Paint, Al_2O_3 and CuO Coating

The absorber surface temperature for different coating materials (black paint, Al_2O_3 and CuO coating) for the solar stills with standard time is shown in Fig. 3 and the details of coating materials is presented in Table 2.

3.2.2 Setup-II:- Solar Distillation Unit with Black Paint, Coal Powder and Activated Carbon Coating

The details of coating materials for setup-II are shown in Table 3.

The absorber surface temperature for different coating materials (black paint, coal powder and CuO) for the solar stills with standard time is shown in Fig. 4.

Table 3 Details of coating materials for setup-II experiments

Number of distillation unit	Three
Coating material of first distillation unit	Only black paint (50 gm.)
Coating material of second distillation unit	Black paint (50 gm.) + coal powder (3 gm.)
Coating material of third distillation unit	Black paint (50 gm.) + activated carbon (3 gm.)

Table 4 Details of coating materials for setup-III experiments

Number of distillation units	Three
Coating material of first distillation unit	Only black paint (50 gm.)
Coating material of second distillation unit	Black paint (50 gm.) + TiO_2 Micro powder (2 gm.)
Coating material of third distillation unit	Black paint (50 gm.) + TiO_2 Nano powder (2 gm.)

3.2.3 Setup-III:- Solar Distillation Unit with Black Paint, TiO_2-Micro, TiO_2-Nanocoating

The details of coating materials for setup-III experiments are shown in Table 4.

With this experimental setup, the observations related to absorber surface temperature with standard time for different coating materials (black paint, TiO_2-micro and TiO_2-nano) for the solar stills are shown in Fig. 5.

3.2.4 Water Quality Measurements

The water output of distillation units was collected and tested for pH and TDS values. Initial value of TDS in water used for distillation was around 2500 mg/L. The result of different distilled water samples is shown in Table 5 and Fig. 6.

Table 5 pH and TDS of water samples obtained using different distillation units

Water samples from different units containing the material	pH	TDS (ppm)
Black paint	6.86	33
Black paint and CuO	7.02	41
Black paint and Al_2O_3	6.98	39
Black paint and coal powder	6.89	90
Black paint and activated carbon	7.01	91
Black paint and TiO_2 (nano)	7.00	38
Black paint and TiO_2 (micro)	6.85	26

Fig. 6 Distilled water samples obtained from solar distillation units with different absorber coatings

4 Results

From Fig. 3, which shows the results of the first experimental setup, it is evident that the highest surface temperature is obtained with black paint and Al_2O_3 combination coating material applied at the absorber base surface of solar distillation unit. Figure 4 shows that the highest surface temperature is obtained with black paint and activated carbon combination coating material at base surface of solar distillation unit with the second experimental setup. During the third experimental setup, TiO_2 micro and nanopowder were mixed with black paint and the comparative temperature profiles are shown in Fig. 5. The highest surface temperature is obtained with black paint and TiO_2 nanopowder combination coating material at base surface of solar distillation unit.

Water quality tests such as pH and TDS measurements were performed on the distilled water samples obtained from the distillation stills as yield. The results show that the TDS considerably reduced with the distillation. With different coatings, the TDS values are in the range of 26–91 ppm.

5 Conclusions

The result of all setups shows that the coating material helps to increase thermal gain from the solar radiation. In setup I, the combination of black paint and aluminum oxide had the highest surface temperature gain as compared with only black paint and black paint with copper oxide. Similarly in setups II and III, black paint with activated carbon and black paint with nanotitanium oxide powder show higher surface temperature results as compared with the other materials. From the study, it can be concluded that Al_2O_3, activated carbon and TiO_2 nanopowder are better choices for coating materials. These materials can be easily mixed with the black paint to improve performance of the system. Water quality tests related to pH and TDS measurement of the distilled water samples show that the water quality of distilled water is improved after distillation. The TDS value is very less in distilled water sample and the pH value is near the standard value. The higher thermal absorptance due to coating materials may help to increase the rate of distillation and improve the yield of the distilled water output from solar distillation unit.

Acknowledgements The authors gratefully acknowledge the support of the Royal Academy of Engineering, UK through Newton-Bhabha Higher Education Partnership Project.

References

1. G.N. Tiwari, H.N. Singh, R. Tripathy, Present status of solar distillation. Sol. Energy **75**, 367–373 (2003)
2. G.N. Garg, J. Prakash, *Solar Energy Fundamentals and Applications* (Tata McGraw Hill Publications, 2010), p.13
3. M. Thirugnanasambandam, S. Iniyan, R. Goic, A review of solar thermal technologies. Renew. Sustain. Energy Rev. (ELSEVIER) **14**, 312–322 (2010)
4. H. Sharon, K.S. Reddy, A review of solar energy driven desalination technologies. Renew. Sustain. Energy Rev. (ELSEVIER) **41**, 1080–1118 (2015)
5. Q. Wang, N. Li, B. Bolto, M. Hoang, Z. Xie, Desalination by pervaporation: a review. Desalination (ELSEVIER) **387**, 46–60 (2016)
6. P. Prakash, V. Velmurugan, Parameters influencing the productivity of solar stills - a review. Renew. Sustain. Energy Rev. (ELSEVIER) **49**, 585–609 (2015)
7. A.A. El-, E. El-Bialy, Advanced designs of solar desalination systems: a review. Renew. Sustain. Energy Rev. (ELSEVIER) **49**, 1198–1212 (2015)
8. V. Belessiotis, S. Kalogirou, E. Delyannis, Solar distillation-solar stills. Therm. Sol. Desalin. **85**, 103–190 (2016)
9. M. Edalatpour, K. Aryana, A. Kianifar, G.N. Tiwari, O. Mahian, S. Wongwises, Solar stills: a review of the latest developments in numerical simulations. Sol. Energy (ELSEVIER) **135**, 897–922 (2016).
10. D.D.W. Rufuss, S. Iniyan, L. Suganthi, P.A. Davies, Solar stills: a comprehensive review of designs, performance and material advances. Renew. Sustain. Energy Rev. (ELSEVIER) **63**, 464–496 (2016)
11. L. Cindrella, S. Prabhus, CuO-PANI nanostructure with tunable spectral selectivity for solar selective coating application. Appl. Surf. Sci. (ELSEVIER) **378** 245–252 (2016)
12. A. Elfasakhany, Performance assessment and productivity of a simple-type solar still. Appl. Energy (ELSEVIER) **183**, 399–407 (2016)
13. M.H. Sellami, S. Guemari, R. Touahir, K. Loudiyi, Solar distillation using a blackened mixture of Portland cement and alluvial sand as a heat storage medium integrated with nanocomposite energy storage system. Desalination (ELSEVIER). **394**, 155–161 (2016)
14. G. Xiao, X. Wang, M. Ni, F. Wang, W. Zhu, Z. Luo, K. Cen, A review on solar stills for brine desalination. Appl. Energy (ELSEVIER) **103**, 642–652 (2013)
15. A. Ahsan, M. Imteaz, R. Dev, H.A. Arafat, Numerical models of solar distillation device: present and previous. Desalination (ELSEVIER) **311**, 173–181 (2013)

A Review on Hardware Implementations of Signal Processing Algorithms

Neelesh Ranjan Srivastava and Vikas Mittal

1 Introduction

The advancements in the Integrated Circuits (ICs) resulted in high-performance and high-speed circuitries remodeled as ASIC and FPGA [1] that enable the user to implement complex signal processing algorithms on hardware platforms [2–4]. With this technology, implementations of complicated algorithms come to reality [5]. The requirement of hardware implementations of signal processing algorithms has abruptly increased due to the large stored data values [6, 7] and the requirement of analyzing and synthesizing data in very less time. This goal cannot be achieved by simple computer software as they run in operating systems having low processing speed [8].

Embedded vision is the vision of a product to enhance its functions in some way. So, some significant challenges occur in developing systems, including a large volume of data to process, quick processing, and limited physical space and weight [9].

FPGAs are increasingly seen as a platform for these applications. They have advantages above conventional computing platforms. Parallel processing enables large volumes of data to be processed at the same time. Pipelined processing [10] enables increased processing throughput and reduced latency.

Despite these advantages, using FPGAs to implement vision systems has other disadvantages such as they are hard to program, etc.

N. R. Srivastava (✉)
KIET Group of Institutions, Ghaziabad, Uttar Pradesh, India
e-mail: nr.srivastava@kiet.edu

V. Mittal
National Institute of Technology, Kurukshetra, Haryana, India
e-mail: vikas_mittal@nitkkr.ac.in

S. Vadhera et al. (eds.), *Latest Trends in Renewable Energy Technologies*, Lecture Notes in Electrical Engineering 760, https://doi.org/10.1007/978-981-16-1186-5_25

For the sake of the above discussion, the GPUs are the best options in almost all the disadvantages of FPGAs that can be illustrated as design time, redundancy, and pipelining [11].

This paper elaborates on the hardware implementation of signal processing algorithms with a stand-alone system that uses a single coding language for both the processes of software simulation and hardware implementations. As per the above discussions, the system can be defined best with GPU.

This stand-alone system can be designed for different algorithms. Therefore, GPUs are a better option compared with FPGAs due to many advantages as GPUs have a wide application field with a mature ecosystem of applications and processes, they also offer an efficient platform for the new applications, they are easily programmed as only a single language can be used for all programming, as the GPUs can be implemented as stand-alone systems, the development time will be very less and above all, they are compatible to floating-point operations [2], which are very much required for the signal processing operations.

2 State of the Art

The literature review for this paper and topic can be easily summarized, year by year, from 2005 to 2020. The research on hardware implementations of signal processing algorithms is getting interesting in recent years. It has been the focus of many surveys and research in recent years. The interest is basically over the need for a specialized and specific hardware platform for the implementations. The stand-alone system must be defined as a cost-effective and open source, in terms of production cost and single coding language, respectively.

In this section, we are going to explore different related surveys and research proposed for the given topic. We are also going to define the achievements of the work and gaps concerning the aim established in this paper.

GPUs are a better choice for video processing applications due to the instruction set, pipelines, and clock rates in comparison to the FPGAs [12–14]. FPGAs can also be defined for the implementations of the industrial control systems as well [15], in terms of the hardware platform. The FPGA systems are diagnosed and defined with some of the best specifications like small device size, high speed, low coast, and a short time to market for efficient implementations [12, 16, 17]. A comparative study for the GPU vs FPGA is surveyed based on computing performance for some of the specific GPUs like Convey HC-1 and Nvidia GTX285 [1]. In the study of hardware platforms, machine learning is done, where different aspects like accuracy, energy, throughput, and cost requirements are to be studied and controlled during the design process [9, 18]. A study proposes a new GPU memory extension solution by solving input/output (I/O) bottlenecks that happen in a multi-GPU system [19]. At last, there is a paper surveyed for the hardware implementation of artificial intelligence algorithms very close to the aspect of this paper [20].

The objective of the survey is to answer multiple research questions concerning GPU applications for the algorithm implementations, these research questions are:

- RQ1: Hardware Selection: What is the main hardware-defined for the algorithm implementations?
- RQ2: Architecture Perspective: What are the architectures required for the implementation purposes and what are the coding language requirements?
- RQ3: Specification and cost perspective: What are the basic specifications defined for the hardware requirements so that the costing of the complete system must be low and affordable to common people?

3 Hardware Implementation Process Flow

See Fig. 1.

The process starts with a top-level block diagram, then the latency and output of each block are defined. For the implementation of each block in terms of state machines, the prototype for the least clear parts of the algorithm is mentioned [21]. Then the rest of the algorithm is implemented by synthesizing the design. At last, the design is simulated, and test cases are developed [22, 23].

4 GPUs

4.1 GPU-Accelerated Computing

The GPU provides better performance for software applications and makes them faster [24]. GPU-accelerated computing is processed by enhancing its computation-intensive parts by the GPU while other sections are being executed in the CPU. A CPU comprises cores designed for sequential serial processing while parallel architecture consists of more efficient and smaller cores that can handle multiple tasks simultaneously. This type of accelerated computing has been rigorously used in various processes [25] such as video editing, medical imaging, fluid dynamics simulations, color grading, and enterprise-level applications. Also, it has been effectively used in complex fields such as artificial intelligence and deep learning [17].

4.2 The architecture of GPUs

Traditionally, a GPU has thousands to lakhs of Arithmetic Logic Units (ALUs) compared with the traditional ASIC or CPU that has approximately 4–8 ALUs [26] as shown in Fig. 2, which is a diagrammatical structural comparison between a GPU

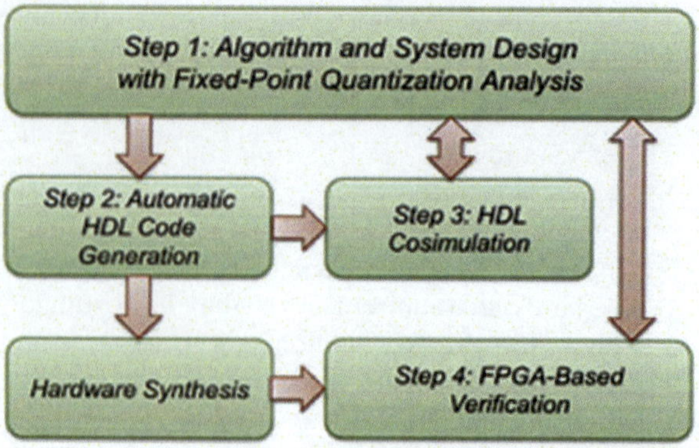

Fig. 1 Process flow of hardware implementation of an algorithm

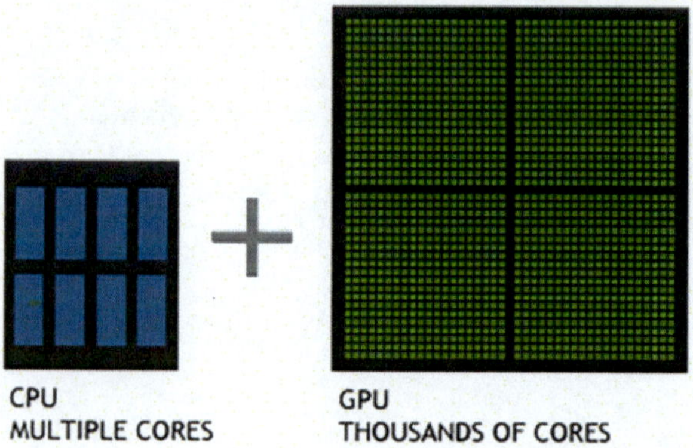

Fig. 2 Structure of CPU and GPU

cell and an FPGA (CPU) cell [27]. The comparison of the floating-point functioning capability between CPU and GPU is that it later is highly specified for the parallel computation of intensive functions and applications [28].

4.3 Languages Required for GPU Programming

In modern times, coding platforms ignore language barriers and focus on high-level computing [29]. There are some prominent and famous coding languages to fulfill all these requirements and conditions:

- **CUDA (Compute Unified Device Architecture)**
  This platform was released by NVIDIA in 2007. It can speed up the computing processes by the best use of GPU power [30]. The programmers can call or fetch CUDA from other languages also like C, C++, FORTRAN, or PYTHON [31].
- **Open CL (Open Computing Language)**

This platform was released by the Khronos Group in 2009. It is a popular open-source and free standard of cross-platform parallel programming. It improves the speed and response timing of many applications like vision processing, neural network, professional creative tools, and scientific and medical software.

- **Open ACC (Open Accelerators)**
  It is the newest standard for parallel computing and was released by a group of companies like Cray, CAPS, NVIDIA, and the Portland Group, in 2015, to simplify parallel programming of different heterogeneous GPU systems.

5 Comparison of different Hardware Platforms

The comparisons and used techniques are described as below:

Performance Parameters:

The selection of appropriate performance parameters is important for evaluating the efficiency of vision kernels of hardware accelerators to provide a proper assessment of different hardware for different applications. Following parameters have been considered in this paper as described below:

Run-time Parameter:

Run-time parameters of vision kernels [32] can be evaluated by calculating the delay time of a kernel's coding. A high-resolution accurate timer is used to measure delays during execution. In this paper, the execution time of CPUs, as well as GPUs, is considered along with the time for configuring data movers in FPGAs.

Energy Parameter:

The energy parameter defines the quantity of energy dissipated by hardware accelerators to execute a kernel's operation [32]. Two components of this consumed power exist in the literature:

(1) Static power:

The power is consumed in the idle condition of the system.

(2) Dynamic power:

The power consumed after the static level of the system while it is working or computing.

Energy-Delay Product (EDP) Parameter:

Run-time parameters cannot represent the real picture of the entire performance. This parameter considers the output of the algorithm along with the energy consumed by it as it consists of the product of energy consumed and delay [32]. It can help in comparing different hardware for selecting appropriate hardware for a specific operation. A lower value of EDP is required for efficient hardware.

In this study, we examined two platforms for implementing vision applications: Nvidia Jetson TX2 and Xilinx ZCU102. These platforms are equipped with power

Table 1 Comparison of Microprocessor, FPGA, ASIC, and GPU architectures based on different parameters

Parameters	Microprocessor	FPGA	ASIC	GPU
Flexibility during development	Medium	High	Very High	Low
Flexibility after development	High	High	Low	High
Parallelism	Low	High	High	Medium
Performance	Low	Medium	High	Medium
Power consumption	High	Medium	Low	High
Cost	Low	Medium	High	Low
Setup cost	None	None	High	None
Unit cost	Medium	High	Low	High
Time to market	Low	Medium	High	Medium

measurement IC enabling the measurement of power for CPU cores and GPU cores on Jetson platform, programmable logic, full power CPU cores, and low-power CPU cores on the FPGA platform. Jetson TX2 shell scripts power trends and log values with the system's timestamp into a text file. This additional measurement task adds to the total power consumption and hence adversely affects the overall performance. Appropriate compensation for this has been considered in this paper for the presented data.

In ZCU102, the Xilinx system controller connects through a UART to separate microcontroller, i.e. MSP430, which controls and reports power data.

First, 1000 frames are processed on a CPU core, and afterward, a similar number of frames are processed on the accelerated hardware. The average frame rate is analyzed by measuring the time interval between any two contiguous vertical lines and is divided by 1000. All the considered frames are in grayscale at 1080p resolution.

FPGA board: The Xilinx Zynq UltraScale + MPSoC ZCU102 board consists of a 16 nm XCZU9EG FPGA and an on-board 4 GB 64bit DDR4 RAM with a bandwidth of 136 Gb/s [32].

GPU board: The Nvidia Jetson TX2 (Pascal 256 CUDA cores (16 nm)) consists of 8GBs of 128bit DDR4 RAM with a bandwidth of 477.6 Gb/s. Both FPGA and GPU consist of on-chip ARM CPU cores with NEON SIMD optimization [32].

A comparison of general Microprocessor, FPGA, ASIC & GPU structures is shown in Table 1.

6 Conclusion and Future Work

A review of hardware implementations of signal processing algorithms has been presented successfully. It has been observed that GPU-based implementations of signal processing algorithms are better in terms of the following.

Due to the cutting-edge design of parallel processing framework, huge number of complex computing units is involved in their architecture, unlike limited numbers as used in ASICs or FPGAs. This complex architecture of the GPU is well suited for signal processing, video processing, or image processing applications.

The architecture of GPUs consists of more processor cores as compared with any of the ASICs or FPGAs, which results in faster processing and complex computations.

The cost of any system designed using GPUs is less as it requires minimum peripherals as compared with FPGA- or ASIC-based systems. Also, in this case, a single coding language (usually an open-source language, which is easily and freely available for the user) is used for cost-effective simulations as well as implementations.

The comparative analysis also suggests that GPUs have the highest number of the desired values for the considered parameters.

Given the above, it can be concluded that it is more beneficial to use GPUs for the hardware implementations of signal processing algorithms as stand-alone systems as compared with ASICs or FPGAs.

References

1. D.H. Jones, A. Powell, C. Bouganis, P.Y.K. Cheung, GPU versus FPGA for high productivity computing, in *2010 International Conference on Field Programmable Logic and Applications, Milano* (2010), pp. 119–124
2. K. Elango, K. Muniandi, Hardware implementation of FFT/IFFT algorithms incorporating efficient computational elements. J. Electr. Eng. Technol. **14**, 1717–1721 (2019)
3. E. Setiawan, T. Adiono, Implementation of systolic co-processor for deep neural network inference based on SoC, in *International SoC Design Conference (ISOCC), 2018*
4. A. Safari, S. Mekhilef, Simulation and hardware implementation of incremental conductance MPPT with direct control method using Cuk converter. IEEE Trans. Industr. Electron. **58**(4), 1154–1161 (2011)
5. M. Han, J.M. Song, H.G. Yang, Y. Kim, Implementation of multi-channel FM repeater using digital signal processing algorithm in FPGA, in *International SoC Design Conference (ISOCC), 2018*
6. E. Onat, FPGA implementation of target detection algorithm at real-time video signal processing using harris corner detector filter, in *26th Signal Processing and Communications Applications Conference (SIU), 2018*
7. T. Rissa, R. Uusikartano, J. Niittylahti, Adaptive FIR filter architectures for run-time reconfigurable FPGAs, in *Proceedings, IEEE International Conference on Field-Programmable Technology, 2002*
8. C. Siddappa, M. Wickert, CAF implementation on FPGA using python tools, in *Proceedings of the 18th Python in Science Conference (SCIPY 2019)*
9. V. Sze, Y.-H. Chen, J. Emer, A. Suleiman, Z. Zhang, Hardware for machine learning: challenges and opportunities, in *2018 IEEE Custom Integrated Circuits Conference (CICC)* (2018)
10. S.P. Mohanty, GPU-CPU multi-core for real-time signal processing, in *2009 Digest of Technical Papers International Conference on Consumer Electronics, Las Vegas, NV, 2009*, pp. 1–2
11. N. Sulaiman, Z. Obaid, M.H. Marhaban, M.N. Hamidon, Design and implementation of FPGA-based systems - a review. Aust. J. Basic Appl. Sci. (2009)

12. B. Cope, P.Y.K. Cheung, W. Luk, S. Witt, Have GPUs made FPGAs redundant in the field of video processing? in *Proceedings. 2005 IEEE International Conference on Field-Programmable Technology, 2005., Singapore* (2005), pp. 111–118
13. L. Shi, X. Gao, X. Yang, Z. Chen, M. Zheng, Algorithm optimization and hardware implementation for merge mode in HEVC. J. Real-Time Image Process. (2018)
14. S.L.Harris, D. MoneyHarris, Digital Design and Computer Architecture, ARM Edition (2016), pp. 172–237
15. E. Monmasson, M.N. Cirstea, FPGA Design Methodology for Industrial Control Systems—A Review. IEEE Trans. Industr. Electron. **54**(4), 1824–1842 (2007)
16. I. Grout, *Digital Systems Design with FPGAs and CPLDs* (Newnes Publication, 2008), pp. 177–216
17. T. Baji, Evolution of the GPU device widely used in AI and massive parallel processing, in *2018 IEEE 2nd Electron Devices Technology and Manufacturing Conference (EDTM)* (2018)
18. J.B. Srivastava, R.K. Pandey, Implementation of digital signal processing algorithm in general purpose graphics processing unit (GPGPU). Int. J. Innov. Res. Comput. Commun. Eng. **1**(4), (2013)
19. Y. Kim, J. Lee, J.-S. Kim, H. Jei, H. Roh, Efficient multi-GPU memory management for deep learning acceleration, in *2018 IEEE 3rd International Workshops on Foundations and Applications of Self* Systems (FAS*W)* (2018)
20. M.A. Talib, S.Majzoub, Q. Nasir, D. Jamal, A systematic literature review on hardware implementation of artificial intelligence algorithms. J. Supercomput. (2020)
21. R. Woods, Y.Yi, *FPGA Based Implementations of Signal Processing Systems* (Wiley, New York, 2017)
22. J. Rodrigues, M. Kamuf, H. Hedberg, V. Owall, A manual on ASIC front to back end design flow. 75–76 (2005)
23. M. Kiran, K.M. War, L.M. Kuan, L.K. Meng, L.W. Kin, Implementing image processing algorithms using 'Hardware in the loop' approach for Xilinx FPGA, 2008, in *International Conference on Electronic Design, Penang, 2008*, pp. 1–6
24. S. Afifi, H. GholamHosseini, R. Sinha, FPGA implementations of SVM classifiers: a review. SN Comput. Sci. **1**(3), 1–7 (2020)
25. N. Raut, A. Gokhale, FPGA implementation for image processing algorithms using xilinx system generator. IOSR J. VLSI Signal Process. (IOSR-JVSP) (2013)
26. D. Sundfeld, G. Teodoro, J. Gorodkin, A.C. Melo, Using GPU to accelerate the pairwise structural RNA alignment with base-pair probabilities, in *Special Issue in Parallel, Distributed, and Network-Based Processing*, 2020.
27. P. Jawandhiya, Hardware design for machine learning. Int. J. Artif. Intell. Appl. **9**(1), 63–84 (2018)
28. M. Zhu, Y. Zhuo, C. Wang, W. Chen, Y. Xie, Performance evaluation and optimization of HBM-enabled GPU for data-intensive applications. IEEE Trans. Very Large Scale Integr. (VLSI) Syst. **26**(5), 831–840 (2018)
29. H. Andrade, I. Crnkovic, A review on software architectures for heterogeneous platforms, in *2018 25th Asia-Pacific Software Engineering Conference (APSEC), Nara, Japan, 2018*, pp. 209–218
30. G. Li, Q. Chang, C. Xi, X. Ma, Acquisition design and implement of Beidou B3I signal based on GPU, in *IOP Conference Series: Materials Science and Engineering, IOP Conference Series: Materials Science and Engineering*, vol. 715, Number 1, (2020)
31. A.C. Crespo, J.M. Dominguez, A. Barreiro, M. Gómez-Gesteira, B.D. Rogers, GPUs, a new tool of acceleration in CFD: efficiency and reliability on smoothed particle hydrodynamics methods. J. PLOS ONE **6**(6), e20685 (2011)
32. M. Qasaimeh, K. Denolf, J. Lo, K. Vissers, J. Zambreno, P.H. Jones, Comparing energy efficiency of CPU, GPU and FPGA implementations for vision kernels, 2019, in *IEEE International Conference on Embedded Software and Systems (ICESS)*, Las Vegas, NV, USA, 2019, pp. 1–8

A Review on Solar PV Cell and Its Evolution

Devesh Jaiswal, Monika Mittal, and Vikas Mittal

1 Introduction

An increase in energy demand and other related factors like enhanced energy costs, limited reserves of conventional resources, and environmental pollution made renewable energy an affordable and dependable energy resource. The US Energy Information Administration (EIA) predicted an increase in world energy consumption by nearly 50% between 2018 and 2050 in its latest "International Energy Outlook 2019 (IEO2019) Reference Case". With the fast growth of power generation, renewable energy (wind, solar, and hydropower) is the rapidly growing energy source during this period [1]. Naturally existing self-replenishing resources (viz. sunlight, tides, rain, wind, biomass, waves, and thermal energy stored in the earth's crust) are used to harness renewable energy. Therefore, this energy is the most tremendous unconventional and the handiest approach to overcome the existing challenges like deficiency of fossil fuel reserves, greenhouse gas emissions, and different environmental issues. Among all renewable energy resources, solar energy is the most sustainable form that would contribute to 4.9 gigatons (Gt) of CO_2 emission reduction, which is 21% of the total emission reduction. It is capable of producing more than 25% of total electricity needs in 2050 [2].

The Solar PV domain is exploring new technologies for enhancing its conversion efficiency. A solar PV cell is the most essential and primitive component of a solar energy system. It directly converts the incident solar energy into electrical energy through photoelectric effect. A number of semiconductor materials and technologies exist presently resulting in low-cost and high-efficiency design of the solar cells. The

D. Jaiswal (✉) · M. Mittal
Electrical Engineering Department, NIT Kurukshetra, Kurukshetra 136119, Haryana, India
e-mail: deveshjaiswal15@gmail.com

V. Mittal
Electronics and Communication Engineering Department, NIT Kurukshetra, Kurukshetra 136119, Haryana, India

S. Vadhera et al. (eds.), *Latest Trends in Renewable Energy Technologies*, Lecture Notes in Electrical Engineering 760, https://doi.org/10.1007/978-981-16-1186-5_26

parameters on which the optimality of its conversion efficiency depends are bandgap utilization, light trapping, other optoelectric properties etc.

The first solar cell was operated at the conversion efficiency of less than 1% [3]. Later it was increased to 4% by using silicon materials [4]. Further advancements were based on the selection of proper materials to achieve cost-effectiveness, reliability, and high efficiency using wafer, thin-film, and polymer technologies. Efficiencies in the range of 20–25% were reported. Traditionally, silicon cell used for the conversion of solar energy had the drawbacks like low efficiency, high cost, and requirement of higher temperature for its surface fabrication [5]. Theoretical efficiency of this solar cell, 16.27%, reduces to 8.52% for cloudy weather and moist surface [6]. Recently, researchers have discovered perovskite solar cells to further improve the conversion efficiency overcoming the drawbacks of silicon solar cells by providing high absorption coefficient, less cost and high-power conversion efficiency of 25.2% [7–9].

This paper is organized as: Section 2 discusses the structure of a typical solar PV cell. Section 3 presents a review of different materials used for solar cells. The effects of insolation and temperature on PV cells are presented in Sect. 4. Section 5 presents the summary and discussion, followed by conclusion at the end.

2 Structure of a Typical Solar Cell

An array of solar cells is arranged in a different manner to form PV modules of different sizes. The incident light on it is reflected, absorbed, or transmitted through it. Out of which only the absorbed light is responsible for electricity generation. Flat-plate and convex lenses are the two types of PV cells. Flat-plate cells are mounted at fixed angles, but the angle can be changed to track the optimal sun exposure. On the other hand, convex lenses require less material for the same power output, but they perform best during a clear sky unlike flat-plate PV cells.

The working principle of a PV cell is similar to a semiconductor diode consisting of P-and N-type material, contacts grid, base, antireflective coating, and glass cover or lens as shown in Fig. 1.

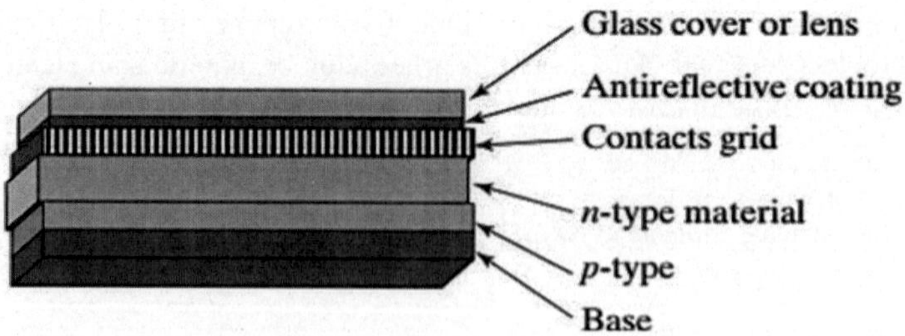

Fig. 1 Parts of PV cell

3 Materials Used in Solar Cells

A PV solar cell is a simple or compound semiconductor device providing conditional and not continuous conduction of electricity. A wide range of solar cell technologies, currently used, differ on the basis of used material, manufacturing process, efficiency, and cost. The evolution of a PV solar cell can be organized into three generations based on different materials used as shown in Fig. 2.

3.1 First-Generation

These cells are most prevalent, traditional, and are based on silicon wafer. Czochralski wafers, which are a kind of silicon wafer, are made using the Czochralski process [10]. This process is very power efficient and, hence, 90% of the manufactured solar cells use silicon. They are further classified into two types: Single/monocrystalline and poly/multicrystalline silicon solar cells [11].

Single/Monocrystalline Silicon Solar Cell

This cell is formed into bars and cut into wafers with the help of cylindrical ingots using Czochralski process. The corners of the cells appear to be clipped like octagon due to being cut from cylindrical ingots. These cells have efficiency in the range of 17–18% [12]. Authors in [13] improved its performance and achieved an efficiency of 19.6% using both the technologies viz. Metal Wrap Through cells, which has the advantage of lower shading area and lower series resistance losses, and Selective

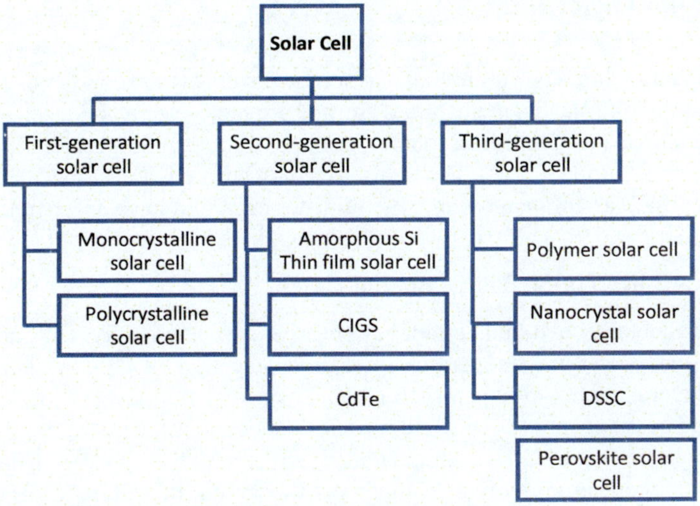

Fig. 2 Classification of the evolution of a solar cell

Emitter cells, which has the advantage of higher blue response and lower contact resistance. The cost of silicon material can be reduced by increasing the cell thickness. Authors in [14] achieved 17% peak efficiency with 50-μm thickness. Authors in [15] experimentally demonstrated the impact of different doping and minority carrier lifetimes on n-type Czochralski UMG-Si (upgraded metallurgical grade silicon) solar cells and confirmed an efficiency of 21.1% from silicon solar cells constructed using 100% UMG-Si.

Polycrystalline Silicon Solar Cell

There are two different methods for producing polycrystalline solar cells. One method is similar to that used for single-crystal cell where polycrystalline cells are produced by slicing the polycrystalline boules of silicon. This technique is not cost-effective. The second method uses glow discharge or sputtering techniques in which polycrystalline film is deposited on a substrate. This technique is cost-effective and suitable for mass production. The efficiency of polycrystalline solar cells ranges from 12 to 14% [16]. There are two reasons for comparatively high efficiency viz. each cell has many crystals and electrons have less freedom to move, the occurrence of "grain boundaries" due to defects in crystals. Therefore, these cells are more economical but less efficient as compared with monocrystalline ones. Authors in [17] compared n-type multicrystalline silicon solar cells with p-type multicrystalline and concluded that the former has higher potential of being more efficient (i.e. 19.6%) for a similar purity level. Authors in [18] achieved the highest efficiency of 21.9% for multicrystalline cells featuring a full-area passivating rear contact (TOPCon) with diffused boron front emitter.

3.2 *Second Generation*

These are manufactured by employing one or more layers of semiconductor materials on the substrate like plastic, metal, glass etc. and are more economical. They include amorphous silicon (a-si) thin-film solar cell, copper indium gallium selenide (CIGS), and Cadmium Telluride (CdTe). The light-absorbing layers of thin-film solar cells are about 1 μm thick as compared with silicon wafer cells that have 350 μm thickness [19].

Amorphous Silicon (a-Si) Thin-Film Solar Cell

These cells generally have small efficiency but are most eco-friendly. They are manufactured at low processing temperatures and are cheaper and widely available. They use triple layered technology in which silicon, as a very thin layer, is deposited on the backside of the substrate. They are made of nanocrystalline (n-Si) or microcrystalline silicon (μc-Si), which has a higher bandgap (1.7 eV) that can absorb the visible part of the solar spectrum more as compared with crystalline silicon (c-Si) (1.1 eV) [20]. The efficiency of a-Si varies in the range of 4–8%. Authors in [21] achieved 8.04% efficiency using hydrogenated a-Si: H. In 2013, using a-Si: H in the top cell, a-SiGe: H in the middle cell, and hydrogenated microcrystalline silicon (μc-Si: H) in the bottom cell stabilized 13.4% efficiency of a-Si: H solar cell achieved in [22].

Copper, Indium, Gallium, and Selenium (CIGS)

These four elements are used in CIGS-based thin-film solar cell technology using sputtering, evaporation, electrochemical coating, printing, and electron beam deposition techniques. CIGS has direct bandgap and provides efficiency in the range of 10–12% [12, 23]. Authors in [24] achieved the highest efficiency of 8.4 $\pm$ 0.8% by integrating CMOS microchips with copper indium gallium (di)selenide (CIGS) thin-film solar cells. The presence of potassium (K) on this module increased the efficiency to 21% in comparison to 20.4% provided by polycrystalline silicon (p-Si) [25]. Authors in [26] reported a record efficiency of 21.7% from a novel thin-film Cu(In,Ga)Se2 (CIGS) solar cell.

Cadmium Telluride (CdTe)

It is a direct bandgap material and II–VI p-type compound semiconductor, which has a large absorption coefficient, ~1.5 eV bandgap, and stable compound. Due to this, CdTe has a higher light absorption capacity and its operating efficiency ranges from 9 to 11% [27]. Its theoretical efficiency is likely to be in the range 28–30% [28]. This material has been considered as an eminently extensible technology whose efficiencies have reached up to 22% and modules 18.6% in the lab environment [29].

3.3 *Third Generation*

These cells employ the newest technologies to achieve high efficiency using organic materials like PSC (polymer solar cell), (NCSC) nanocrystal solar cell, (DSSC) dye-sensitized solar cells, and perovskite solar cell [30].

Polymer Solar Cell (PSC)

This cell is formed when a polymer foil or ribbon is coated with thin serially connected functional layers. They have higher flexibility, low material cost, and good processing ability. Authors in [31] analyzed that heptane/o-dichlorobenzene (ODCB) mixture has a good combination with (P3HT)/C-70 [poly(3-hexylthiophene)] film for high device performance and the optimized ratio of P3HT donor to C-70 accepter to achieve 2.24% conversion efficiency. Another author in [32] enhanced the performance by reporting 30% conversion efficiency when organic optoelectronic devices are combined with Ni NPs (nickel nanoparticles) by thermal evaporation. Authors in [33] demonstrated that the highest conversion efficiency of 10.3% is possible when PM6 (fluorinated donor polymer) is combined with all polymer acceptor-based PSC [PFBDT-IDTIC].

Nanocrystal Solar Cell

These cells are also called Quantum dots (QD). They are formed when a layer of nanocrystals is deposited on Si- or organic material-based substrate. With the advancement of nanotechnology, semiconductor materials like Si, CdTe or CIGS are targeted to be replaced by nanocrystal semiconductor materials. The best nanocrystal solar cells are formed by PbS QD or PbSe QD as an active layer [34]. It is possible to increase efficiency by up to 60% using nanocrystals in solar cells.

Dye-sensitized Solar Cells (DSSC)

These are liquid-typed cells. DSSC consists of four components: (1) the semiconductor electrode (NiO p-type and TiO_2 n-type), (2) a dye sensitizer, (3) redox mediator, and (4) a counter electrode. Authors in [35] investigated that the efficiency increases by more than 10%, when photosensitization of nano-grained TiO_2 coatings is combined with the optically active dyes. Authors in [36] analyzed DSSC PV behavior experimentally in the sense that the composite feature of ND (nanodiamond) with TiO_2 improves the conversion efficiency by 18.7%.

Perovskite Solar Cell

In recent years, these cells have remarkable discovery, which have combined features of different elements and provide the advantage of high absorption coefficient. Another advantage is that it can be used in single-junction technology to increase the efficiency by 23% as well as used in tandem junction technology to achieve the efficiency of 27%. Perovskite is a crystal structure represented by formula ABX3, where A and B are site locations of different sizes and X is halogen (Cl^-; Br^-; I^-) [12]. Authors in [37] reported that they enhanced their performance by 29.6% using tandem junction technology combining perovskite materials with crystalline silicon cell/CIGS/CZTSSe. These solar cells can have an efficiency up to 31% [38]. However, they have drawbacks such as less durability and stability, degradation of material etc.

4 Effects of Solar Insolation and Temperature

The performance of solar cells also depends upon the environmental conditions such as insolation and temperature. The effect of increased insolation is more current and voltage at the output resulting in more power at the output of a solar cell. This is due to the fact that electrons in the cell material get more excitation energy causing an increase in their mobility. The effect of insolation on individual material is not available in the literature to the best of authors' knowledge. The increase in temperature has a similar effect due to a decrease in the bandgap of material for a given insolation, resulting in the increase in generated electron-hole pairs, but here this results in slightly rising in output current and a significant drop in output voltage, thereby reducing the overall output power and reduced efficiency. The effects of high temperatures on different solar cell materials are shown in Table 1.

Table 1 Highlights of different generations of a solar cell [12]

Parameters		Efficiency (%)	Energy bandgap (eV)	High temp. effect	Merits	Demerits	Cost
First generation	Mono-crystalline Silicon	14–17	1.1	Not stable performance	Ecofriendly, reliable, stable, efficient and abundant availability of raw materials	Time- consuming and complicated manufacturing process	Expensive
	Poly-crystalline Silicon	12–14	1.1	Not stable performance	Inexpensive compared to mono Si	Less efficiency	Expensive
Second generation	a-Si Thin Film	4–8	1.1–1.5	Stable performance on both low and high temp	Economical and high absorption capacity	Reliability is less when it is used for outdoor applications	Required only 50% of the expense of traditional Si-based cell
	CIGS	10–12	>1.2	Stable performance on both low and high temp	Flexible, durable, high eficiency	Absorbing capaity is less	Required only 50% of the expense of traditional Si-based cell
	CdTe	9–11	1.5	Stable performance on both low and high temp	Flexible, durable	Poisonous due to Cd	Required only 50% of the expense of traditional Si-based cell

(continued)

Table 1 (continued)

Parameters		Efficiency (%)	Energy bandgap (eV)	High temp. effect	Merits	Demerits	Cost
Third generation	Polymer	3–10	2	Not stable performance	Flexible, durable	Requirement of installation duration and space is high	Required only 50% of the expense of traditional Si-based cell
	Nanocrystal	7–8	3	Excellent stable performance	Flexible, durable	Requirement of installation duration and space is high	Required only 50% of the expense of traditional Si-based cell
	Dye-sensitized	10	3	Not stable performance	Flexible, durable	Requirement of installation duration and space is high	Required only 50% of the expense of traditional Si-based cell
	Perovskite	31	1	Excellent stable performance	Flexible, durable, high efficiency	Requirement of installation duration and space is high	Required only 50% of the expense of traditional Si-based cell

5 Summary and Discussion

It has been observed from the literature that many materials and structures can be used in the manufacturing of solar cells to increase their efficiency. Based on different technologies and materials, the solar cells are summarized as below:

- First-generation solar cells (monocrystalline and polycrystalline) are based on the silicon wafer. In general, they have high conversion efficiency in the range of 12–18%. They have advantages of abundant availability of raw material and eco-friendly, but the downsides of these cells are that they have complexities in the fabrication of silicon and are expensive.
- Second-generation solar cells (a-Si, CIGS, and CdTe) are based on thin film. They have a lower efficiency (4–11%). Recently, the highest efficiency of 21.7% has been recorded for CIGS. They have advantages of high absorption capacity and lower cost in comparison to first-generation cells. The downsides of these cells are that they contaminate the environment while being fabricated and consist of hard-to-find materials.
- Third-generation cells (PSC, nanocrystal, DSSC, and perovskite) are fabricated using the latest solution-processable technology. They have high efficiency at high temperatures and are mechanically robust. The performance of these cells has been increased from 4 to 20% using perovskite material.
- Fourth-generation cells based on carbon nanotubes (CNT), graphene(G), and its derivatives have exhibited good efficiency and they are still under the initial state of research. Japan's Gifu University and the Tokyo Institute of Technology have predicted that a maximum of 38.7% conversion efficiency can be achieved using chalcogenide perovskite material and using tandem technology.

6 Conclusion

A brief review on solar cell, its working, structure, evolution, and effect of solar insolation and temperature has been successfully presented in this paper. The type of the used materials and technologies have a great bearing on its conversion efficiency. Different generations are discussed along with their highlights. Presently, fourth-generation cells are in the early research stage. However, there are some other factors such as PV panel location, temperature, shading, orientation, maximum power point tracking, fault finding etc., which significantly affect the overall performance of a PV system. Therefore, the overall performance of a solar PV system can be further improved by focusing on these factors individually in the future.

References

1. (EIA), U.S.E.I.A.: International Energy Outlook 2019 (IEO2019). https://www.eia.gov/outlooks/ieo/pdf/ieo2019.pdf. Accessed 24 Sept 2019
2. IRENA: Future of Solar Photovoltaic: Deployment, investment, technology, grid integration and socio-economic aspects (A Global Energy Transformation: paper). In. International Renewable Energy Agency, Abu Dhabi, (2019)
3. C.W. Tang, Two-layer organic photovoltaic cell. Appl. Phys. Lett. ((United States), Medium: X; Size:) 183–185 (1986)
4. J. Rostalski, D. Meissner, Monochromatic versus solar efficiencies of organic solar cells. Sol. Energy Mater. Sol. Cells **61**(1), 87–95 (2000). https://doi.org/10.1016/s0927-0248(99)00099-9
5. M. Abdulkadir, A.S. Samosir, A.H. Yatim, modeling and simulation based approach of photovoltaic system in simulink model, 2012
6. Y.H. Mahmood, F.S. Atallah, A.F. Youssef, Studying the weather condition affecting on solar panel efficiency. Tikrit J. Pure Sci. **25**(3), 98–102 (2020). https://doi.org/10.25130/j.v25i3.995
7. N. Abdullahi, C. Saha, R. Jinks, Modelling and performance analysis of a silicon PV module. J. Renew. Sustain. Energy **9**, 033501 (2017). https://doi.org/10.1063/1.4982744
8. F. Adamo, F. Attivissimo, A.D. Nisio, M. Spadavecchia, Characterization and testing of a tool for photovoltaic panel modeling. IEEE Trans. Instrum. Meas. **60**(5), 1613–1622 (2011). https://doi.org/10.1109/TIM.2011.2105051
9. N.-G. Park, efficiency perovskite solar cells: materials and devices engineering. Trans. Electr. Electron. Mater. **21**(1), 1–15 (2020). https://doi.org/10.1007/s42341-019-00156-0
10. P. Würfel, U. Würfel, *Physics of Solar Cells: From Basic Principles to Advanced Concepts* (Wiley, New York, 2016)
11. A. McEvoy, L. Castaner, T. Markvart, *Solar Cells: Materials, Manufacture and Operation* (Academic Press, 2012)
12. S. Sharma, K. Jain, A. Sharma, Solar cells: in research and applications—a review. Mater. Sci. Appl. **06**, 1145–1155 (2015). doi:https://doi.org/10.4236/msa.2015.612113
13. Y. Weiwei, W. Xusheng, Z. Feng, Z. Lingjun, 19.6% cast mono-MWT solar cells and 268 W modules, in *2012 IEEE 38th Photovoltaic Specialists Conference (PVSC) PART 2, 3–8 June 2012* (2012), pp. 1–5
14. S. Chih-Tang, K.A. Yamakawa, R. Lutwack, Effect of thickness on silicon solar cell efficiency. IEEE Trans. Electron. Devices **29**(5), 903–908 (1982). https://doi.org/10.1109/T-ED.1982.20797
15. P. Zheng, F.E. Rougieux, X. Zhang, J. Degoulange, R. Einhaus, P. Rivat, D.H.Macdonald, 21.1% UMG Silicon Solar Cells. IEEE J. Photovolt. **7**(1), 58–61 (2017). doi:https://doi.org/10.1109/JPHOTOV.2016.2616192
16. P. Jayakumar, Resource assessment handbook. Asia and Pacific Center for Transfer of Technology of the United Nations (2009)
17. F. Schindler, J. Schön , B. Michl, S. Riepe, P. Krenckel, J. Benick, F. Feldmann, M. Hermle, S.W. Glunz, W. Warta, M.C. Schubert, High efficiency multicrystalline silicon solar cells: potential of n-type doping, in *2015 IEEE 42nd Photovoltaic Specialist Conference (PVSC), 14–19 June 2015* (2015), pp. 1–3
18. J. Benick, A. Richter, R. Müller, H. Hauser, F. Feldmann, P. Krenckel, S. Riepe, F. Schindler, M.C. Schubert, M. Hermle, A.W. Bett, S.W. Glunz, High-efficiency n-type HP mc silicon solar cells. IEEE J. Photovolt. **7**(5), 1171–1175 (2017). https://doi.org/10.1109/JPHOTOV.2017.2714139
19. K. Chopra, P. Paulson, V. Dutta, Thin-film solar cells: an overview. Prog. Photovolt. **12**, 69–92 (2004). https://doi.org/10.1002/pip.541
20. D. Shi, Z. Guo, N. Bedford, 10 - Nanoenergy Materials, in *Nanomaterials and Devices*. ed. by D. Shi, Z. Guo, N. Bedford (William Andrew Publishing, Oxford, 2015), pp. 255–291
21. Y. Tawada, M. Kondo, H. Okamoto, Y. Hamakawa, Hydrogenated amorphous silicon carbide as a window material for high efficiency a-Si solar cells. Sol. Energy Mater.s **6**(3), 299–315 (1982)

22. S. Kim, J.-W. Chung, H. Lee, J. Park, Y. Heo, H.-M. Lee, Remarkable progress in thin- film silicon solar cells using high-efficiency triple-junction technology. Sol. Energy Mater. Sol. Cells **119**, 26–35 (2013). https://doi.org/10.1016/j.solmat.2013.04.016
23. B. Srinivas, S. Balaji, M. Nagendra Babu, Y. Reddy, Review on present and advance materials for solar cells. Int. J. Eng. Res.-Online **3**, 178–182 (2015)
24. J. Lu, W. Liu, A. Lu, Y. Sun, J. Lu, Integration of solar cells on top of CMOS chips-Part II: CIGS solar cells. IEEE Trans. Electron Devices **58**, 2620–2627 (2011). https://doi.org/10.1109/TED.2011.2156799
25. P. Reinhard, F. Pianezzi, B. Bissig, A. Chirila, P. Blösch, S. Nishiwaki, S. Buecheler, A. Tiwari, Cu(In, Ga)Se_2 thin-film solar cells and modules—a boost in efficiency due to potassium. IEEE J. Photovolt. **5**, 656–663 (2015). https://doi.org/10.1109/JPHOTOV.2014.2377516
26. P. Jackson, D. Hariskos, R. Wuerz, O. Kiowski, A. Bauer, T. Magorian Friedlmeier, M. Powalla, Properties of Cu(In,Ga)Se2 solar cells with new record efficiencies up to 21.7%. physica status solidi (RRL) - Rapid Res. Lett. **9999** (2014). doi:https://doi.org/10.1002/pssr.201409520
27. J.J. Becker, C.M. Campbell, Y. Zhao, M. Boccard, D. Mohanty, M. Lassise, E. Suarez, I. Bhat, Z.C. Holman, Y. Zhang, Monocrystalline CdTe/MgCdTe double-heterostructure solar cells with ZnTe hole contacts. IEEE J. Photovolt. **7**(1), 307–312 (2017). https://doi.org/10.1109/JPHOTOV.2016.2626139
28. A. Bosio, N. Romeo, S. Mazzamuto, V. Canevari, Polycrystalline CdTe thin films for photovoltaic applications. Prog. Cryst. Growth Charact. Mater. **52**(4), 247–279 (2006). https://doi.org/10.1016/j.pcrysgrow.2006.09.001
29. T. Baines, T.P. Shalvey, J.D. Major, 10 - CdTe solar cells, in *A Comprehensive Guide to Solar Energy Systems*, ed,. by Letcher, T.M., Fthenakis, V.M. (Academic Press, 2018), pp. 215–232
30. A. Gaur, G.N. Tiwari, Performance of photovoltaic modules of different solar cells. J. Sol. Energy **2013**, 734581 (2013). https://doi.org/10.1155/2013/734581
31. H. Tang, G. Lu, X. Yang, The role of morphology control in determining the performance of P3HT/C-70 bulk heterojunction polymer solar cells. IEEE J. Sel. Top. Quantum Electron. **16**(6), 1725–1731 (2010). https://doi.org/10.1109/JSTQE.2010.2042034
32. P. Bi, F. Zheng, H. Jin, W. Xu, L. Feng, X. Hao, Performance enhancement in polymer-based organic optoelectronic devices enabled by discontinuous metal interlayer. IEEE J. Photovolt. **6**(6), 1522–1529 (2016). https://doi.org/10.1109/JPHOTOV.2016.2598257
33. H. Yao, F. Bai, H. Hu, L. Arunagiri, J. Zhang, Y. Chen, H. Yu, S. Chen, T. Liu, J.Y.L. Lai, Y. Zou, H. Ade, H. Yan, Efficient all-polymer solar cells based on a new polymer acceptor achieving 10.3% power conversion efficiency. ACS Energy Lett. **4**(2), 417–422 (2019). doi:https://doi.org/10.1021/acsenergylett.8b02114
34. R.J. Ellingson, M.C. Beard, J.C. Johnson, P. Yu, O.I. Micic, A.J. Nozik, A. Shabaev, A.L. Efros, Highly efficient multiple exciton generation in colloidal PbSe and PbS quantum dots. Nano Lett. **5**(5), 865–871 (2005). https://doi.org/10.1021/nl0502672
35. M. Liang, W. Xu, F. Cai, P. Chen, B. Peng, J. Chen, Z. Li, New triphenylamine-based organic dyes for efficient dye-sensitized solar cells. J. Phys. Chem. C **111**(11), 4465–4472 (2007)
36. M.H.K. Tafti, S.M. Sadeghzadeh, Dye sensitized solar cell efficiency improvement using TiO 2/nanodiamond nano composite. Sādhanā **43**(7), 113 (2018)
37. C. Ubani, M. Ibrahim, M. Teridi, Moving into the domain of perovskite sensitized solar cell. Renew. Sustain. Energy Rev. **72**, 907–915 (2017)
38. D. Shi, Y. Zeng, W. Shen, Perovskite/c-Si tandem solar cell with inverted nanopyramids: realizing high efficiency by controllable light trapping. Sci. Rep. **5**, 16504 (2015)

Evolutionary Progress of the Electric Car Market with Future Directions

Rishabh Bhardwaj and Sandeep Gupta

1 Introduction

In the early 90s, we all required some other fuels rather than petrol and diesel because the vehicles operated by these fuels are very noisy and producing pollution that causes many harmful effects on the environment. So, to overcome this problem, many people give their efforts from the 90s to the present day [1, 2]. Most of them are focused only on electric vehicles. The motive of this paper is to tell about the people that which type of technology is used to make the electric vehicle and described that electric engines are far better than the gasoline engines.

The report comprises the purpose that why electric vehicles get famous very quickly and the reason is that they improve the environment condition of today. The final effect about the electric vehicles on the globe and on the all-human cast is very good and beneficial. As compared with the gasoline-powered engine, the electric vehicles are very much efficient, silent and a big point they are 97% cleaner than the gasoline-powered vehicles [3]. The other competitions of the electric vehicles are gas-powered vehicles, but they cause very harmful injections of the gases in the environment, which cause asthma and respiratory system problems, whereas the electric vehicles do not produce such gases in the environment that cause any type of disease [4].

This paper starts with some basic knowledge of the past of the electric vehicles, especially the peak and bottom of the production and the purpose for the modification. The forward section tells about some technical information of the electric vehicles, which includes its parts and functions of operation. Then the following parts explain to us the modification and pros and cons of electric vehicles.

R. Bhardwaj · S. Gupta (✉)
JECRC University, Jaipur, India
e-mail: jecsandeep@gmail.com

R. Bhardwaj
e-mail: ashoksharma7975@gmail.com

S. Vadhera et al. (eds.), *Latest Trends in Renewable Energy Technologies*, Lecture Notes in Electrical Engineering 760, https://doi.org/10.1007/978-981-16-1186-5_27

2 Evaluation of Electric Cars

This is very difficult to explain the first electric car and its inventor. So based on previous literatures, it is a step-by-step invention of the batteries to the electric motors. This leads to the first invention of the electric vehicle on road in the eighteenth century [5, 6]. In the very first part of the century, scientists from many countries like Hungary Netherlands America, etc. start making toy cars with a great idea of battery powered and getting success in the development of small-scale electric cars [7]. Despite all these, Robert Anderson, a British inventor, achieved success in the making of the first crude electric carriage, whereas, America gave big news by launching its first electric car in 1890, which was a great discovery of William Morrison. The capacity of the vehicle is carrying six passengers with a top speed of 22.53 kilometers per hour, which is also helpful in Spark interest in vehicles [8].

At the starting of the eighteenth century, the great scientist Richard Trevithick invented the steam-powered carry age, which initiates the horseless transportation. After this invention over 30 years of glamorous and dirty engines of steam, the first cell-operated electrical vehicle was invented in 1834. After 50 years, the first invention of gasoline-powered internal combustion engine vehicle (ICEVs) was closed in 1885, fully stopped, hence the electric vehicles are now not new for us and they are 50 years older than ICEVs. In 1828, a Hungarian scientist discovers a very small prototype of a car that is filled by an e-motor [1].

The scientist, Robert Anderson, of Scotland discovers the first electric carriage. The time is not known, but it is between the years of 1832 and 1839. Another scientist would discover an electrical phaeton that is much more powerful than electric carriage. The highest range of this vehicle is 18 miles and its top speed is 14 miles per hour, which costs $2000. The recession and decline in the making of electric vehicles take place in 1920. Reasons for this lesson production include a good road network, high manufacturing of ICEVs compared with of EVs, and the main reason is the price of EVs is much more costly than the ICEVs. CEO at the end of 1935, the EVS gets completely vanished from the automobile market. The years 1960 and 1970 give rebirth to electric vehicles because the gasoline-powered vehicles are creating very harmful and poisonous gases in the environment for the people in all over the world [9].

Refuelable batteries are not available at the time, so we can recharge and use them till 1840. The scientists Thomas Deven and Scotsman Robert Davidson finally launch some better and fortunate electrical vehicles on the road in 1842. Both of these are using the latest technology sales in their vehicles, but the batteries are till now not rechargeable. After some time, a scientist, named Frenchman Gaston Plante, discovers a very good battery with enhancing capacity in 1865. And also his fellow in 1881, named Cameli Faure, upgrade the storage of the batteries in 1881. At the end of the eighteenth century, some countries like France and England became the first countries to give reinforcements in the widespread expansion of electric vehicles. Gustave Trouve, who is the French scientist, displays a three-wheeler car in November 1881 at France. As time passed, there are a lot of discoveries related

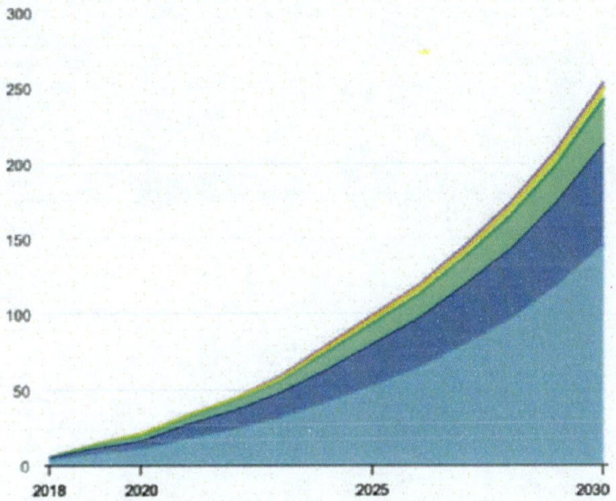

Fig.1 Electric vehicle supply situation in years 2018–2030 [11]

to electric vehicles take place and increased in the late eighteenth century and early nineteenth century [10].

In the upcoming years, electric vehicles from different companies will come out or be launched. Some cities like New York have some Group E taxis groups. At the end of 1900, the electric vehicles are at their composition. Every third vehicle on road is now an electrical vehicle, this type of strong sale continues till the next 10 years. The ancient pages of history have been discussed above. We also see that in the basic needs of human society, our food, shelter, clothing and transportation, transportation is the most important after food because this really gives us freedom of life. Figures 1, 2, 3 represent the future and growth of electric vehicle systems.

3 EV's Parts Details

3.1 Data Preprocessing

The main part of the electric vehicle is its motor because the vehicle is operated by the working motor. This motor is supplied by the rechargeable cells (or battery packs). By looking at the infrastructure, no one points out the difference between the internal combustion engine vehicles and electric vehicles, the only difference between them is the electric vehicles are very much silent (Fig. 4).

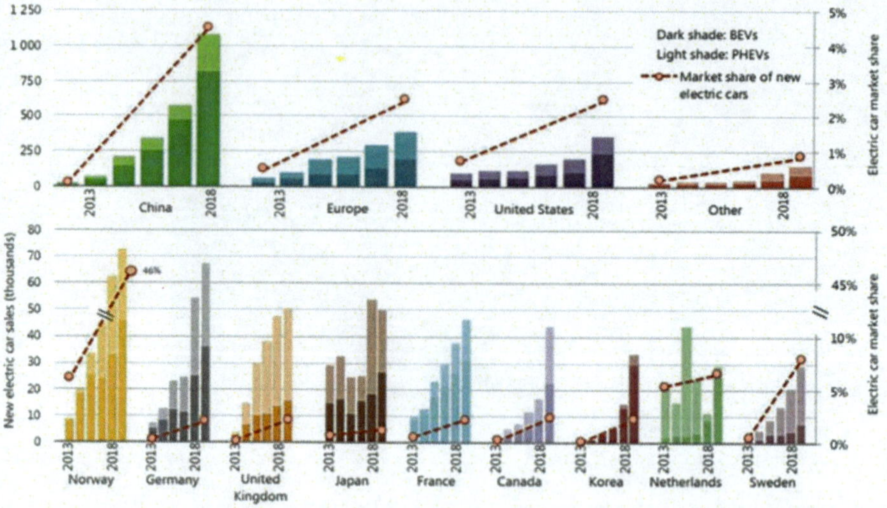

Fig. 2 Global electric car sales and market share in past years [12]

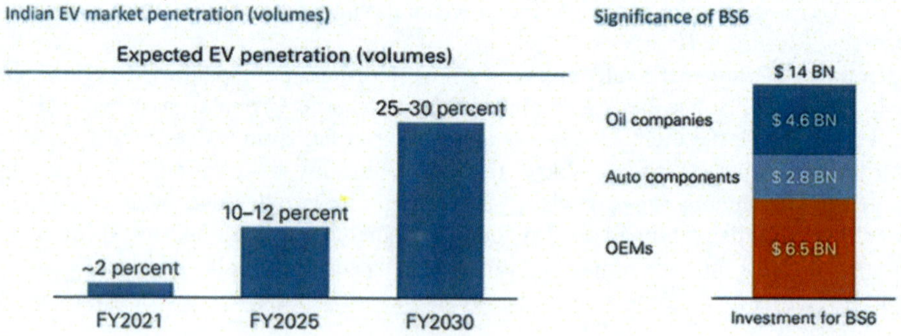

Fig. 3 Overview of Indian EV market penetration [13]

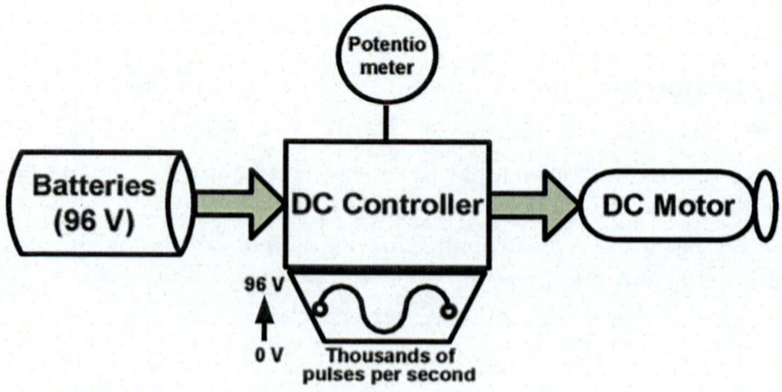

Fig. 4 Parts of an EV system [14]

The components of the electric car are [15]:

1. Battery (which is rechargeable)
2. Control unit generally called controller
3. Motor.

There is a step-by-step transformation of power takes place that first controller feeds the power from the rechargeable cells or batteries and supply it to the electric motor. The main use of the cells is to provide a continuous supply of voltage to the motor. After the receiving voltage from the batteries, the motor then starts to rotate the transmission, which finally results in the rotation of wheels that cause moment in the vehicle. Despite all these, some more main components of the EVS are

1. Potentiometer
2. Motor
3. DC controller
4. Batteries.

4 Working of the EVs

The most interesting part of the electric vehicles is its working. In this process, the rider applies force on the accelerator, then there is an instant activation of the potentiometer takes place. The work of the potentiometer is to receive and send signals to the controller and check how much voltage is getting to supply for operation at this level. after checking all the condition and treats the potentiometer readings were the controller balance the voltage, then collect voltage from the batteries and directly send it to the motor for further operation. After receiving the voltage from the controller motor starts transmission and then transmission rotates the wheels, which finally causes the movement of the car. Figure 5 shows the battery charging process in the EVs system. The complete workflow in the electric vehicles area is representing in Fig. 6.

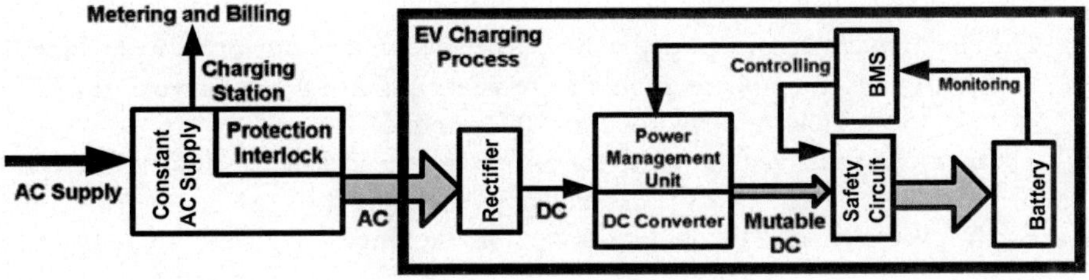

Fig. 5 Battery charging system in EV

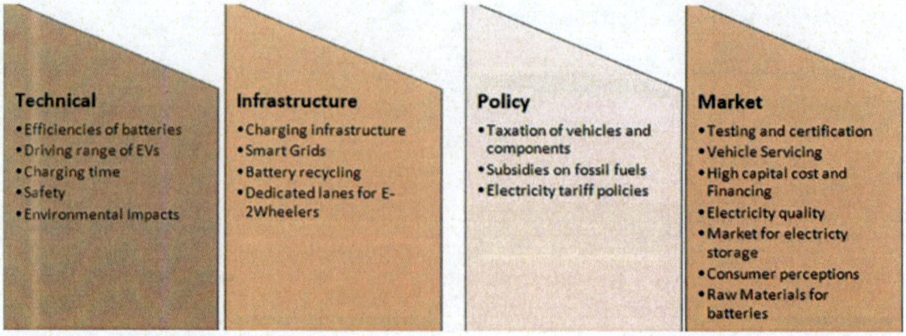

Fig. 6 Ecosystem for electric vehicles

5 Advantages of the Electric Vehicles with Modification

The electric vehicles are run on electricity instead of gasoline. The movement in electric vehicles is taking place by the working of electric motor, whereas internal combustion engine vehicles are engine based; the electric vehicles store their fuel in a form of rechargeable batteries as compared with the fuel tank. The electric vehicles are easy to plug in and plug out anywhere for refueling in comparison of petrol pumps. Nowadays some companies offer hybrid cars with automatic charging. These types of cars charge themselves without any plugging [10]. So the future cars offer with only electric vehicles, not hybrid vehicles, and also with automatic charging capacities. This reduces charging station dependencies.

There are some other options that also take place for automatic charging like apply the solar film on the top of the roof of the car, so when the car drives in sunlight, it gets automatically charged. And in the future, there will also be seen that electric vehicles get recharged by roads itself by the process of wireless charging and by many other advanced methods. Main advantages of the electric vehicles are shown following:

1. The maintenance of electric vehicles is very low as compared with the internal combustion engine vehicles.
2. The electric vehicles have very stable parts because they do not have any bulky parts or frames as compared with the internal combustion engine vehicles.
3. The electric vehicles are environmentally friendly.
4. The method of solar technologies can be used for the free recharging of batteries.
5. Electric vehicles decrease the dependency on crude oil.
6. Electric vehicles have the facility of flexible fuelling.

6 Conclusion

In the latest outlook, e-mobility in India will be an important part for staying alive because the air pollution level is very terrific high and the only solving this problem is with the help of green sources and energy transmission. Therefore, EVs can be used in this factor. A distributed storage in the metropolitan energy system is possible by the use of EVs with the smart grid implementation and intermittent renewable energy sources. However, appropriate preparation with reference to supervising and commanding charging infrastructure is needed. The automotive industry players and charging infrastructure, batteries and mobility service providers have taken various actions to ramp up industry action. So now the middle-class family is also getting some luxury instead of the two-wheeler on which the carrying three to four passengers, without spending a lot of money on petrol and diesel and also do not harm the environment and themselves also. Companies are also designing and testing products fit for the Indian market with the main focus on two-wheelers and three-wheelers.

References

1. S. Beggs, S. Cardell, J. Hausman, Assessing the potential demand for electric cars. J. Econ. **17**(1), 1–19 (1981)
2. S. Secinaro, V. Brescia, D. Calandra, P. Biancone, Employing bibliometric analysis to identify suitable business models for electric cars. J. Clean. Prod. 121503.(2020)
3. T. Randall, *Here's How Electric Cars will Cause the Next Oil Crisis* (Bloomberg, New York, 2016). Accessed 25 Mar 2016
4. E.G. Durney, "Truly electric car." U.S. Patent Application 12/509,427, filed January 27 (2011)
5. T. Wilberforce, Z. El-Hassan, F.N. Khatib, A. Al Makky, A. Baroutaji, J.G. Carton, A.G. Olabi, Developments of electric cars and fuel cell hydrogen electric cars. Int. J. Hydrog. Energy **42**(40), 25695–25734 (2017)
6. A. Grenier, S. Page, The impact of electrified transport on local grid infrastructure: a comparison between electric cars and light rail. Energy Policy **49**, 355–364 (2012)
7. C.C. Chan, The state of the art of electric, hybrid, and fuel cell vehicles. Proc. IEEE **95**(4), 704–718 (2007)
8. C.C. Chan, The state of the art of electric and hybrid vehicles. Proc. IEEE **90**(2), 247–275 (2002)
9. https://www.consultancy.in/news/901/indian-electric-vehicles-market-to-take-off-in-2020-says-arthur-d-little
10. T.A. Skouras, P.K. Gkonis, C.N. Ilias, P.T. Trakadas, E.G. Tsampasis, T.V. Zahariadis, Electrical vehicles: current state of the art, future challenges, and perspec- tives. Clean Technol. **2**(1), 1–16 (2020)
11. https://www.iea.org/data-and-statistics/charts/electric-vehicle-stock-in-the-ev3030- scenario-2018–2030
12. https://www.weforum.org/agenda/2019/10/how-can-india-transition-to-electric-vehicles-heres-a-roadmap/
13. A. Holms, R. Argueta, A Technical Research Report: The Electric Vehicle. Argueta–6–7, March, 11 (2010)
14. J.Y. Yong, V.K. Ramachandaramurthy, K.M. Tan, N. Mithulananthan, A review on the state-of-the-art technologies of electric vehicle, its impacts and prospects. Re New. Sustain. Energy Rev. **49**, 365–385 (2015)
15. M. Brain, How electric cars work (2002). Accessed 29 Jan 2010

Modelling and Simulation of Grid-Connected Renewable Energy Systems

Ankush Sinha and Shashi Bhushan Singh

1 Introduction

Integration of unconventional sources to the grid has increased by leaps and bounds in the last few years. Because of a number of merits such as uncomplicated and reasonable designs, efficient execution etc., grid connection has become one of the favoured modes of utilising renewable energy resources, whether it may be wind or solar. It is quite advantageous at the time of disaster by providing reinforcements during emergency situations. What is more in most cases of small-scale DGs, transformer is also, not required that makes system more accessible, cost-effective and modest [1, 2]. Also, it engages net-metering by which if SPV system produces more power than consumption during a sunny day, this excess solar power will be returned to the power grid to rotate your electric meter in retrograded direction, and when this happens, the local power provider must usually provide credits for excess generated power [3]. In Sect. 2, components for integrating DGs with grid with its advantages have been discussed. In Sect. 3, challenges and solutions related to this topic have been discussed. Section 4 explains the simulation and work done. Results and discussions are shown in Sect. 5 and conclusions and future scope are discussed in Sect. 6.

A. Sinha (✉)
SREE, NIT Kurukshetra, Kurukshetra, India
e-mail: ankush.nitkkr@yahoo.com

S. B. Singh
Electrical Engineering Department, NIT Kurukshetra, Kurukshetra, India
e-mail: sbsingh@nitkkr.ac.in

S. Vadhera et al. (eds.), *Latest Trends in Renewable Energy Technologies*, Lecture Notes in Electrical Engineering 760, https://doi.org/10.1007/978-981-16-1186-5_28

2 Interfacing Small-Scale DGS with Grid

Power conversion and conditioning are indispensable, if one needs to homogenise the output from RES and grid. DC–DC converters and inverters are employed for performing the conversion job, whereas filters are deployed for conditioning [1, 4].

2.1 *DC–DC Converters*

DC–DC Converters are integral part of integrating RES with the grid. Normally boast converters are used as the output voltage from these sources that are not good enough in magnitude to be fed directly to the load or to the grid. Hence, they step up the voltage to the desired levels [1]. In most cases, power from RES is intermittent as it is heavily dependent on external environmental conditions. Hence, they not only augment the voltage but also mingle with the MPPT circuit to feed maximum power at the output [4, 5].

2.2 *Three Phase Inverter*

The acquired DC voltage from the PV system is connected to the DC/AC converter via a DC/DC converter, which intensifies the voltage from PV system and then an inverter converts the DC voltage to AC voltage and current and connects it to the power grid for grid applications and individual loading for stand-alone applications and the output voltage and frequency, which is regulated by inverter and has to be synchronised with grid [6, 7].

2.3 *Filters*

Power output from inverter cannot be fed directly to the grid or load because of its substandard and low-grade power quality of the output. Normally on the grid side, passive filters are employed to compensate reactive power, improve power factor, and reduce harmonics and other power quality and related issues in the system.

A combination of inductors and capacitor banks is designed and implemented to combat a number of quality-related issues [8]. In modern times, active power filters and discrete filters are also used, may not be directly connected to the grid, but commingle with the control schemes that are conceived to enhance the power quality [7].

3 Challenges in Grid-Integrated RES

There are a number of confrontations that one has to face before integrating RES to the grid, for instance, protection issues, islanding, voltage and frequency variations, grid desynchronisation, harmonics and flicker etc., and many other hurdles that have to be passed before integrating DGs to the grid [9–11]. This paper deals with solving the issues of harmonics and grid synchronisation; hence, they have been discussed in this section.

3.1 Grid Synchronisation

The voltage fluctuation due to DG's synchronisation with an Electric Power System (EPS) region does not exceed 5 per cent of the PCC prevailing voltage point. With synchronism loss protection, the disconnection of 250 kW or greater DGs from the EPS region can be rendered without deliberate time delay [4]. DG at all times has to maintain system frequency and should cease feeding to utility if the frequency swerves from the specified range for at least that time [4].

3.2 Harmonics and Flicker

Harmonic adulteration is a major concern for power system professionals due to increased losses, unnecessary heating in rotating machinery, and acute interference with communication circuits that share common AC power lines and produce noise leading to misleading results. Harmonic frequencies are the integral multiple of fundamental frequency that arise due to a number of reasons, the chief reasons behind that is non-linearity of modern loads like furnaces, VFDs, UPS etc. [12]. Only fundamental frequency constitutes to useful AC power and other frequencies only make their contribution in losses and a parameter has been devised to quantify harmonics, which is known as Total Harmonic Distortion (THD) [5].

$$THD = \sqrt{\frac{V_{RMS}^2}{V_1^2} - 1}$$

where

VRMS is the RMS value of the whole signal

V1 is the RMS value of fundamental component of signal

The importance of the problems arising through harmonics from converters has prompted some agencies to impose restrictions on those converters [13]. IEEE standard 519 imposes a restriction on both current and voltage harmonics from the converters or from non-linear loads. IEEE Std. 519 first came into existence in 1981 and last revised and updated in 2014. THD limits as considered by IEEE that must not exceed 5% at any times [14]. For safety and protection issues, IEEE introduced Std–1547–2003, which was revised in 2018 for DR interconnection with the main electrical power systems, which is accepted worldwide by most electrical utilities and they consider it as a guiding light while interfacing DGs with the grid. One of the main features of this standard is, if due to faults or other disturbances, the voltage and frequency deviate from the recommended base values, DGs need to be disconnected from the main grid [15].

3.3 Grid Synchronisation Algorithms

To obey grid codes and implement guidelines issued by the authorities, several algorithms and control schemes have been devised and a number of articles were presented in the literature. For example, LST/ILST algorithm, SRF theory and many others have been utilised for grid synchronisation and power quality enhancement [16, 17]. In this paper, SRF theory was employed; hence, it has been briefly discussed below.

3.4 SRF Theory for Synchronisation

SRF Synchronous Reference Frame in coalition with PLL is another algorithm for grid synchronisation. It is also called dqPLL algorithm [4].

DqPLL uses equations of Park's transformations. This transform converts the time-domain components of a three-phase system in an *abc* reference frame to direct, quadrature, and zero (dq0) components in a rotating reference frame. This can preserve the active and reactive powers with the powers of the system in the *abc* reference frame by implementing an invariant version of the Park transform. For a balanced system, the zero component is equal to zero [18]. Considering d axis voltage rotating reference frame leading the a axis in three-phase reference frame by angle at t = 0 and d axis is rotating at rad/sec, hence = and q axis is considered in quadrature, that is, 90° phase apart from d axis as shown in Fig. 1 [4].

Equation 2 delineates the relationship between dq and abc reference frames.

$$\begin{bmatrix} V_d \\ V_q \end{bmatrix} = \frac{2}{3}\begin{bmatrix} \cos\theta & \cos(120-\theta) & \cos(240-\theta) \\ -\sin\theta & -\sin(120-\theta) & -\sin(240-\theta) \end{bmatrix}\begin{bmatrix} V_a \\ V_b \\ V_c \end{bmatrix} \quad (2)$$

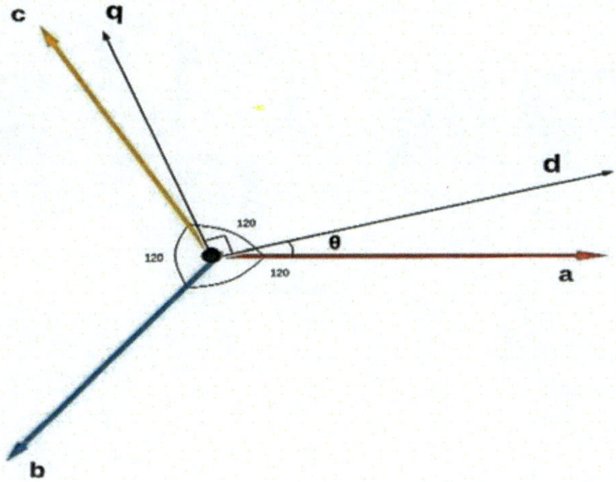

Fig. 1 Phasor diagram-abc to dq

4 Simulation and Work Done

The main part of the simulation is based on SRF theory for fundamental extraction and grid synchronisation. The basis of this work is taken from A.K. Verma, B. Singh, and D.T. Sahani [19] and J. Saroha, G. Pandove and M. Singh [20] and they have worked on the same, grid integration of SPV formulating control schemes to ensure admirable power quality at load and to the grid [19, 20]. In this chapter, the role of filters in insuring power quality and the comparison between the results in the presence and absence of filters has also been discussed. Along with that filter, the characteristic response is also shown which the papers in references [19, 20] have not discussed.

4.1 Schematic Representation with Main Control Scheme

Diagrammatic representation of grid-connected solar photovoltaic system is shown in Fig. 2. First, DC output of 400 V, 15KW is considered from Ref. [21]. Then the DC output from the RES is amplified to the desired level by DC–DC boast converter, parameters of the converter. After that the amplified DC output is inverted to AC by DC–AC converter comprising of MOSFET AC switching devices and triggering pulses/signals to these switches is controlled by the control schemes, which is further discussed in this chapter. The converted AC output is not fit to be directly fed to the load or grid as the output from the inverter is rich in harmonics and also devaluates the power factor at load. Hence, passive filters, i.e. combination of inductors and capacitor banks are required to keep a check of the reactive power.

Here we have considered two types of load and have evaluated the results. In the first case, we considered linear RL load and have analysed the results, and in the

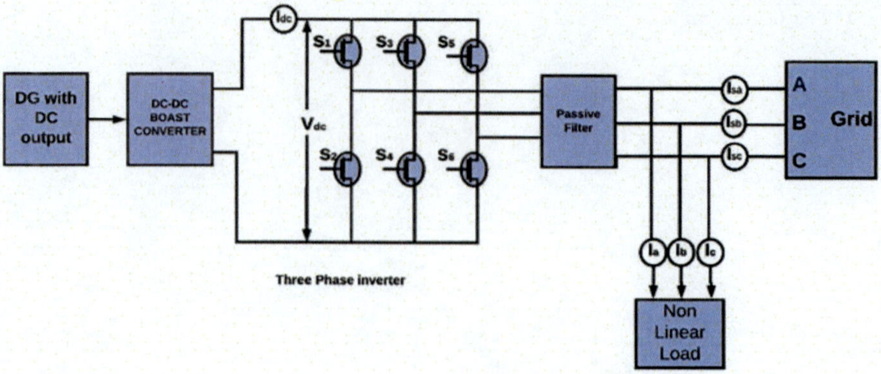

Fig. 2 Diagrammatic representation of grid-connected solar photovoltaic system

second case, we have considered three-phase non-linear dynamic load and visualised if system is working in satisfactory conditions.

Main control scheme is described in Fig. 3. The DC output from the boast converter is compared with the reference DC voltage Vref*. The voltage from boast converter contains ripples. These ripples need to be separated to have admirable power quality at AC side. The error is then reduced by PID controller with values Kpd = 2 and Kid = 1.5 [19].

The output of PID controller, i.e. Id* is then compared with DC current output of boast converter and the resultant is considered as Id, i.e. d phase current to be fed to dq0 to abc converter (Inverse Park's Transform), which generates three-phase currents, i.e. Ia*, Ib* and Ic*, which is then compared with reference three-phase current in hysteresis current controller, which generates triggering signals for respective MOSFETs in the inverter circuit.

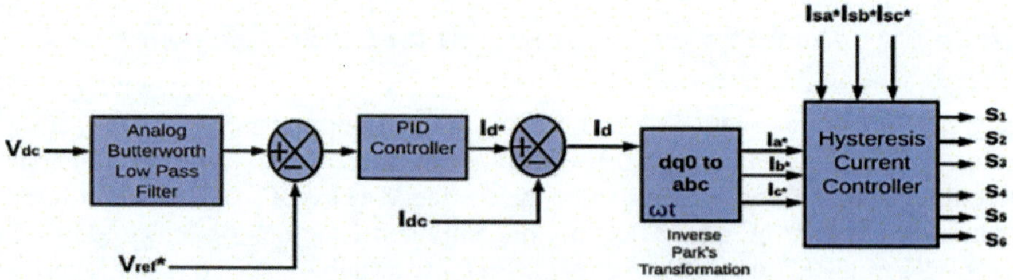

Fig. 3 Main control scheme

4.2 Auxiliary Control Schemes

The work also includes two control schemes, first for reference DC voltage extraction and next for three-phase reference current extraction.

The block diagram of reference voltage extraction is shown in Fig. 4. The three-phase grid voltage is considered as alluded voltages, which is transformed into αβ0 by Clarke transformation and value of Vref* is given by Eq. 3

$$V_{ref}^{*} = \sqrt{\frac{2}{3}\left(V_{\alpha}^{2} + V_{\beta}^{2}\right)}$$

Figure 5 shows the control scheme for reference current extraction. In this scheme, three-phase load currents, i.e. Ia, Ib, and Ic are transformed into αβ0 using Eq. 6. It is then transformed back to reference three-phase current by Inverse Park transformation. The three-phase load current contains number of harmonic currents other than fundamental. Hence, the transformed DC from Clarke transformation contains a DC ripples, hence generating erroneous reference and hence randomistic AC side outputs. Hence analogue butter-worth second-order filter is added to separate out

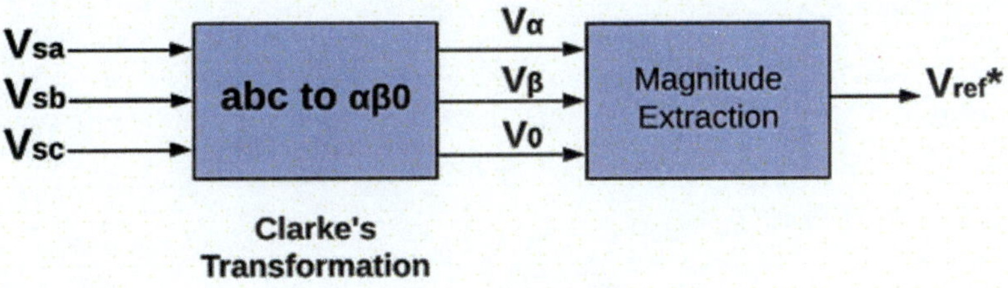

Fig. 4 Auxiliary control scheme for reference voltage extraction

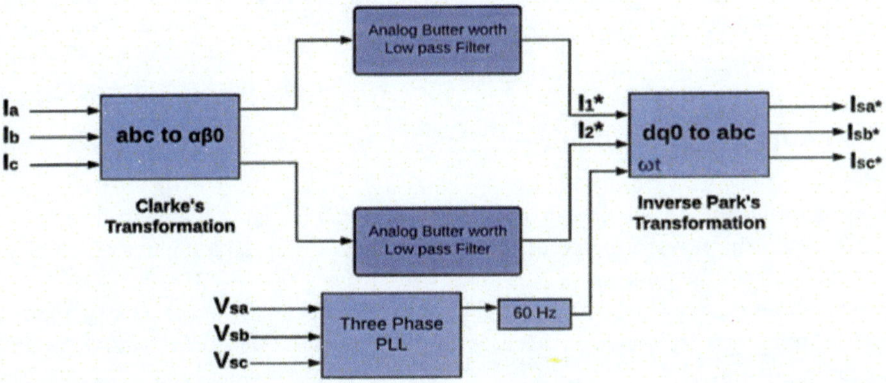

Fig. 5 Auxiliary control scheme for reference current extraction

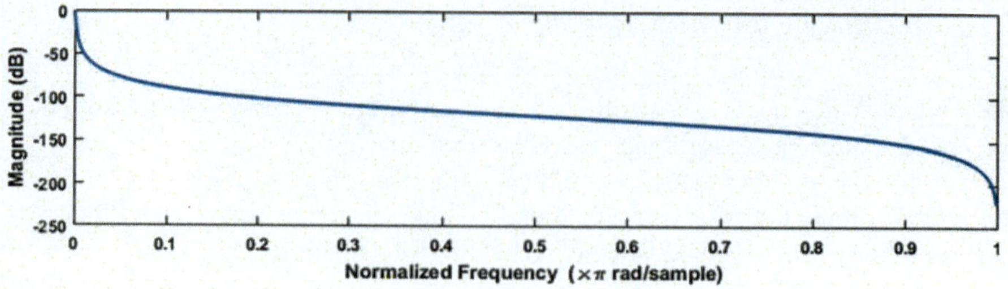

Fig. 6 Magnitude response of second-order butter-worth Filter

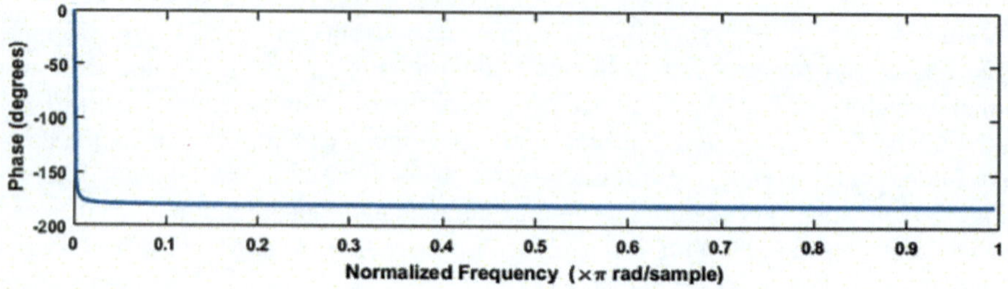

Fig. 7 Phase response of second-order butter-worth Filter

ripples from converted DC output and then inverse park transformation block enables these signals to be considered as reference signals for HCC block.

4.3 Filter Characteristics

Figures 6 and 7 show the magnitude and phase response of a second-order low pass butter-worth filter.

5 Results and Discussion

Figure 8 shows the AC side results when filters are not added in the control scheme for RL load. The waveform depicted in this figure is quite randomistic, which can be depicted from the figures.

When filter is added in the schemes, the control voltage in DC side is nearly ripple-free and reference three-phase current with filter is nearly sinusoidal as is depicted from Figs. 10 and 11.

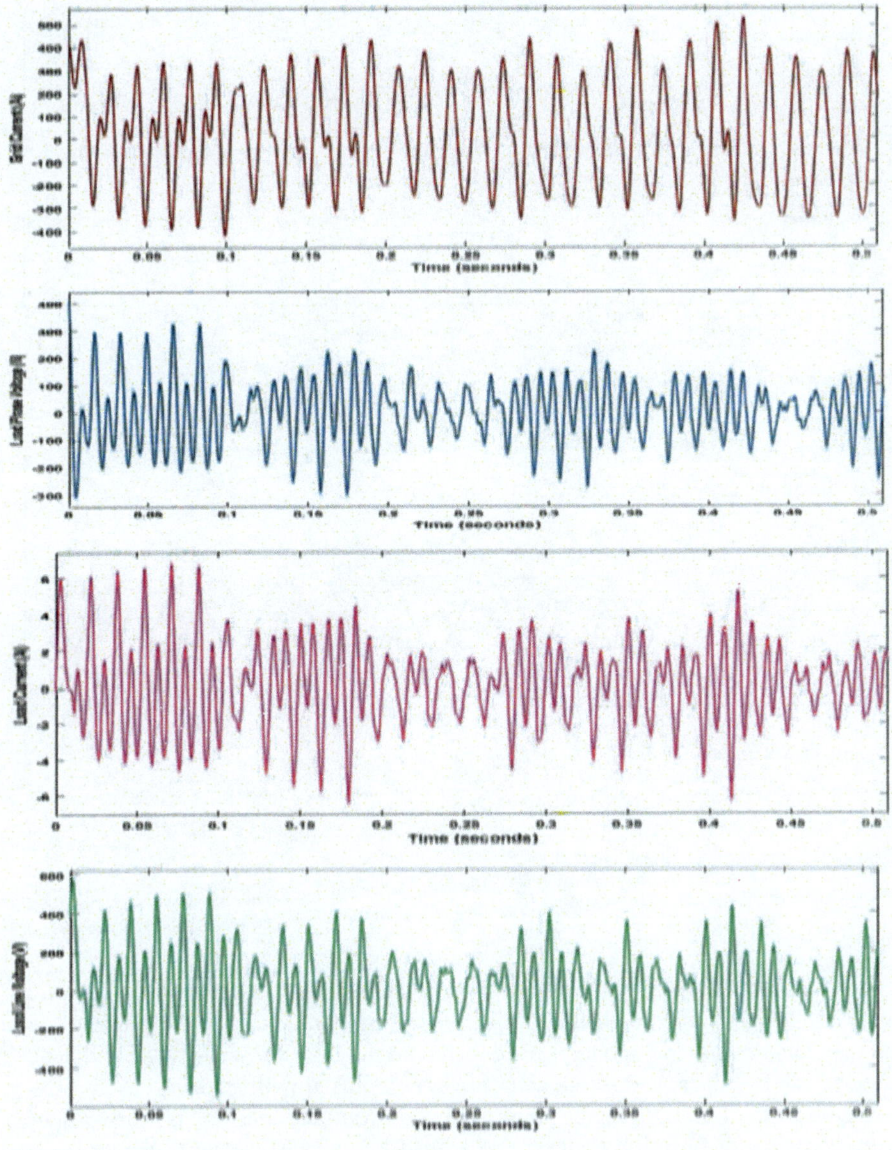

Fig. 8 AC side results with RL load without filter

Here we have considered the system with two kinds of load. In the first case, the credibility of system is tested with linear RL load, and in the second case, three-phase dynamic load was added and results were verified.

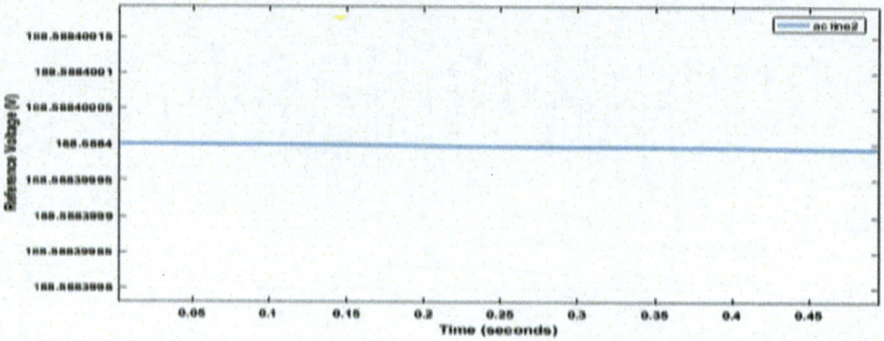

Fig. 10 Control voltage

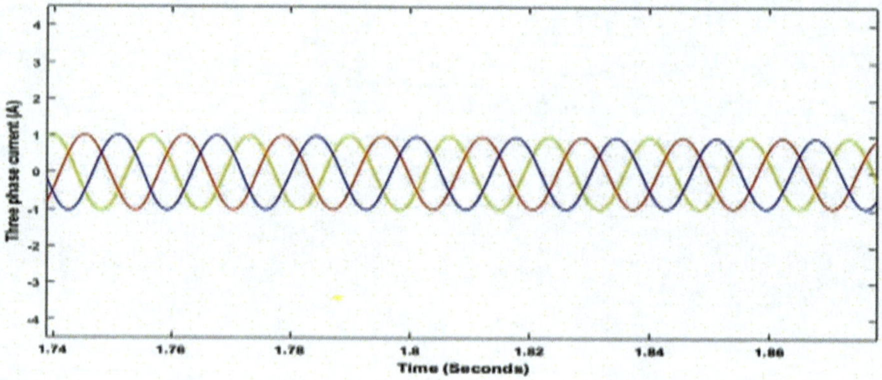

Fig. 11 Reference three-phase currents

Case 1

Figure 12 shows the AC side results of the system with RL linear load. With this figure, it is clear that in presence of filter, AC side parameters that are load phase and line voltage and grid and load currents have a much better wave shape and hence have low THD values, and frequencies are synchronised with grid.

Figures 13, 14 and 15 depict the grid, inverter output, and load powers, respectively. Power transferred to load or grid from RES is nearly at unity power factor. Also after initial transients, power is nearly constant, hence keeping system stable.

Case 2

In this case, three-phase dynamic load was connected and results were obtained. Figure 16 shows the AC side results with this load. Figures 17, 18 and 19 depict the load, inverter output, and grid powers, respectively.

With Table 1 and all the above figures, it is clear that why filter is necessary in the system.

Case 1

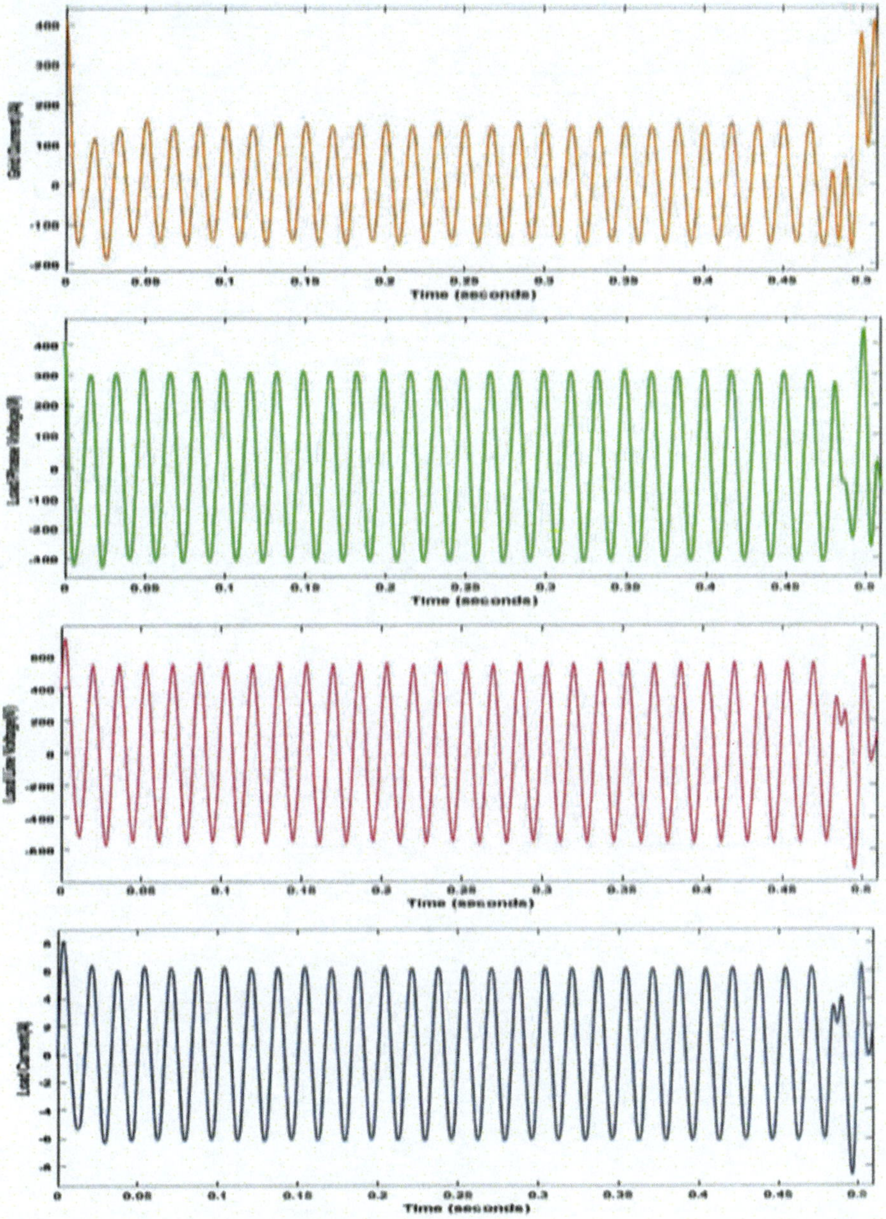

Fig. 12 AC side results with RL load

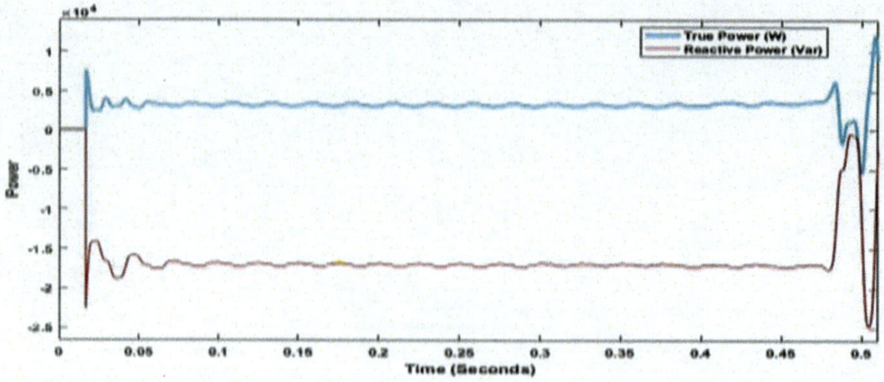

Fig. 13 Grid power with RL load

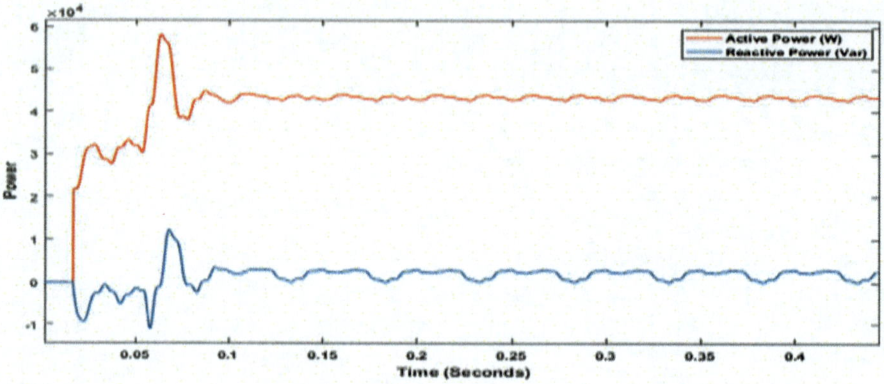

Fig. 14 Inverter output power with RL load

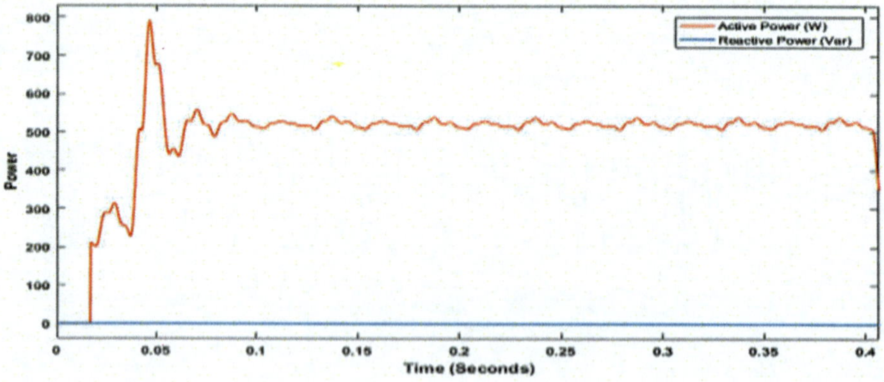

Fig. 15 Load power with RL load

Case 2

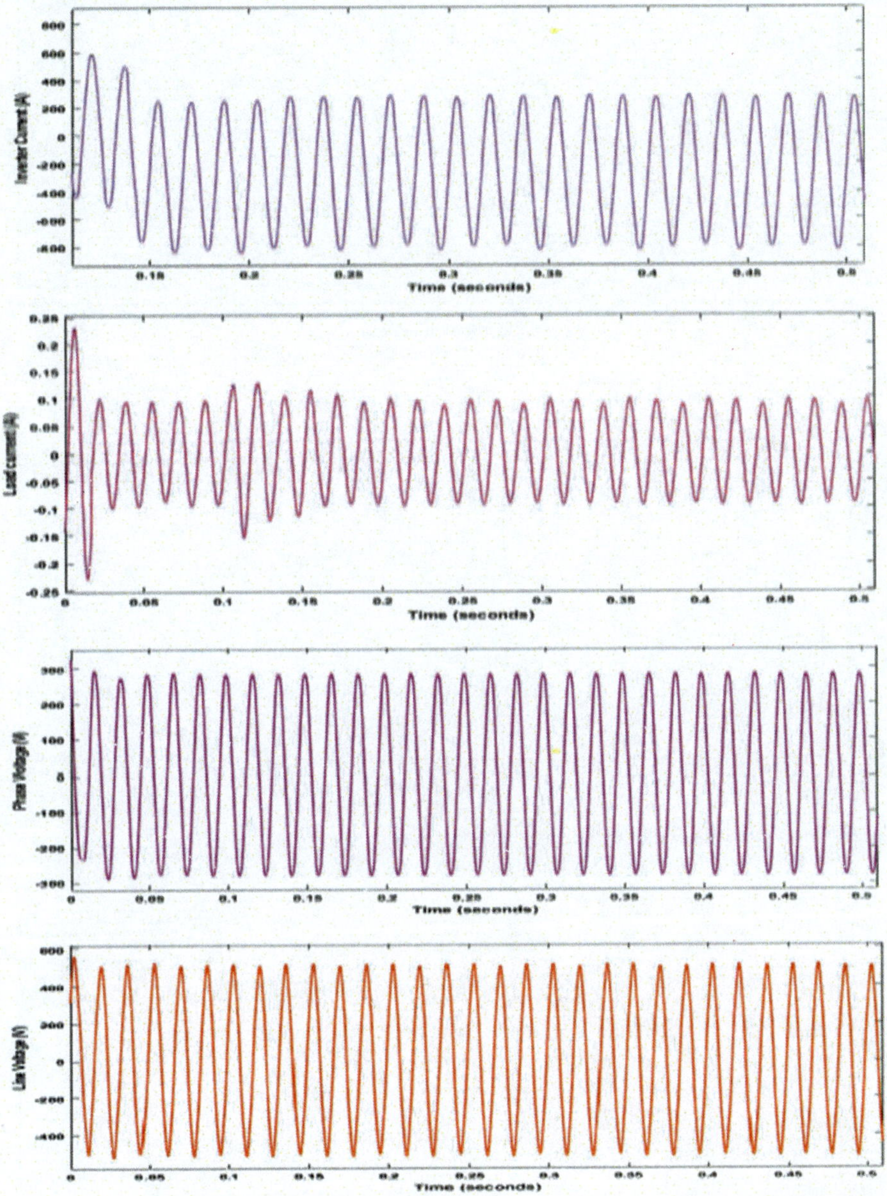

Fig. 16 AC side results with three-phase dynamic load

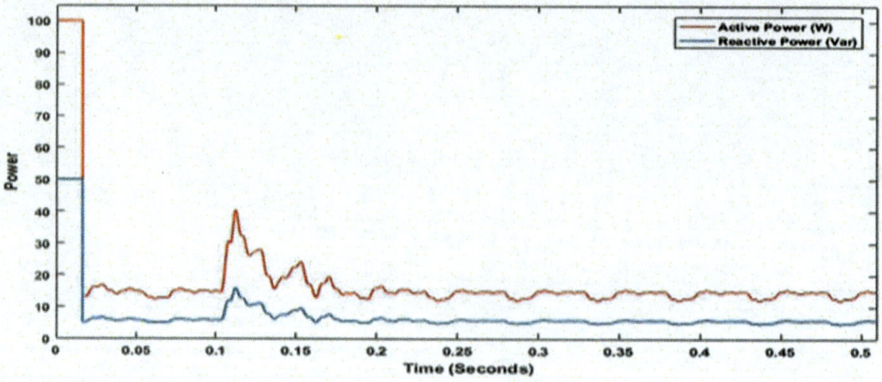

Fig. 17 Three-phase dynamic load

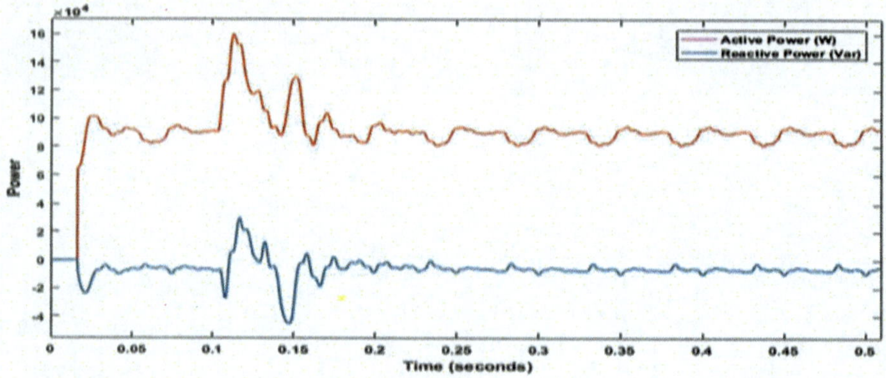

Fig. 18 Inverter output with three-phase dynamic load

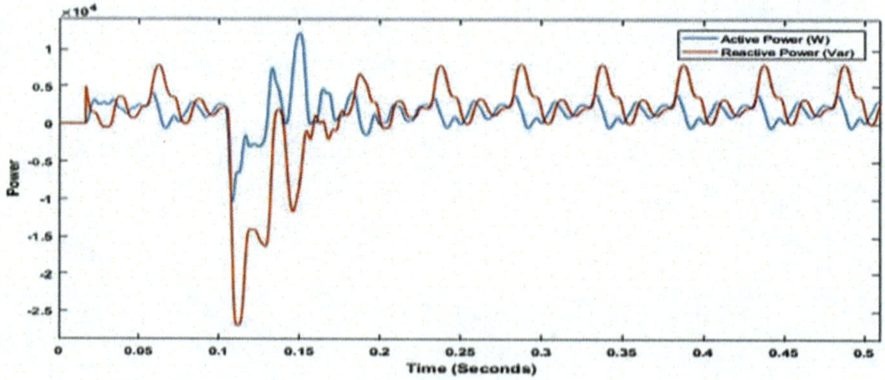

Fig. 19 Grid power with three-phase dynamic load

Table 1 THD results with and without filter in RL load

Circuit parameters	THD without filter	THD with filter
Grid current	45.6%	3.65%
Load current	54.3%	2.79%
Load line voltage	47.2%	2.72%
Load phase voltage	54.9%	2.74%

6 Conclusions and Future Scope

A lot of thought processes have been done and with analysis of a number of literatures that were available in the direction of research has been done in this paper and results have been discussed and inspected by developing the simulations using MATLAB SIMULINK. A software prototype of grid-connected RES system has been worked upon and analysed. We have tried to address some of the issues related to power quality at the grid as well as at the load side. The results at the AC side are more than par for the course as far as IEEE std. 519 is concerned for both type of loads and would be suitable for most of the grid codes with slight alterations in magnitudes. This is a general control scheme, which can be used for any small-scale grid-connected RES system, which may be SPV, fuel cell or any other RES system that produces DC output. Also, if worked upon more in this direction, with improvements in DC side with some modifications, or use of bidirectional converters, this can be used for charging and discharging of electrical vehicle whether may be Vehicle to Grid or Grid to Vehicle. With some further modifications, in place of analogue filter use of digital filter may be taken into consideration because digital filter has a number of advantages like they remain uninfluenced by ageing and other environmental or external conditions like noise, temperatures, pressures etc.; expeditious response; low cost; modifiability and many other advantages over other filters.

References

1. J. Singh, S. Ganguly, Study and design of grid connected photovoltaic system, thesis report electrical and instrumentation dept., Thapar University, Patiala
2. A. Nottrott, J. Kleissl, B.Washom, Energy dispatch schedule optimization and cost benefit analysis for grid-connected, photovoltaic-battery storage systems An international journal on renewable energy- Elsevier Vol. 55, July 2013, pp. 230–240
3. C. Schelly, E.P. Louie, J.M. Pearce, Examining interconnection and net metering policy for distributed generation in the United States. Renewable Energy Focus- Elsevier Volumes **22–23**, 10–19 (2017). December
4. S. Sumathi, L.A. Kumar, P. Surekha, *Solar PV and Wind energy conversion systems- An introduction to theory, modelling with MATLAB/SIMULINK, and the role of soft computing techniques* (Springer International Publishing, Switzerland, 2015)
5. P.S. Bimbra, "Power Electronics", sixth edition, 2018

6. I. Bhattacharya, Y. Deng, Simon Y. Foo, Active filters for harmonics elimination in solar photovoltaic grid-connected and stand-alone systems, in *IEEE 2nd Asia Symposium on Quality Electronic Design*, Penang, Malaysia, 3–4 August 2010
7. P.C. Sen, Power Electronics, 46th reprint edition, (2017)
8. M. Prodanovic, T.C. Green, Control and filter design of three-phase inverters for high power quality grid connection. IEEE Trans. Power Electron. **18**(1), 373–380 (2003)
9. V. Telukunta, J. Pradhan, A. Agrawal, M. Singh, S.G. Srivani, Protection challenges under bulk penetration of renewable energy resources in power systems: a review. CSEE J. Power and Energy Syst. **3**(4), 365–379 (2017)
10. H.R. Baghaee, D. Mlakić, S. Nikolovski, T. Dragicčvić, Anti-islanding protection of PV-based microgrids consisting of PHEVs using SVMs. IEEE Trans. Smart Grid **11**(1), 483–500 (2020)
11. V. Kumar, A. S. Pandey; S.K. Sinha, Grid integration and power quality issues of wind and solar energy system: a review, in *2016 International Conference on Emerging Trends in Electrical Electronics & Sustainable Energy Systems (ICETEESES)*, Sultanpur, India, 11–12 March 2016
12. R.D. Henderson, P.J. Rose, Harmonics: the effects on power quality and transformers. IEEE Trans. Ind. Appl. **30**(3), 528–532 (2007)
13. J.R. Rodríguez, J.W. Dixon, J.R. Espinoza, J. Pontt, P. Lezana, PWM regenerative rectifiers: state of the art. IEEE Trans. Industr. Electron. **52**(1), 5–22 (2005)
14. IEEE Std. 519-2014, IEEE recommended practice and requirements for harmonic control in electric power systems, New York, NY
15. IEEE Std. 1547–2003, IEEE Standards for Interconnecting Distributed Recourses
16. B. Singh, S.R. Arya, Adaptive theory-based improved linear sinusoidal tracer control algorithm for DSTATCOM. IEEE Trans. Power Electron. **28**(8), 3768–3778 (2013)
17. B. Singh, C. Jain, S. Goel, ILST control algorithm of single-stage dual purpose grid connected solar PV system. IEEE Trans. Power Electron. 29(10), October, 5347–5357 (2014)
18. D. Mondal, A. Chakrabarti, A. Sengupta, Power System small Signal Stability Analysis and Control", First Edition,2014, Elsevier Inc
19. A.K. Verma, B. Singh, D.T. Sahani, Grid interfaced solar photovoltaic power generating system with power quality improvement at AC mains, IEEE 3rd Int. Conf. Sustainable Energy Technology, Kathmandu, Nepal, October 2012
20. J. Saroha, G. Pandove, M. Singh, Modelling and simulation of grid connected SPV system with active power filtering features. J. Inst. Eng. India Ser. B **99**(1), 25–35 (2018)
21. Manisha Joshi, G.A. Vaidya, Modeling and simulation of single phase grid connected solar photovoltaic system, in *2014 Annual IEEE India Conference (INDICON)*, Pune, India, 11–13 December 2014

Wireless Sensor Network Node-Based Locusts' Protection for Agricultural Fields

Kshitij Shinghal, Amit Saxena, Rajul Misra, and Vikas Kumar

1 Introduction

The year 2020 has been full of challenges for complete human kind. We can see in Fig. 1 that, in early 2020, there was a break of novel Corona Virus, Covid 19, which had a deep impact all over the world.

The world was trying to recover from the shock where the nature started giving shocks one after another in form of storms, floods, conflicts, social conflicts, depression etc. as shown in Fig. 1. While everyone was trying to cope with the challenges, the locusts swarm/hopper bands started raiding agricultural fields. These locust swarms entered India towards several districts of Rajasthan from Pakistan's Sindh province and are affecting the north and central Indian states as shown in Fig. 2.

Figure 2 depicts the movement of locusts indicated by red arrows, the areas affected are shown with red, orange and yellow dots in order of severity of locusts' attack, respectively. These locust swarms keep on covering the area at an average pace of about 30–35 Km/hr with the flow of wind [1–3]. They can travel 150 Km to 250 Km per day on average. This speed is much greater than the average speed of human, bird etc. as shown in Fig. 3.

K. Shinghal (✉) · A. Saxena · R. Misra
Department of Electronics & Communication Engineering, Moradabad Institute of Technology, Moradabad, UP, India
e-mail: kshinghal@gmail.com

A. Saxena
e-mail: amitssaksena@gmail.com

R. Misra
e-mail: rajulmisra71@gmail.com

V. Kumar
Dr. A.P.J. Abdul Kalam Technical University, Lucknow, UP, India
e-mail: vikas_mittal_in@rediffmail.com

S. Vadhera et al. (eds.), *Latest Trends in Renewable Energy Technologies*, Lecture Notes in Electrical Engineering 760, https://doi.org/10.1007/978-981-16-1186-5_29

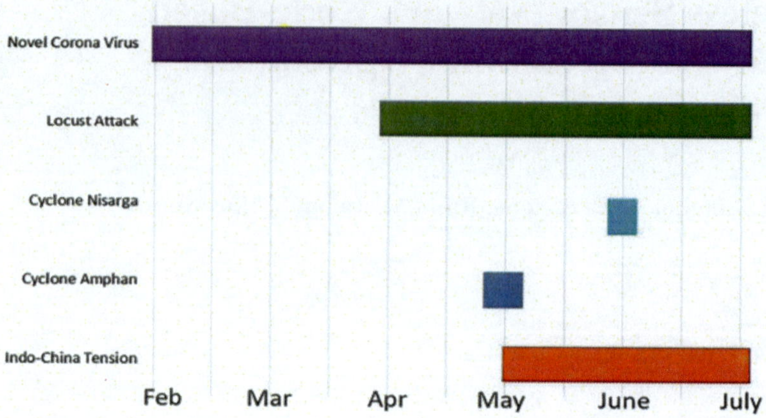

Fig. 1 Challenges in the year 2020

Figure 3 helps to get an idea how fast locusts travel, covering large areas, ravaging and destroying everything that is green and edible in their path. This was a damaging blow to the people who were already striving hard to fight other calamities. Locusts commonly known as grasshoppers have big hind legs for hopping. The classification shown in fig. 4a, b, c represents grasshopper, a locust and a cicada, respectively.

The major difference between a grasshopper and a locust is that grasshopper does not change its behaviour or appearance, whereas a locusts change their behaviour and appearance. A locust is green and prefers to live alone when it is in solitary form (when plenty of food is available) and becomes larger, brown in colour with bigger eyes and tends to form swarms or bands of adults or hopping nymphs when in Gregorian form (when food becomes scarse or scanty). Locusts in solitary as well as Gregorian form are shown in Fig. 5 [4].

The locust eats everything and anything that is green bar a few exceptions. A desert adult locust will consume roughly in terms of its own weight in food per day, which is 2 grams on a daily basis. An awfully tiny part of a mean swarm (or regarding one metric ton of locusts) consumes fare identical quantity of food per day as regarding 10 elephants or 25 camels or 2500 people as shown in Fig. 6.

During the rainy season or when food is available, the locusts remain in solitary form and as food becomes scanty, it takes Gregorian form and starts migrating over large distances in search of food. Generally, the locust lives about 3 to 5 months, but their life depends on many factors such as weather, availability of food, climate etc. In the next section, the authors discussed the proposed method, implementation details, result and discussion followed by conclusion.

Fig. 2 Route and impact of locust on Indian Subcontinent

2 Proposed Methodology

The locusts devour everything that is edible such as fruits, vegetables, leaves, plants etc. that come into their path. The whole body of locusts is covered with biological sensors and as soon as they touch anything, that is, food, they start eating it. The only exception to this is 'Neem Tree Leaves'. Locusts do not eat neem tree leaves and are repelled from anything that has been sprayed with neem tree leaf water solution. This fact is utilized in the present work to repel and change the route of locust swarms. In present paper, WSN nodes are deployed in the vicinity of agricultural field as shown in Fig. 7 for early detection of the movement of locust swarms. The WSN nodes are

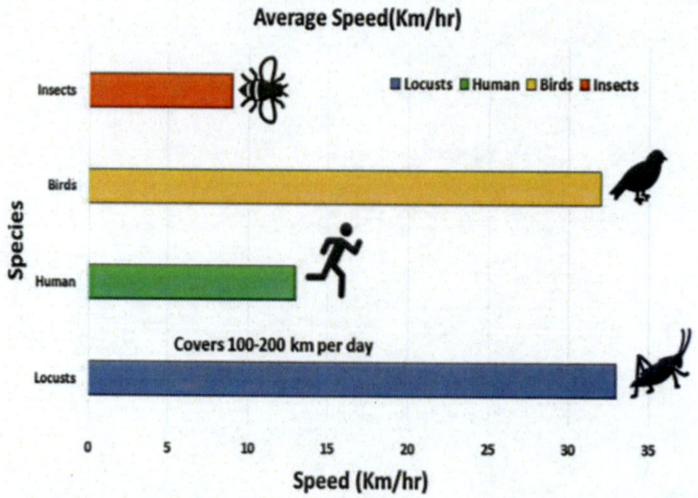

Fig. 3 Comparative average speed of locust

Fig. 4 Classification of **a** grasshopper, **b** cicadas, **c** locusts

Fig. 5 Locusts **a** Solitary mode, **b** Gregorian mode [4]

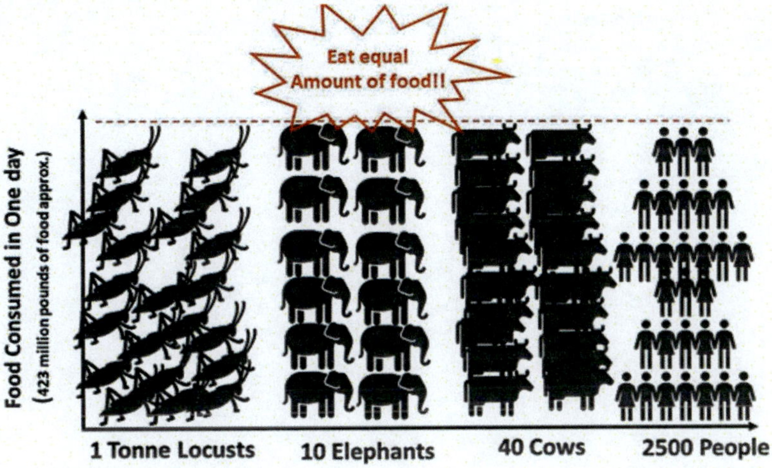

Fig. 6 Average food consumed by locusts in 1 day

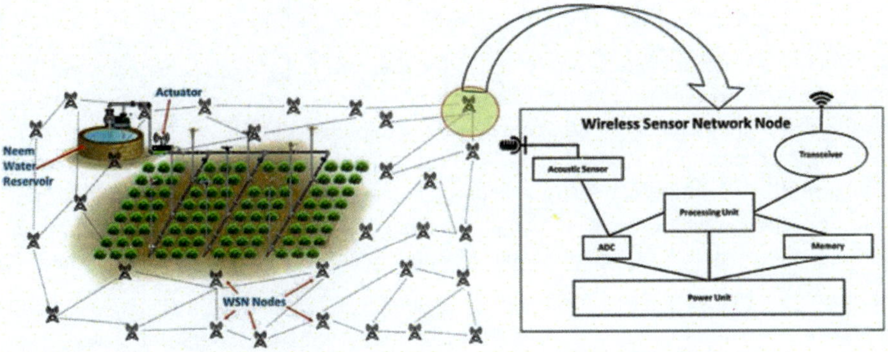

Fig. 7 Wireless sensor network node deployed in field

deployed in near vicinity of the field. The complete block diagram of typical WSN node used in the present work is shown in Fig. 7.

The arrangement to sprinkle neem water on detection [5, 6] of arrival of locusts swarm is shown in Fig. 8.

The system shown in Fig. 8 consists of a neem water reservoir which in turn is connected with pump. The pump supplies neem water solution to distribution pipes that are connected through riser to the sprinklers. As the nodes detect an approaching locust swarm, it sends signal to the actuator in vicinity to pump [7–9]. The actuator turns ON the sprinkler system spraying neem water on the crop and fine mist in the atmosphere. The WSN nodes have two sensors, namely, acoustic sensor and scent sensor to detect the arrival of locust swarms or hopper bands. When locust swarms move, they make a buzzing sound, which can be detected by acoustic sensor and they

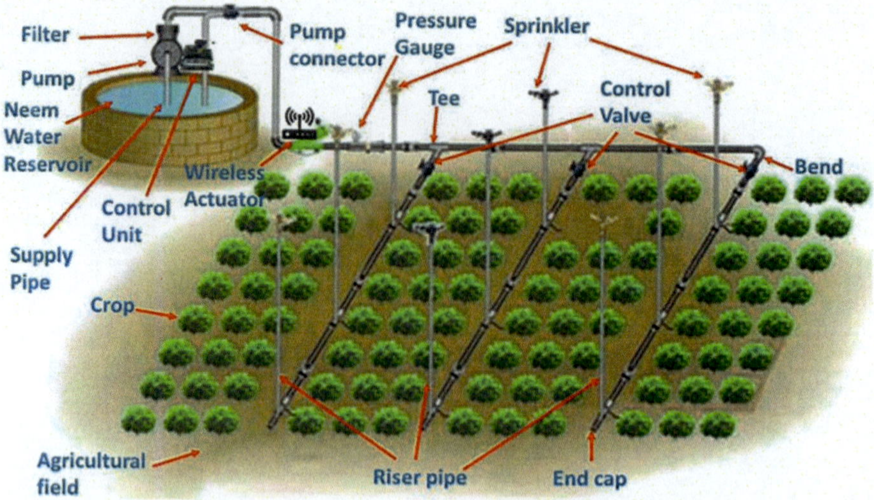

Fig. 8 Layout of agricultural field equipped with the proposed system

also release a smell for other locusts to follow their trail, the scent sensor detects this smell, the combined output from these sensors confirms arrival of locust swarm. The interface of acoustic sensor and scent sensor is depicted Fig. 9.

Further to remain in groups, they give off a scent that helps them to keep track of one another. A small sensor (gas sensor) is used to detect this scent. The output from these two sensors is sent to detect the early presence of locust swarms/hopper bands. If the presence of locust swarms/hopper bands is detected, the WSN nodes

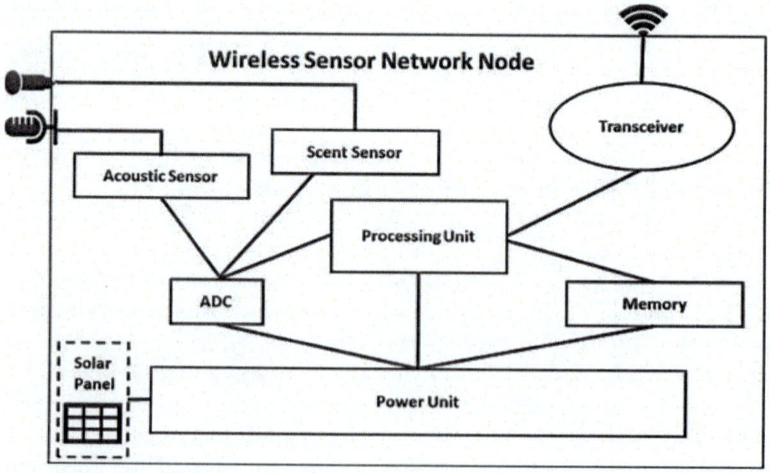

Fig. 9 Block diagram of WSN node

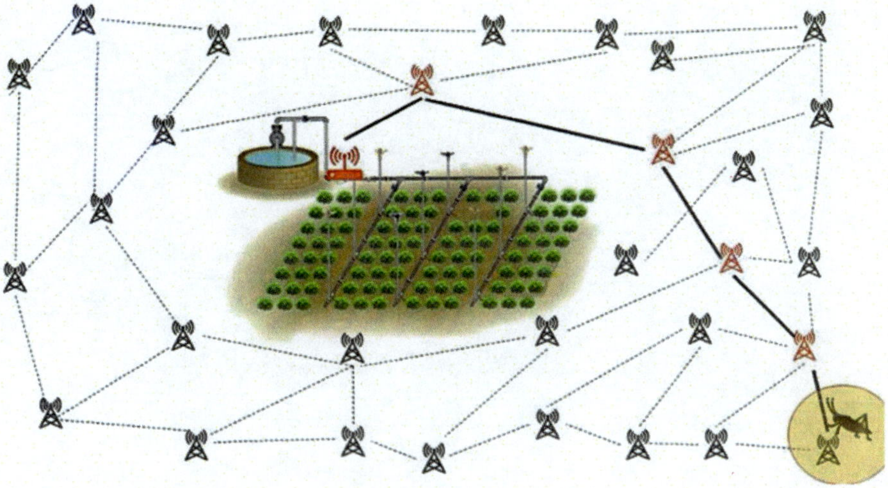

Fig. 10 WSN node and actuator system on detection of locusts

send message to the field using LEECH Protocol [10, 11] as shown in Fig. 10. The actuator situated at the field becomes activated as shown in Fig. 10 and starts the sprinkler system.

The sprinkler system is connected to neem water reservoir as shown in Fig. 8. It starts spraying neem water on the crops and agricultural field. The smell of neem water forces the locust swarms to change their path leaving the crop of the field unharmed.

3 Implementation Details

To implement the proposed system, a neem water reservoir has to be constructed in the field as shown in Fig. 8. Furthermore, the sprinkler arrangement to spray neem water has to be fixed in the field as depicted in Fig. 11. The actual deployment of WSN nodes in the vicinity of the agricultural field is shown in Figs. 12 and 13.

The WSN node is fixed with two sensors: (1) Acoustic sensor and (2) Gas sensor (as shown in Fig. 9, to detect the buzzing sound of the swarm and to detect scent released from the swarm. The combined output of the sensors helps the processing unit to take the decision whether the swarm is present or not. The power source shown is battery,however, the additional power source in form of solar panel can be added to prolong the battery life. Here, the size of WSN node is not important as has to be deployed in remote areas, though timely detection and early warning is an important parameter while implementing the present system. As early warning will ensure that sprinkler sprays the neem water solution on time producing enough smell to divert the path of locusts swarms/hopper bands.

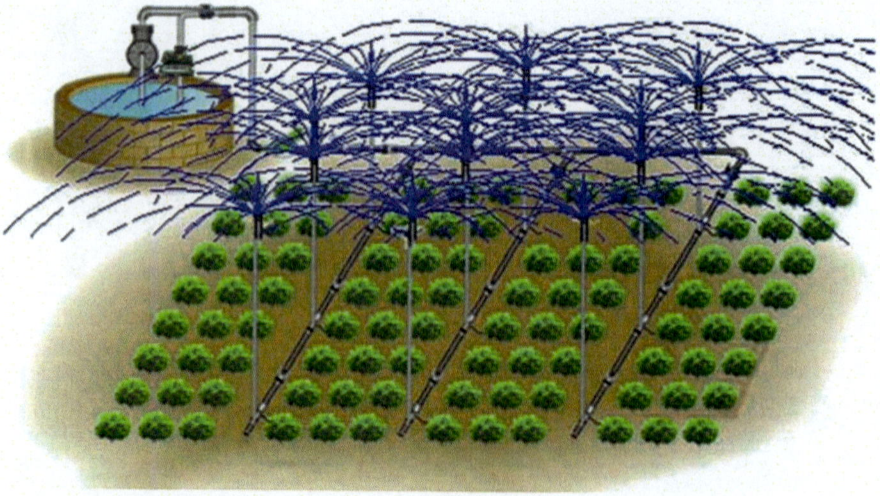

Fig. 11 Activation of sprinklers system on detection of the locust swarm

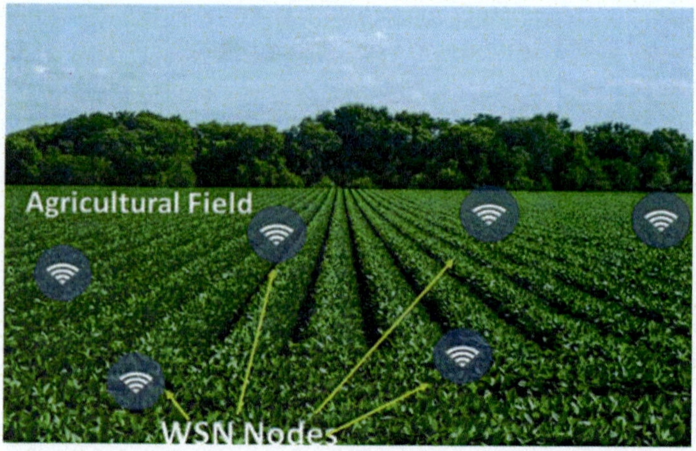

Fig. 12 Interconnected WSN nodes

3.1 Hardware Description

The WSN node shown in Fig. 9 consists of the following components:

1. Acoustic Sensor
2. Smell Sensor
3. Transceiver
4. Processor Board

Fig. 13 Actual deployed of WSN nodes

The features and specifications of the above components are given in the appendix.

3.1.1 Acoustic Sensor

SKU: 321988 shown in Fig. 14 is an analogue sound sensor microphone module for Arduino. It is suitable for present applications, i.e. detecting sound of approaching locust swarm.

Fig. 14 Acoustic sensor SKU321988 [12]

Fig. 15 Smell Sensor MICS-5524 [13]

3.1.2 Smell Sensor

MICS-5524: Smell Detection Sensor Module shown in Fig. 15 is a robust MEMS sensor for smell detection. It is suitable for present applications, i.e. detecting smell of approaching locust swarm. However, it cannot tell you which gas it has detected but can effectively detect the presence of locusts on basis of smell released by them. When smell released by locusts is detected, the analogue voltage will increase in the proportion of detected gas, which can also give an approximate idea about the size of the swarm.

3.1.3 Transceiver

SKU:319016RF shown in Fig. 16 is RF transmitter receiver module working at 315 MHz and is a wireless link kit for Arduino. This Radio Transmitter and Receiver pair is perfectly matched for the present application and allows controlling WSN nodes from a distance up to 500 feet wireless.

3.1.4 Processor Board

WeMos ESP8266 D1 R2 V2.1.0 WiFi development board shown in Fig. 17 is an ESP8266 WiFi-based board that uses the Arduino layout with an operating voltage of 3.3V [15]. This processor board is perfectly matched for the present application.

Fig. 16 Transceiver SKU:319016RF [14]

Fig. 17 WeMos ESP8266 processor board [15]

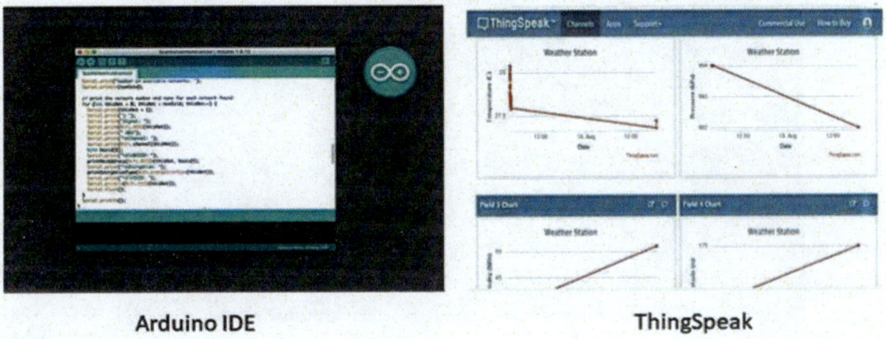

Fig. 18 Arduino open [16]. *Source*: IDE and ThingSpeak interface

3.2 *Software Description*

Arduino open source IDE and ThingSpeak shown in Fig. 18 are open application platform, which were used to establish communication protocol between nodes and server. ThingSpeak offered an open source API that was used to interface nodes with actuator. ThingSpeak is used to read data from sensors and actuate sprinklers. The licence for ThingSpeak is under GPLv3 for open source use and can be licensed from ioBridge for closed source applications.

4 Results and Discussions

The proposed system can be used effectively to:

(1) Detect the arrival of locusts swarms/hopper bands
(2) It can easily divert the path of locust swarms or hopper bands.

Figures 19 and 20 show the actual implementation of WSN node and field testing of the node, respectively.

An important trade-off during the design is speed because early warning is necessary for timely starting of sprinkler system, whereas size and power supply can be compromised to get optimum speed.

Another important parameter is that we set the threshold of processing unit at lower side, i.e. it will check the following conditions:

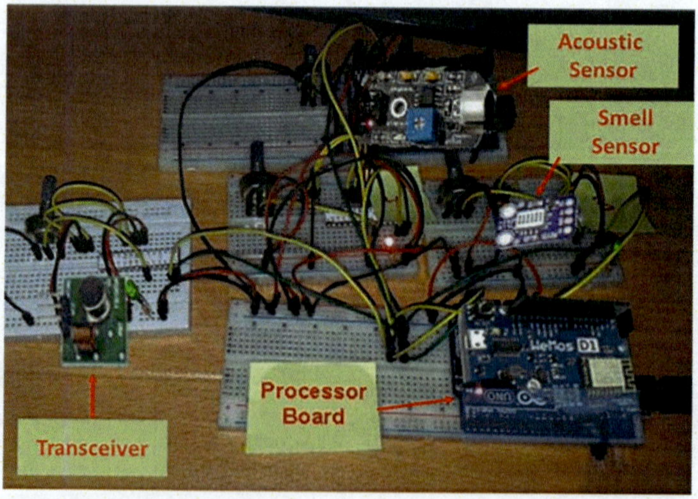

Fig. 19 Actual implementation of WSN node

Fig. 20 Field testing of WSN node

(1) There is a high probability that swarm/hopper bands present. O/p: Sprinkler started
(2) There are chances that swarm/hopper bands present. O/p: Sprinkler started
(3) There are medium chances that swarm/hopper bands present. O/p: Sprinkler started.
(4) No chances of hopper band/swarm. O/p: Sprinkler not started.

Setting the threshold to lower level may start sprinkler in case of false event detection. But in the present solution, false event detecting poses no loss as neem water is in no way harmful for crop or agricultural field. In no case there should not be a miss of locust swarm, however, to ensure that a few false triggering are bearable.

5 Conclusion and Future Work

In the present scenario, when the world is facing several challenges, the locusts pose another threat of devastating fields, destroying crops. The proposed solution not only solves the problem of locusts destroying the crops but also helps to divert the paths of locusts swarms/hopper bands towards the barren lands. Once diverted towards such direction, the swarm will eventually die due to scarcity of food without the use of any harmful chemicals, pesticides etc. In the present work, the WSN nodes and sprinklers are deployed in the vicinity of agricultural field, but in near future by developing the models and calculating the exact paths/routes of locust swarms, the present system can be deployed on the borders to divert the path of locust swarms at the entry point itself. Though doing so will require developing models and expenditure on establishing the network and equipment on the border.

Table 1 Specification of Acoustic sensor [12]

IC chip	LM386
Operating voltage(VDC)	5 V
Signal amplitude	Vcc/2
PCB size (mm)	35 × 18
Mounting hole (mm)	3
Module weight	0.085 kg
Module dimensions	5 × 3 × 2 cm

Acknowledgements We owe our sincere feelings of gratitude to the management of MIT Group of Institutions, Moradabad. We are also thankful to Prof. Rohit Garg, Director MIT, for his support, guidance and suggestions, which helped us a lot to write the paper.

Appendix

1 Features of Acoustic Sensor

The features and characteristics of sensor module match the requirement of acoustic sensor for the present application. The acoustic sensor SKU: 321988 has the following features [12]:

1. Detectable sound signal size
2. Built-in filter–rectifier circuit, DC signal output
3. Good sensitivity, built-in amplifier circuit, adjustable gain
4. Voltage signal for sound intensity can be obtained by AD conversion
5. Analogue voltage signal output, signal amplitude VCC/2
6. Compatible for Arduino sensor interface.

2 Specifications of Acoustic Sensor

The specifications of Acoustic sensor are listed in table 1.

3 Features of Smell Sensor

This sensor is sensitive to [13]:

1. CO (~1 to 1000 ppm).
2. Ammonia (~1 to 500 ppm).

Table 2 Specification of smell sensor [13]

Model	MICS5524
Input voltage (Volt)	4.9–5.1
Maximum heater power dissipation (mW)	88
Relative humidity range (%RH)	5–95
Operating temperature range (Celsius)	−30~85
Dimensions (mm)	20 × 13 × 3.4
Weight (gm)	1
Module weight	0.077 kg
Module dimensions	8 × 6 × 2 cm

3. Ethanol (~10 to 500 ppm).
4. H2 (~1 – 1000 ppm).
5. Methane / Propane / Iso-Butane (~1,000++ ppm).

4 Specifications of Smell Sensor

The specifications of smell sensor are listed in Table 2.

5 Features of Transceiver

This wireless transmitter and receiver pair operates at 315Mhz. They can easily fit into a breadboard and work well with microcontrollers to create a very simple wireless receiving frequency (MHz) data link. Since these are only transmitters, they will only work communicating data one way, you would need two pairs (of different frequencies) to act as a transmitter/receiver pair [14].

6 Specifications of Transceiver

The specifications of receiver module and transmitter module are listed in Tables 3 and 4, respectively.

7 Features of Processor Board

The board is controlled by the ESP8266 chip (a 32-Bit processor) and has a larger flash memory compared with an Arduino Uno. It consists of 11 digital I/O pins and 1

Table 3 Specification of receiver module [14]

Model	XY-MK-5 V
Operating voltage (VDC)	5
Quiescent current (mA)	4
Receiving frequency (MHz)	315
Receiver sensitivity	−105 dB
Dimensions (mm)	30 × 14 × 7
Communication distance (m):	20–200 m

Table 4 Specification of receiver module [14]

Launch distance (m)	20–200
Operating voltage (V)	3.5−12
Operating mode	AM
Transfer rate (KB/sec)	4
Transmitting power (mW)	10
Transmitting frequency (MHz)	315

analogue (input) pin. The board can be connected using a Micro-B-type USB cable. The D1 R2 is a WiFi capable ESP8266EX-based development board in the form of the Arduino UNO board format. This board is compatible with the Arduino IDE and with NodeMCU.

On-Board Switching Power Supply:

- Input Voltage Range: 9 V to 12 V
- Output: 5 V at 1A Max

8 Specifications of the Processor Board

The specifications of processor board are listed in Table 5.

Table 5 Specification of processor board [15]

Microcontroller	ESP8266
Digital I/O pins	11
Operating voltage(VDC)	3.3
Analogue I/O pins	1
Flash memory	4 MB
Dimensions in mm (LxWxH)	68 × 54 x 12
Weight (gm)	20
Module weight	0.105 kg
Module dimensions	6 × 6 × 2 cm

References

1. K.H.J. Huang, Remote sensing of locust and grasshopper plague in China: a review, in *2016 Fifth International Conference on Agro-Geoinformatics (Agro-Geoinformatics)* (Tianjin, 2016), pp. 1–6. https://doi.org/10.1109/Agro-Geoinformatics.2016.7577686.
2. H. Wehn, B. Rabus, D. Wood, A. McCardle, Prediction of locust outbreaks from RADARSAT-1 multi-angle data, in *IGARSS 2004. 2004 IEEE International Geoscience and Remote Sensing Symposium*, vol. 5 (Anchorage, AK, 2004), pp. 3543–3546. https://doi.org/10.1109/IGARSS.2004.1370475
3. T. Starner, D. Kirsch, S. Assefa, The locust swarm: an environmentally- powered, networkless location and messaging system, in *Digest of Papers. First International Symposium on Wearable Computers* (Cambridge, MA, USA, 1997), pp. 169–170. https://doi.org/10.1109/ISWC.1997.629938
4. S.D. Gordon, J.C. Jackson, S.M. Rogers, J.F.C. Windmill, Listening to the environment: hearing differences from an epigenetic effect in solitarious and gregarious locusts. Proc. R. Soc. B. 281 (1795). https://doi.org/10.1098/rspb.2014.1693
5. K. Nishio, N. Ihara, T. Yamasaki, Simple analog-digital circuit for detection of approaching object based on visual systems of pigeon and locust, in *2010 11th International Conference on Control Automation Robotics and Vision* (Singapore, 2010), pp. 1714–1718. https://doi.org/10.1109/ICARCV.2010.5707233.
6. M.R.M. Kassim, A.N. Harun, Applications of WSN in agricultural environment monitoring systems, in *2016 International Conference on Information and Communication Technology Convergence (ICTC)* (Jeju, 2016), pp. 344–349. https://doi.org/10.1109/ICTC.2016.7763493.
7. J. Brinkhoff, J. Hornbuckle, Characterization of WiFi signal range for agricultural WSNs, in *2017 23rd Asia-Pacific Conference on Communications (APCC)* (Perth, WA, 2017), pp. 1–6. https://doi.org/10.23919/APCC.2017.8304043.
8. J.B. Patel, C.B. Bhatt, B. Patel, K. Parwani, C. Sohaliya, Field irrigation management system using wireless sensor network, in *2011 Nirma University International Conference on Engineering* (Ahmedabad, Gujarat, 2011), pp. 1–4. https://doi.org/10.1109/NUiConE.2011.6153317
9. J. AdelineSneha, R. Chakravarthi, J.A. Glenn, A review on energy efficient image feature transmission in WSN for micro region pest control, in *2016 International Conference on Electrical, Electronics, and Optimization Techniques (ICEEOT)* (Chennai, 2016), pp. 4859–4862. https://doi.org/10.1109/ICEEOT.2016.7755643
10. Y. Li, L. Ding, F. Liu, The improvement of LEACH protocol in WSN, in *Proceedings of 2011 International Conference on Computer Science and Network Technology* (Harbin, 2011), pp. 1345–1348. https://doi.org/10.1109/ICCSNT.2011.6182209
11. R. Regmi, P.W.C. Prasad, A. Alsadoon, A. Elchouemi, S. Sreedharan, Modified LEACH algorithm for wireless sensor networks in agricultural field, in *2017 IEEE International Conference on Power, Control, Signals and Instrumentation Engineering (ICPCSI)* (Chennai, 2017), pp. 3100–3104. https://doi.org/10.1109/ICPCSI.2017.8392296
12. Data sheet of SKU321988 Acoustic sensor available online at https://test.robu.in/product/analog-sound-sensor-microphone-module-for-arduino/
13. Datasheet of Smell sensor MICS5524 available online at https://www.mouser.in/datasheet/2/18/1084_Datasheet-MiCS-5524-rev-8-1144838.pdf
14. Datasheet of Transceiver SKU:319016RF available online at https://robu.in/product/rf-transmitter-receiver-module-315mhz-wireless-link-kit-for-arduino/
15. Datasheet of WeMos ESP8266 processor board available online at https://www.rhydolabz.com/arduino-arduino-boards-c-152_123/wemos-d1-r2-wifi-esp8266-development-board-arduino-compatible-p-2431.html
16. Arduino IDE https://www.arduino.cc/pro/arduino-pro-ide
17. ThingSpeak https://thingspeak.com/

Current Scenario of Solar Power and Various Schemes for Stimulation and Expansion of Solar Energy Sector in India

Sunny Vaish, Ravneet Kaur, Deepika Bhalla, and Naveen Kumar Sharma

1 Introduction

Power and energy are among the key factors that have a significant contribution towards the economy of a nation. Research projects that are linked to the utilities for defining, formulating, implementing and then redefining for improvement finally result in the national development along with the benefit of it reaching the consumer [1]. During the last decade of the previous millennium, most of the utilities were forces to their operations, structure and the ways the business was done. Earlier the functioning of the utilities was vertical and closely held and the demand changes it to an open system. The fast depletion of reserves of fossil resources for meeting the energy demand of high-energy intense industries and the resulting visible change, the use of these reserves caused to the environment caught the much-needed attention for decelerating the dependence on fossil fuel. The industrial activity and demand for development result in dumping carbon into the atmosphere; of the total 8 billion tonnes of carbon emissions, 81.25% comes from fossil fuels and the remaining from deforestation [2]. Renewable Energy Sources (RESs) such as pico and micro hydro, solar, wind, biomass, fuel-cell, etc. are non-polluting, clean and can easily substitute fossil fuels provided technology and economics are achieved.

The power sector in India is extremely diversified, and it also has a considerable growth rate. For both capacity addition and energy security, the power sector of India

---

S. Vaish · R. Kaur (✉) · D. Bhalla · N. K. Sharma
Department of Electrical Engineering, IKGPTU Jalandhar, Punjab 144 603, India
e-mail: ravneetasr2506@gmail.com

S. Vaish
e-mail: sunnyvaish3@gmail.com

D. Bhalla
e-mail: deepika.bhalla89@gmail.com

N. K. Sharma
e-mail: naveen31.sharma@gmail.com

S. Vadhera et al. (eds.), *Latest Trends in Renewable Energy Technologies*, Lecture Notes in Electrical Engineering 760, https://doi.org/10.1007/978-981-16-1186-5_30

is amongst the fastest growing along in the world, and along with it, it's diversified. Now solar power is an integral part of India's energy expansion plan; it is not only for capacity addition but also for energy security. With the restructuring in the power sector, India looked towards solar power as an alternate source of energy. The utilities started taking much interest in this energy that is free of cost and also environment friendly for generation purpose. Since the last decade, the cost of generating power has been lessening due to technology advances resulting in the decline in the cost of equipment used in Solar Power Plants (SPP) along with efficient solar panels. The need for skilled labour exists, however, the manpower required is lesser; hence, the cost of operating a solar power plant is less in comparison to its overall cost. Moreover, with the advancement of the automation of solar power plants, there is a reduction in the cost of labour along with the cost of operation. Hence, the cost of commissioning a solar power plant is bound to come down in the near future [1].

There was a change at the global level in the regulations and laws that govern the power sector and India also has trailed this global change. In June 2003, Electricity Act 2003 came into force in India with the objective for introduction of competition for the much-needed power for all and protection of the consumer interest. The provisions of the Act were National Electricity Policy, Open access in transmission, power trading, mandatory SERC rural electrification, open access in distribution, licence-free generation and distribution, mandatory metering and in its absence stringent penalties for power theft. The Electricity Act of 2003 replaced the Electricity Act 1910, Electricity Regulatory Commission Act 1998 and Electricity Supply Act 1948. The aim of the Act and its subsequent amendments have been pushing the power sector onto a route of robust commercial growth and enabled all the States and Centre to move in synchronization and harmonization of their operations and management for national development through reliable and cheap quality of power and energy [3, 4].

As of 31 March, 2020, the grid-connected installed total capacity from all sources is 370106.46 MW, which includes power from renewable and non-renewable sources. More than half of the total installed from all sources is from fossil fuel, and coal contributing to most of it, which amounts to be 55.43%. As per data of the Central Electricity Authority (CEA) of India, the country is highly depended on coal for meeting the demand for power and energy. The breakup of total installed capacity from all the sources in India: of the 62.8% total thermal is 54.2% from coal, 6.7% from gas, 1.7% from lignite and 0.1% from diesel. Figure 1 represents the graphical bar representation of total generation including RES in BU (Billion Units) from all sources. The RES data from Ministry for New and Renewable Energy include Small Hydro Project, Solar and Wind Energy, Biomass Power, Biomass Gasifier and Urban & Industrial Waste Power.

The above date of installed capacity is not what is needed for achieving a sustainable energy target for the future, the 71% dependence on fossil fuels is to be done away with, which can be achieved by the exploitation of the potential available RES, which would be an assured reduction of the environmental and ecological impacts of power generation [5]. Greenhouse gas (GHG) emissions and air pollution are due to the extensive use of lignite and coal. Growing fluctuation in the oil prices and its

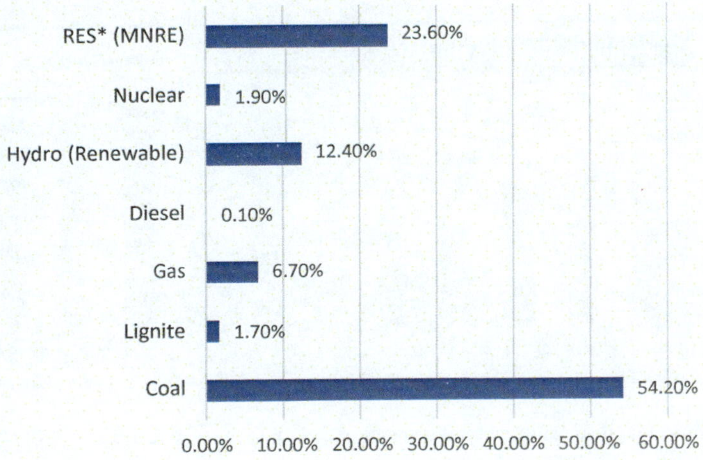

Fig. 1 Installed Capacityfrom all the sources in India [4]

import dependency, international bilateral issues among the oil-producing nations are contributing to the additional economic burden and uncertainty of diesel-based thermal units in the country. High dependency on import has resulted in greater energy insecurity and the same could continue in future. The country needs to decrease dependence on both coal and oil, and along with it focuses on RES to get a secure and stable energy system [5].

In the recent past, India has been among the fastest-growing economy, this has levied a massive burden on power sector for supplying adequate and reliable power for both industrial and urban demands. This growing economy has made India the third-largest consumer of electricity in the world [6]. As on date, the power generation is still insufficient to meet the demand, although there has been a considerable annual addition in the generation capacity. Over the past decade, there have been challenges such as fuel shortage for the operation of thermal units, infrastructure issues for the hydro projects, high price for the import of technology for the renewable sector along with the subsidy for promoting it; the consequence is that the tariff is increasing. Considering the irradiation level and the number of sunny days in India, solar energy can offer a reasonably priced solution to the issue of power deficiency and carbon footprints. The graphical representation of total generation including renewable sources in BU is as shown in Fig. 2.

Solar radiation is plentifully available along with a guarantee of no rise in the cost. Hence, moving towards renewable energy generation can provide relief from increasing tariff and energy security for consumers. The potential of solar power in Saudi Arabia has been extensively discussed in reference [7] along with the economic aspects of power and energy from the sun for that country. Solar energy technologies provide an excellent opportunity for mitigation of GHG emission, thus reducing pollution and retarding the global warming [8]. Suitable attention is required for better development and utilization of RESs in India [9]. Sharma et al. presented

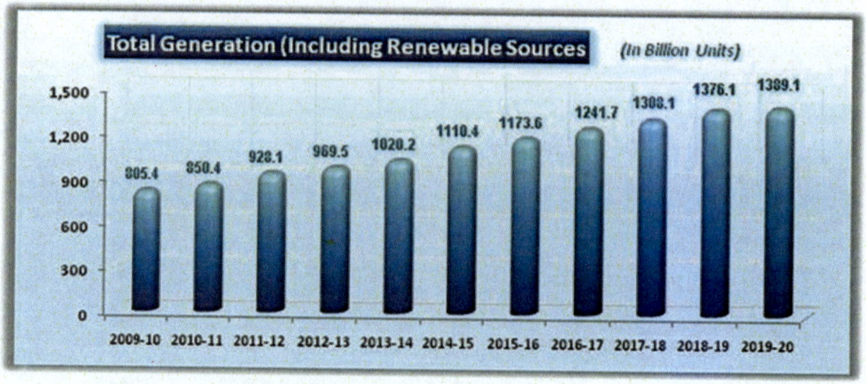

Fig. 2 Graphical representation of total generation including renewable sources in BU [4]

the analysis of technology available and then economic paybacks of solar energy in context with India until the year 2011 [10].

Today both the economic potential and technical up-gradation of solar energy are changing very fast. Hence, recent statistics is indispensable for a precise approximation of power through solar energy. This work mollifies the objective by stating the planning for the development of solar energy, setting the target, achievement of the target and assessment of government initiatives. This paper covers the inclusive study made by the authors of the existing electrical power in India with emphasis on solar energy in terms of its availability, present status, targets laid and the attainments, government initiatives, updated installed capacity along with future strategies. It is intended to provide the recent information regarding solar energy sector in India in regard with the deployment of technologies so as to harness solar energy for future development of remote areas and reduction in fossil fuel dependence.

2 The Electrical Power Installed Capacity and Demand Projection of India

Once in a few years, the Central Electricity Authority carries out Electricity Power Survey (EPS) with the objective of planning of generation capacity, infrastructure for transmission and distribution of power to load centres, by both the state/provincial and the central/federal government in India. The past vs future of electricity demand in India is shown in Fig. 3. The recently released 19th EPS projects the electrical power demand of 1743 TWh (Tera watt hour), which is 6.59% of Compound Annual Growth Rate (CAGR) from 2017. The peak load is 299GW (Giga watt), which is 6.32% CAGR by 2027 (CEA, 2017) [11].

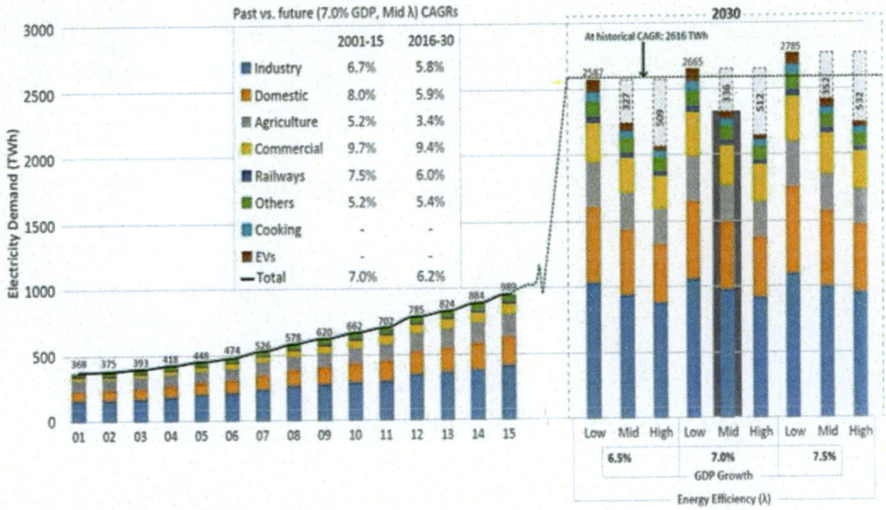

Fig. 3 Past vs Future of Electricity Demand [11]

2.1 *Electricity Generation and Growth*

The target of electricity generation in India for the year 2020–21 from conventional energy sources has been aimed to be 1330 Billion Unit (BU), which is an ambitious growth of 6.33% of the actual conventional generation of the previous year. For the year 2019–2020, generation from conventional sources was 1250.784 BU, which is a growth rate of about 0.12% from the preceding year. During 2019–2020, 1252.6 BU were generated as shown in Table 1 [4].

The target for electrical power generation from conventional sources for the year 2020–2021 comprises of 1138.533 BU from thermal; 140.357 BU from hydro; 43.880 from nuclear and 7.230 BU import from Bhutan; this totals to 1330BU [4].

2.2 *Future Projection of Power Requirement in India*

By the end of the next decade, it is expected that the power demand will increase by 128%. The estimation of the power demand by the industrial sector in 2014–2025 is 550GW, and in the year 2029–2030, it will be 760 GW; which would be an increase of 38.18% [12]. Looking at these projections, for sustainable development, the solution to the increase in electrical power demand lies in the optimal utilization of the available solar days [13].

Table 1 Annual conventional power generation and its growth in India [4]

Year	Contribution of conventional sources towards power generation (billion units)	Percentage growth of generation
2009–2010	771.6	6.6
2010–2011	811.1	5.56
2011–2012	876.9	8.11
2012–2013	912.0	4.01
2013–2014	967.2	6.04
2014–2015	1048.7	8.43
2015–2016	1107.8	5.64
2016–2017	1160.1	4.72
2017–2018	1206.3	3.98
2018–2019	1249.3	3.57
2019–2020	1252.6	0.26
2020–2021	91.913	22.85

3 Renewable Energy Generation Capacity in India

India ranks amongst first few of the countries that produce substantial amount of power from renewable sources [14]. As per statistics available on 31 March, 2020, about 35.86% of India's installed generation capacity is from new and renewable sources; generating approximately 21.22% of total utility electricity in the country [15]. The National Electricity Plan for the year 2018 National Electricity Plan sets out ambitions to achieve 275 GW of renewable by the next 10 years, which would increase contribution from the renewable sources to an estimated 24% in electricity generation and 44% of installed capacity [16]. India has certain prominent locations from where geothermal energy can be harnessed up for meeting the local demand. With geothermal energy, another option after efforts on solar, wind, biomass has shown promising results, now the Ministry of New and Renewable Energy (MNRE) is targeting 1 GW of geothermal capacity by 2022.

The target of stand-alone renewable energy is comparatively easy to achieve; however, there are many technical challenges for connecting power from renewable sources to the grid. In the next 2 years (by March 2022), the government has a target to achieve 175 GW of grid-connected renewable electricity. The breakup of which is 57.14%, from solar, 34.28% from wind, 5.71% from biomass and 2.87% from small hydropower. As of 30 April, 2020, India has 45699.22 MW of installed large hydro capacity. Excluding this contribution from the large hydropower, the installed grid-interactive renewable power capacity as of 30 June, 2020 is depicted in Fig. 4. Assessing the statistics, it can be concluded that new and fast-developing renewable energy sources, managed by the MNRE, are gaining much-needed momentum.

India's renewable installed power capacity target for 2020 is 175 GWalready, 50.09% of it has been achieved. The total installed grid interactive renewable power

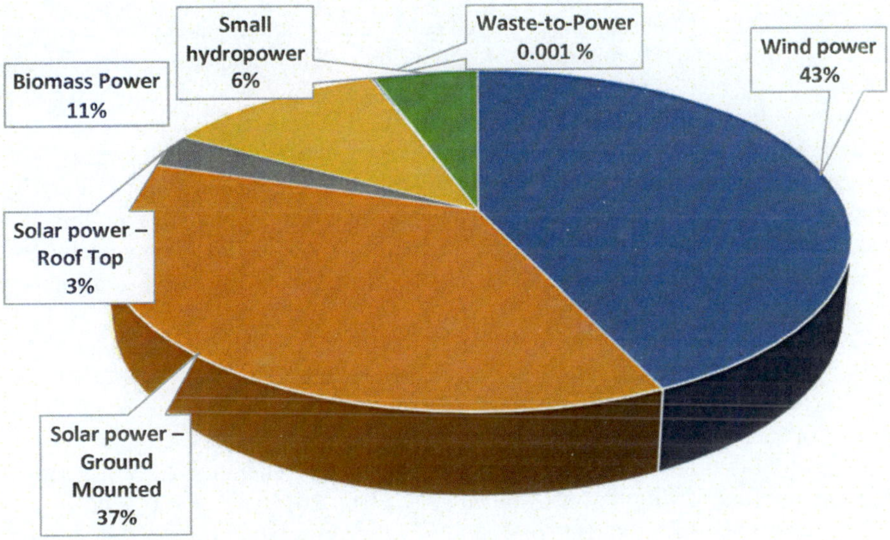

Fig. 4 Installed grid-interactive renewable power capacity [15, 15]

capacity (excluding large hydropower) is 87669.16 MW as of 30 June 2020. For meeting the ambitious target of installed capacity, as on 30 June 2020, as far as the solar power is concerned, nearly half of the target has been achieved, which is around 35122.30 MW. The target attainment is being achieved on daily basis from both roof and ground- mounted panels. Even the installation of power from wind is more than halfway towards achieving the target, the total installed grid interactive wind power is almost 37829.55 MW. The eminent among achieving the target is small hydropower that is above 90%. The contribution of both Biomass bagasse and non-bagasse cogeneration has already reached 95% of the target. The pace of achieving the target is considerable despite the lockdown due to the pandemic 2020.

3.1 Renewable Energy Potential, Target and Regulatory Support

Regulatory provisions play a vital role in encouraging the investment in solar energy. The GOI has a renewable energy target of 175 GW, this will result in accelerated investments in the area. As per the market survey of the Deutsche bank conducted earlier, it was expected that by the year 2019–2020, the annual capital investment in the solar energy sector will surpass that of coal, which is due commitment by global investors worth US$35 billion [18]. This report also estimated that starting from the financial years 2016–2017, the dependence on fossil fuel (primarily coal) can be cut down by 8% by having 5 GW of solar capacity addition per year for 5 years. The results include savings through reduced coal imports along with a significant reduction in GHG emissions [19]. To promote renewable energy penetration, state electricity regulatory commissions are required to aggressively emphasize the installation and use of RES within their jurisdiction. The State Government can promote renewable energy by providing preferential tariffs for installing renewable generators, and the Central Government can provide additional support by allowing open access to the power grid for renewable energy consumers and generators. As part of its 12th 5-year plan, the government aimed to add 30 GW of new renewable energy capacity by 2017 [23]. In 2008, a target was set of achieving 15% solar energy penetration in India by 2020.

As per the requirement of the National Action Plan on Climate Change (NAPCC) under the Electricity Act, so as to strengthen the existing Renewable Purchase Obligations (RPOs), the RPOs set state-wise purchase targets for both the electrical distribution companies and the consumers, for sourcing renewable electricity from third-party generators through the grid [19]. The Electricity Generation (GWh) through RES against Total Utility Power Installed since 2014–2015 is given in Table 2.

The renewable energy targets of 175 GW, which consist of 100 GW generation by solar, and 40% of which will be from rooftop solar [22]. This is a quantum leap from the solar target of 20 GW by 2022 set under the Jawaharlal Nehru National Solar Mission (JNNSM). The new targets include 60 GW of wind power, 10 GW of biomass and 5 GW of small hydro. To achieve these targets, India needs to invest about US$200 billion [24] in the renewable sector. The targets contribute towards the commitment by India submitted to the United Nations Framework Convention on Climate Change (UNFCCC) in December 2015. The commitment was made ahead of the Global Conference of the Parties (COP21) climate talks in Paris.

Table 2 Renewable Electricity Generation among Total Utility Power Installed Year-wise renewable energy generation (GWh) [20]

Source	2014–2015	2015–2016	2016–2017	2017–18	2018–2019	2019–2020
Large Hydro	129,244	121,377	122,313	126,134	135,040	155,970
Small Hydro	8,060	8,355	7,673	5,056	8,703	9,366
Solar	4,600	7,450	12,086	25,871	39,268	50,103
Wind	28,214	28,604	46,011	52,666	62,036	64,639
Bio mass	14,944	16,681	14,159	15,252	16,325	13,843
Other	414	269	213	358	425	366
Total	191,025	187,158	204,182	227,973	261,797	294,288 [21]
Total utility power	1,105,446	1,168,359	1,236,392	1,302,904	1,371,517	1,385,114
Percentage renewable power	17.28%	16.02%	16.52%	17.50%	19.1%	21.25%

3.2 *Scheme to Promote Solar Energy*

The National Action Plan for Climate Change (NAPCC) in June 2008 initiated the development of solar technologies in India. To take it further, in November 2009, National Solar Mission was initiated. Jawaharlal Nehru National Solar Mission" (JNNSM) aimed to reach parity with the grid power tariff by 2022. This targets the development and deployment of solar energy technologies in the country [16, 16].

4 Current Scenario of Solar Power in India

In recent years, as the need for more and more power for development is increasing, the role of solar energy sources has also been increasing with minimum ecological impact.

4.1 *Grid-Interactive Solar Power*

Solar energy is free, does not produce any waste during the process of power generation, it is secure and simple to maintain. Figure 5 details the installed capacity of Grid Interactive Solar Photovoltaic (SPV) power for all the states, which is 35122.30 MW in India as of 31 June 2020. Karnataka ranks first as the solar state; it had a 5,000MW installed capacity by the end of the financial year 2017–2018. This includes Pavagarh solar Park of installed capacity of 2050 MW, which was the world second largest photovoltaic solar park. The second-largest solar power state of India is Rajasthan; total photovoltaic capacity by the end of 2020, it will be of capacity 5222.86 MW. Bhadla photovoltaic solar park is the world's largest solar park as of March 2020, located in Bhadla, Phalodi tehsil, Jodhpur district, Rajasthan. The solar park has a total capacity of 2245 MW by the end of March 2020. Tamil Nadu has the third highest operating solar power capacity in India, and as on June 2020, the total operating capacity is 3918.10 MW.

Telangana ranks fourth with the solar power installed capacity of 3689.36MW. Installed photovoltaic solar capacity in Andhra Pradesh is 3618.77 MW, at the fifth position. Andhra Pradesh has three photovoltaic solar stations, these are: Ananthapuramusolar park, Kurnool solar park and Kadapa solar park, having an approved solar capacity of 1500 MW, 1000 MW and 1000 MW, respectively. Gujrat is in sixth position in developing solar power, and by the end of June 2020 its total photovoltaic capacity reaching 3053.69 MW. Madhya Pradesh is at seventh position with its total photovoltaic capacity reaching 2311.80 MW. Madhya Pradesh has two solar power stations; the approved solar power capacity for Neemuch-Mandsor solar park and Rewa Solar Park is 750MW. Maharashtra, Uttar Pradesh and Punjab are next in order they have a total installed solar power capacity of 1869.97 MW, 1174.10 MW and

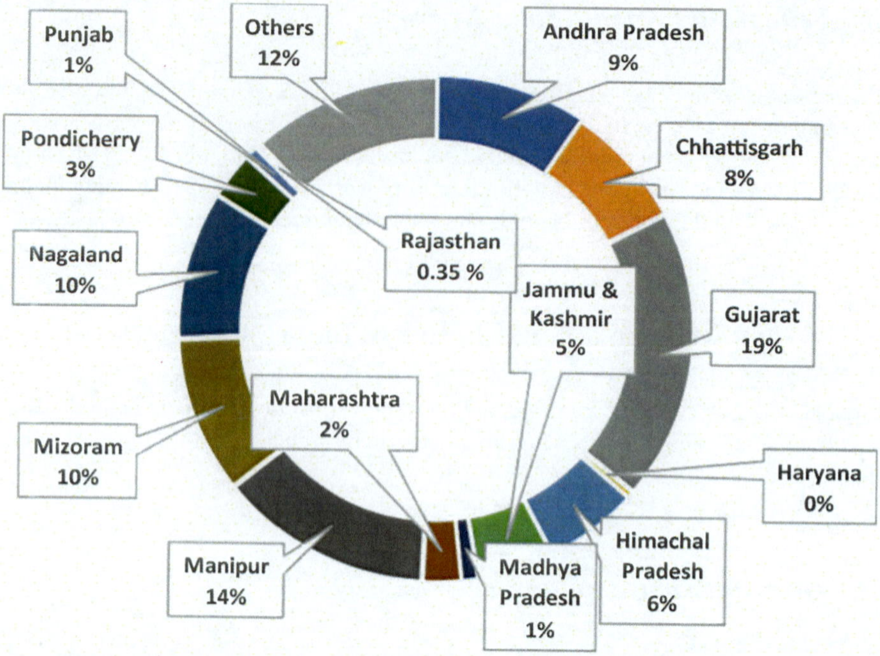

Fig. 5 State wise installed capacity of Grid-Interactive Photovoltaic Solar Power [27]

947.10 MW, respectively. These top 10 solar power states contribute 90% of the total installed capacity of Grid-Interactive photovoltaic solar power in the country. Other 10% installed capacity of Grid-Interactive photovoltaic solar power comes from the remaining left states in the country.

4.2 *Off-Grid Installed Solar Power*

Solar energy in its various forms has been successfully harnessed by India, by the end of June 2020, 17.21 lakhs Home Lighting System (HLS), 7.31 lakhs Street Lighting System (SLS) and 75.49 lakhs Solar Lantern (SL) have been installed and are operational. There are 256156 solar photovoltaic (SPV) pump systems and Solar Power Plants (SPP) reached 215767-Kilowatt peak (kWp). Figure 6 gives the Decentralised/Off-Grid solar energy devices/systems in India.

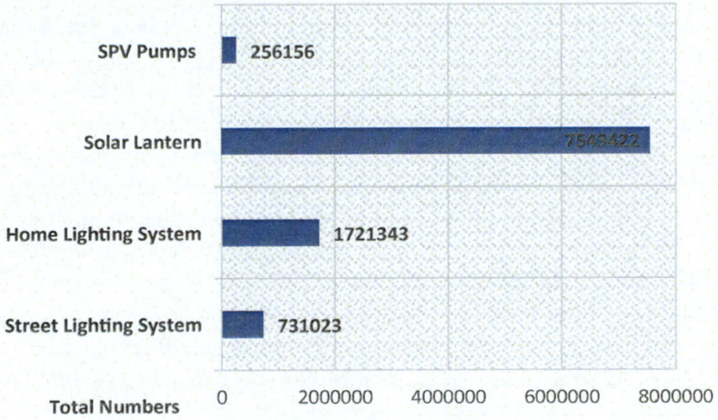

Fig. 6 Cumulative systems installed up to 30 June 2020 [26]

5 Ranking of India's Solar Capacity

There has been an exponential rise in the last decade in the renewal energy installation across the globe. Figure 7 shows the cumulative solar power installation capacity of countries across the globe.

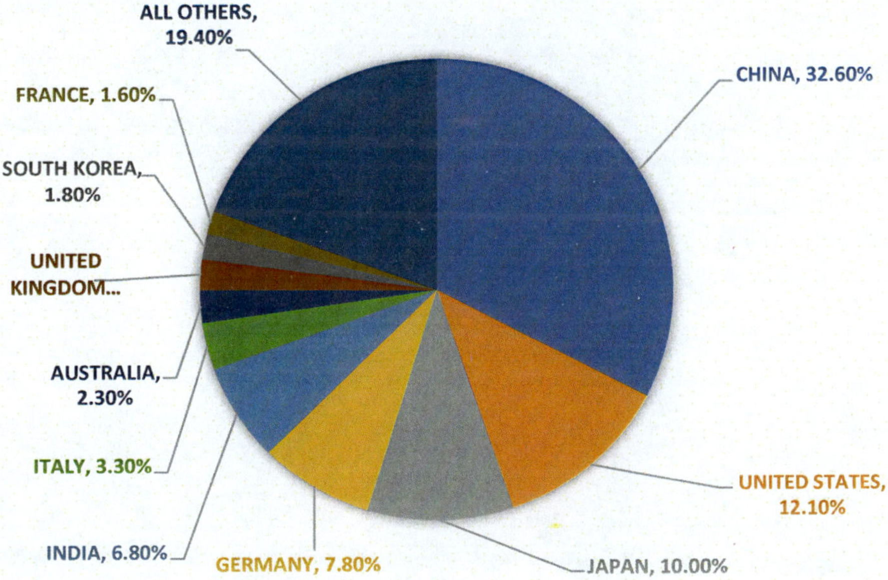

Fig. 7 Country-wise solar power installed capacity [29]

Worlds' global solar installed capacity is 673000MW, where India ranks fifth. India's total Solar Photovoltaic Capacity contributes approximately 6.8% to global renewable solar energy installed capacity. China contributes 32.6% of the global renewable energy installed capacity; its contribution is with 204700 MW. The United States, Japan, Germany, Italy, Australia, United Kingdom, South Korea, France and all other countries stand at second, third, fourth, sixth, seventh, eighth, ninth, tenth and so on, respectively, positions for solar power installed capacity in the world with 75900 MW, 63000 MW, 49200 MW, 20800 MW, 14600 MW, 13300 MW, 11200 MW, 9900 MW and 121600 MW, respectively.

India has been making consistent and sincere efforts to meet the ambitious 2022 targets. The installed capacity of solar power is close to half of its target; it stands at 35122.30 MW as of 30 June 2020. The target is expected to be attained by the ongoing installation and expansion programmes.

Most of the Indian states have significant solar insolation throughout the whole year, which include Karnataka, Andhra Pradesh, Gujarat, Tamil Nadu, Rajasthan, Maharashtra, Madhya Pradesh, Punjab, Telangana, Uttar Pradesh, etc. Figure 8 shows the installed solar projects state wise. Both the central and state governments are

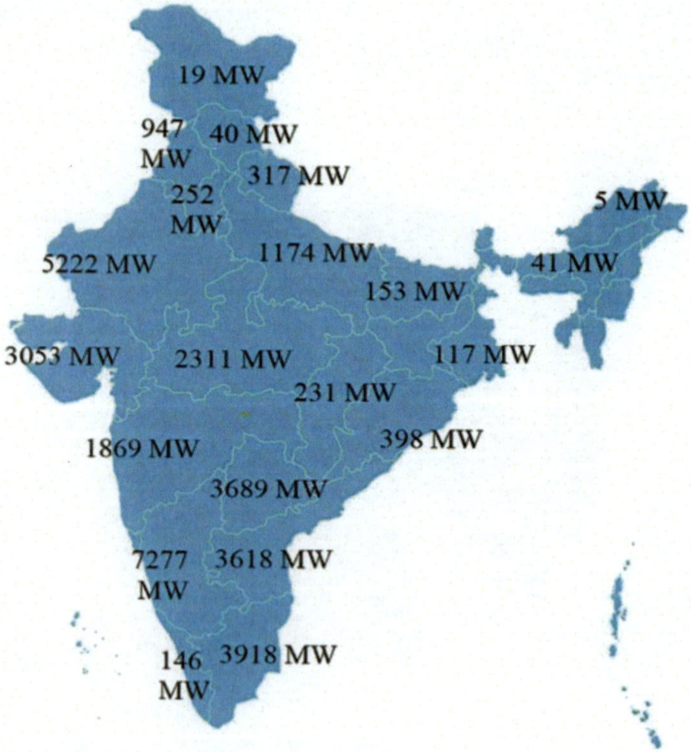

Fig. 8 State-wise status of Solar power installed capacity in India [27]

working towards installing more and more solar projects and providing subsidy where necessary.

5.1 Solar Power Progress of India in 2020–2021

As per records of MNRE, scheme-wise solar progress is given in Table 3. Grid interactive or Grid-connected and off-grid are the two types of solar power productions.

Grid interactive solar power is further categorised as Solar Power–ground mounted and Solar Power–roof top. The capacity of the ground mounted and rooftop type is 32305.15 MW and 2817.15 MW, respectively. Off-grid SPV systems contribute 980.84 MW.

6 Initiatives by Government of India for and Promotion of Renewable Energy

To cater for all the aspects of new and renewable energy, the Ministry of Power of Government of India(GOI) created a new department in the year 1982, naming it as the Department of Non-Conventional Energy Sources (DNES). Looking into the requirement for more efforts in the direction of renewable energy, in 1992, this department was upgraded into a separate Ministry and was called the Ministry of Non-Conventional Energy Sources (MNES). In October 2006, it was renamed as Ministry of New and Renewable Energy (MNRE).

Table 3 Scheme wise solar physical progress in 2020–2021 [17]

Type of solar power scheme	Target	Achievement in FY- 2020–21	
		During April-June 2020	Cumulative as on 30.6.2020)
I. Grid interactive solar power (Capacities in MWp)			
Ground Mounted	9000.00	192.66	32,305.15
Roof Top	2000.00	301.85	2817.15
II. Off-grid/captive solar power (Capacities in MWEQ)			
SPV Systems	500.00	2.45	980.84
TOTAL	11,500	496.96	36,105.14

6.1 Institutions Under the Ministry of New and Renewable Energy

The MNRE set up five institutions of excellence to promote renewable energy, three of these institutions are autonomous bodies and two public sector undertakings. The Autonomous institutes are National Institute of Solar Energy (NISE) at Gurugram in Haryana, National Institute of Wind Energy (NIWE) at Chennai in Tamil Nadu and National Institute of Bio Energy (NIBE) in Kapurthala in Punjab. The two public sector undertakings (PSU) are Indian Renewable Energy Development Agency (IREDA) and Solar Energy Corporation of India (SECI) in Delhi. The autonomous institutes undertake activities related to research and development, standardization, certification, testing, skill development, resource assessment and awareness in their respective energy source and associated technologies. The Ministry established a non-banking financial institution under the administrative control named IREDA, the objective of this PSU is to provide term loans for renewable energy and energy efficiency projects for all the renewable energy sources. SECI comes under the Companies Act—Sect. 3, it is the executing branch of the Ministry for implementation of solar and wind energy projects [26].

6.2 Schemes Under GOI to Boost RE Sector

For sustained industrial growth in specific and national development in general, the power and energy sector are the key sectors for any country. GOI has also recognised the power sector for sustained industrial growth. The initiatives taken by GOI for the improvement of renewable energy contribution to the power sector have been mentioned below:

A. **National Thermal Power Corporation Limited (NTPC)**

Through NTPC, VidyutVyapar Nigam Limited (NVVN) under Phase-I implemented a total of 1GW solar power, this power comes under projects that are grid interactive. These large solar power plants are connected to grid at 33 kV and above voltages. These comprise 500 MW capacity Solar Photovoltaic (SPV) and Solar Thermal (ST) technologies. The National Solar Mission (NSM) is amongst the GOI initiatives under National Action Plan on Climate Change, which has three stages. The first stage is Migration Scheme, the second is NSM Phase-I, Batch-I and the third is NSM Phase-I, Batch-II.

So as to give a quick start to the NSM, and also to have a swift and smooth implementation of the then on-going projects of solar power that were in the implementation stage in different States, in Feb 2010, the government made an initiative to allow the migration of those ongoing projects into NSM. From the then ongoing projects, 16 projects of 54 MW SPV and 30 MW ST capacity were mitigated for long-term procurement of power by NVVN. The Central

Electricity Regulatory Commission (CERC) notified tariff for 2010–11 for the mitigation scheme, which was decided to be 17.91₹ per unit for SPV and 15.31₹ per unit for ST.

B. **National Solar Mission**

The status of NSM Phase-I, Batch-I & Batch-II, provisions, allotment, tariff and commissioning is:

(1) Reverse bidding method is used for the allotment of projects, bids were invited in two batches, namely, NSM Batch-I and nSMBatch-II. Bids for the same were invited in two batches. In August 2010, bids were invited for SPV (150 MW) and ST (470 MW) projects under Batch-I, and in August 2011, bids were invited for SPV (350 MW) projects.

(2) The project capacities for Batch-I were for both SPV and ST, with capacity up to 5 MW and 100 MW, respectively. Allotment was made to 28 SPV projects having a total capacity of 140MW and 7 ST projects with a total capacity of 470MW. The aggregate of combined SPC and ST projects was 610MW. The bids of the power tariff for SPV were at an average of 12.12 ₹ per unit, in the range of 0.95₹ per unit to 2.76 ₹ per unit, For ST projects, bids of the power tariff were at an average of 11.48₹ per unit in the range of 10.49₹ per unit to 12.24₹ per unit. As for the commissioning status of NSM Phase-I, Batch-I, it stands at three ST projects of aggregate 200 MW capacity and 28 SPV projects of aggregate 140 MW capacity.

(3) The project capacities for Batch-II were 5MW to 20MW for SPV. Allotment was made to 27 SPV projects having a total capacity of 340MW. The bid tariff was at an average of 8.77₹ per unit, in the range of 7.49₹ per unit to 9.44₹ per unit. As for the commissioning of the projects, the status of NSM Phase-I, Batch-II is 330MW capacity.

(4) Under the NSM Phase-I, two additional projects were allocated. One of the SPV projects was of 10MW capacity set up by the Solar Corporation of India (SECI). The other SPV project was of 5MW capacity set up by Delhi Mumbai Industrial Corridor Development Corporation Limited.

(5) Under the unbundling scheme of NSM Phase-I, SPV projects of 533MW and SP projects of 200MW have been commissioned.

(6) NVVN purchases power at the decided tariffs from the commissioned plants and sells it to the State Utilities and the distribution companies (DISCOMs). This works under a mechanism of bundling power from unallocated quota of power coal-based stations of NTPC on equal capacity basis. There are 17 thermal plants that are a part of this bundling. The bundling was done so as to effectively reduce the average per unit cost of solar power that the State Utilities and DISCOMs are bound to purchase. A revolving fund of Rs.486 crore has been established as a Payment Security Mechanism ensure timely payments to developers in the event of delays or defaults in payments by State Utilities and DISCOMs in the payments to be made to NVVN.

III. **NTPC State-Specific Bundling Scheme**

In the year 2015, a scheme, namely, State-Specific Bundling Scheme was introduced. This scheme permitted solar power projects to bundle with coal-based power projects. The ratio of solar power to coal-based thermal power for dandling is 2:1. The sole objective to bundle solar and coal projects was to bring down the tariff. The target capacity to be bundled was 3000MW under the scheme. In the year 2018–19, 2750 MW, and in 2019–20, 200 MW has been commissioned. The Kadapa Ultra Mega Solar Park in Andhra Pradesh was commissioned in 2020. The remaining 50MW is ready and will be commissioned in 2020–2021.

IV. **Development of Solar Parks and Ultra Mega Solar Power Projects**

(i) In December 2012, the Scheme for Development of Solar Parks and Ultra Mega Solar Power Projects was rolled out. This scheme has an aggregate capacity of 20,000 MW. Later the capacity of the Solar Park Scheme was doubled and it stood at 40,000 MW in March 2017. The target of this enhancement was to be met with by setting up 50 solar parks by the year 2021–22.

(ii) In general, the capacity of a solar park is 500 MW or above. The projects are implemented only in non-agricultural land. To set up a solar park for each MW capacity roughly 4 to 5 acres of land is required. However, smaller parks with a capacity up to 20 MW are also considered in States/UTs, where there is insufficient land is available. The total central grants approved under the Scheme for Development of Solar Parks and Ultra Mega Solar Power Projects is Rs.8100 crore.

(iii) There is the provision of financial assistance for the solar parks by the SECI. The financial support provided by GOI through the Ministry for the detailed project report of a solar park project is Rs.25 lakh. There is an additional provision of Rs. 20 lakh, of which 60% is earmarked for the development of the internal infrastructure and the remaining 40% is for the development of external power evacuation infrastructure of solar park, otherwise 30% of the project cost, including grid connectivity cost, is provided. The lower amount of the two is given on achieving the milestones prescribed in the scheme.

(iv) Solar Park Scheme has a target to develop at least 50 solar parks. These parks should have an aggregate installed capacity of 40,000 MW of solar power and these must be developed by 2021–22. Up till December 2019, based on the proposals received, an aggregate capacity 22,879 MW has been approved, these are 39 solar parks distributed across 17 States. They are parks that are at different stages of development in different states. As per the status of land for the development of solar parks, 82,600 acres out of 1,31,000 lakh acres of land identified have been acquired. Solar projects of an aggregate capacity of 7767 MW have been commissioned and the details are given in Table 4.

Table 4 Solar Projects commissioned inside Solar Parks till 31 December 2019 [26]

Sr No	Solar Park	Capacity Approved (MW)	Capacity Commissioned (MW)
1	Pavagada SP, KA	2000	2000
2	Kurnool SP, AP	1000	1000
3	Bhadla-III SP, Raj	1000	900
4	Ananthapuamu SP, AP	1500	887
5	Rewa Solar Park, MP	750	735
6	Bhadla-II SP, Raj	680	680
7	Bhadla-IV SP, Raj	500	500
8	Ananthapuamu-II SP, AP	500	400
9	Neemuch-MandsorSP,MP	750	250
10	Kadapa SP, AP	1000	200
11	UP Solar Park, UP	440	165
12	Kasargod SP, Kerala	200	50
	Total	**10,320**	**7767**

E. Scheme for SPV projects under Ministry of Defence

Under Phase-II/III of NSM, with the Viability Gap Funding (VGF), there is a provision of setting up Grid-Connected Solar PV Power Projects by defence establishments. The establishments under Ministry of Defence and Para Military Forces come under this scheme. The proposed renewable energy capacity of 300MW under Ministry of Defence was approved by GOI in January, 2015. Approval for 241 MW has been given to different Defence Organisations, out of which, 53.11% is already commissioned and balance capacities are under implementation stage. Table 5 gives the details of the defence schemes.

Table 5 Present Status of Defence Scheme (as of 31 December 2019) [26]

Sl No	Ministry	Org	Capacity Approved (MW)	Capacity Commissioned (MW)
1	Department of Defence Production (116.5 MW)	OFB, Kolkata	7	7
2		BEL	75.5	62.5
3		BDL	10	10
4		HAL, Nashik	15	15
5		Of, Kanpur	5	5
6		MIDHANI	4	4
7	Department of Defence	DOS/MES	125.45	25
		Total	241.95	128.5

Table 6 Year wise targets under National Solar Mission [26]

Year	Tendering target (MW)
2019–20	30,000
2020–21	30,000

F. **Focus Area under Phase-II of NSM**

Grid interfaced projects come under Phase-II of the NSM. The aim is for accomplishing significantly higher scales of the target of 100 GW by 2022, the details of which are in Table 6. GOI has finalized tendering trajectory so as to achieve the target. Selection of capacity for Phase-II, grid interfaced projects is being done via different schemes such as Viability Gap Funding (VGF), Generation-Based Incentive (GBI) and bundling. GOI has considered all probable choices for enactment; hence, there is provision that the allocation of target capacity may be changed depending upon the resource availability.

G. **Solar Energy Potential and Achievements**

Based upon solar radiation and availability of land, in the country, the potential of solar power has been assessed to be around 750 GWp. In December, 2019, 33,730 MW was the total solar power capacity installed. Tenders of around 22,839 MW were in pipeline for which letter of intent(LOI) has been issued but not commissioned, and for around 28,578 MW, tender issued but LOI yet to be issued.

7 Conclusion

The current scenario up till March 2020, of the solar power sector of India has been assessed. The much distributed, clean and free of cost solar energy has a very low cost of generation and has a potential of reducing the import of fossil fuel for power generation and reducing import dependence. India aims to achieve solar capacity 100000 MW by 2022. The successful achievements of the ambitious targets are through various projects and solar parks for which many new institutes and organisations have been established. This target is being achieved through various initiatives and schemes of both the state and central government. The renewable energy would be contributing about 35% of the energy demand. After making continuous strides towards sustainable development through solar energy, wind energy and small hydropower, now India has identified geothermal energy and the next option for electrical power from renewable energy sources.

References

1. R. Singh, Y.R. Sood, Transmission tariff for restructured Indian power sector with special consideration to promotion of renewable energy sources, in *Proceedings of the IEEE TENCON*, pp. 1–7 (2009)
2. N.K. Sharma, P.K. Tiwari, Y.R. Sood, Environmental friendly solar energy in restructured Indian power sector, in *IET International Conference on Sustainable Energy and Intelligent Systems*, pp. 104–109 (2011)
3. A. Singh, Power sector reform in India: current issues and prospectus. Energy Policy **34**, 2480–2490 (2006)
4. Ministry of power. Available:https://powermin.nic.in/en/content/power-sector-glance- all-india.
5. S.C. Bhattacharya, Chinmoy Jana, "Renewable energy in India: Historicaldevelopments and prospects. ELSEVIER, Energy **34**(8), 981–991 (2009)
6. Electricity domestic consumption, Global Energy Statistical (2015). Available: https://yearbook.enerdata.net/electricity-domestic-consumption-data-byregion.html
7. A. Sulaiman, M.A. Irfan, The techno-economic potential of Saudi Arabia's solar industry. Renew. Sustain. Energy Rev. **55**, 697–702 (2016)
8. N.L. Panwar, S.C. Kaushik, S. Kothari, Role of renewable energy sources in environmental protection: A review. Renew. Sustain. Energy Rev. **15**(3), 1513–1524 (2011)
9. V. Khare, S. Nema, P. Baredar, Status of solar wind renewable energy in India. Renew. Sustain. Energy Rev. **27**, 1–10 (2011)
10. N.K. Sharma, P.K. Tiwari, Y.R. Sood, Solar energy in India: strategies, policies, perspectives and future potential. Renew. Sustain. Energy Rev. **16**, 933–941 (2012)
11. The future of Indian electricity demand-Brookings India (book)-Sahil Ali, (2018). Available:https://www.brookings.edu/wp-content/uploads/2018/10/The-future-of-Indianelectricity-demand.pdf
12. Central Electricity Authority (CEA), 18th EPS (2015)
13. A. Digambar Singh, B. Yog Raj Sood, C. Deepak, Recent Techno-Economic Potential and Development of Solar Energy Sector in India, in *IEEE Region 10 Humanitarian Technology Conference (R10-HTC)*, pp. 246–257 (2019)
14. Central Electricity Authority (CEA) India. Available:https://www.cea.nic.in/reports/others/planning/rpm/Plant-wise%20details%20of%20RE%20Installed%20Capacity-merged.pdf
15. Central Electricity Authority (CEA) India. Available:https://cea.nic.in/reports/monthly/installedcapacity/2020/installed_capacity-03.pdf
16. India 2020–Energy Policy Review by International Energy Agency (IEA). Available: https://niti.gov.in/sites/default/files/2020-01/IEA-India-In-depth-review2020.pdf
17. Ministry of New and Renewable Energy (MNRE), "Scheme wise Physical Progress in 2020–21 & Cumulative upto June, 2020". Available: https://mnre.gov.in/the- ministry/physical-progress.
18. Soft Bank, Bharti and Foxconn to Form "Joint Venture for Renewable Energy in India". Available: https://www.softbank.jp/en/corp/news/press/sb/2015/20150622_01/.June2009
19. Ministry ofNewand Renewable Energy (MNRE), "RenewableEnergy Regulatory Framework". Available:https://mnre.gov.in/information/renewable-energy-regulatoryframework.
20. Central Electricity Authority (CEA) India. Available:https://www.cea.nic.in/reports/monthly/renewable/2020/renewable-03.pdf
21. MNRE, "Loan for Installation of Grid Interactive RooftopSolar PV Plants," MNRE Press-release. Available: https://mnre.gov.in/file-manager/UserFiles/Press-Release-Grid-Interactive-Solar-Rooftop.pdf
22. International Energy Agency (IEA), "Twelfth Five YearPlan (2012–2017) in Planning Commission Governmentof India". Available: https://www.iea.org/policiesandmeasures/pams/india/name-42436-en.php.2013.
23. R. Singh, A. Upadhyay, "Cheap Power or CleanEnergy, India's $200 Billion Dilemma," Bloomberg Business. Available:https://www.bloomberg.com/news/articles/2015-07-29/cheap-power-or-clean- energy-india-s-200-billion-dilemma.(2015).

24. Central Electricity Authority (CEA) India. Available: https://niti.gov.in/sites/default/files/2020-01/IEA-India-In-depth-review2020.pdf
25. Government of India, "Union Budget 2015–16". Available: https://www.pppinindia.com/sector-power.php.
26. C.S. Bhattacharyya, An overview of problems and prospects for the Indian power sector. Energy **19**, 795–803 (1999)
27. Ministry of New and Renewable Energy (MNRE), "Annual Report 2019–2020". Available: https://mnre.gov.in/img/documents/uploads/file_f-1585710569965.pdf.
28. Ministry of New and Renewable Energy (MNRE), "Scheme wise Physical Progress in 2020–21 & Cumulative upto June, 2020". Available: https://mnre.gov.in/img/documents/uploads/file_s-1594347424972.xlsx.
29. Ministry of New and Renewable Energy (MNRE), "Scheme wise Physical Progress in 2020–21 June, (2020). Available: https://mnre.gov.in/img/documents/uploads/file_s-1594347482065.xlsx.
30. Country wise cumulative solar power installed capacity, "International Energy Agency". Available:https://iea-pvps.org/wpcontent/uploads/2020/04/IEA_PVPS_Snapshot_2020.pdf

A Short Review and Investigate Study on Performance of Magneto-Hydrodynamic Using High Reynolds Numbers

Kiran Kumar Namala and V. Bala Murali Krishna

1 Introduction

Magneto-hydrodynamic generator is one of the best options in the electric sector to convert the heat energy and kinetic energy into electricity due to advantages like, operate at high temperatures without wearing the components of the system, unlike conventional generators. The MHD generator has developed to increase overall efficiency, as an auxiliary cycle in thermal and nuclear power stations. The basic block diagram of MHD power generation is shown in Fig. 1 [1]. The classical plasma MHD system is an open-cycle process, and the working fluid passes through the generator and an exhaust [2–4]. The major limitation to work with the plasma MHD system is, it requires a high working temperature, and the working gases need to be ionized at temperatures above 2,000 K [5–7]. High temperatures are not suitable for small-scale and domestic power applications. Moreover, the plasma gases easily contaminate the environment, and the radiation from the plasma is a threat to the health [8–13]. Applications of MHD in areas like marine wave energy, fusion reactors, pump applications, space applications, etc. are well reported in the literature [14–26].

The liquid metal MHD systems work at lower temperature gradients [9] as compared with the plasma systems. The Lead–Bismuth alloy system and other like systems operate at lower temperatures with the conductivities of electric components of few liquid metals that have magnitude more significant than the plasma MHD system. The solar-based thermal systems, fossil fuels, and geothermal energy, etc., are used as resources for making compact liquid metal MHD generators for low magnetic field systems. Harada et al. [15] have proposed a small-scale MHD power

K. K. Namala
Department of Mechanical Engineering, V.N.R.V.J.I.E.T, 500090 Hyderabad, India
e-mail: kirankumar.iitd@gmail.com

V. B. M. Krishna (✉)
Department of Electrical Engineering, Central University of Karnataka, 585367 Gulbarga, India
e-mail: muralikrishna.cuk@gmail.com

S. Vadhera et al. (eds.), *Latest Trends in Renewable Energy Technologies*, Lecture Notes in Electrical Engineering 760, https://doi.org/10.1007/978-981-16-1186-5_31

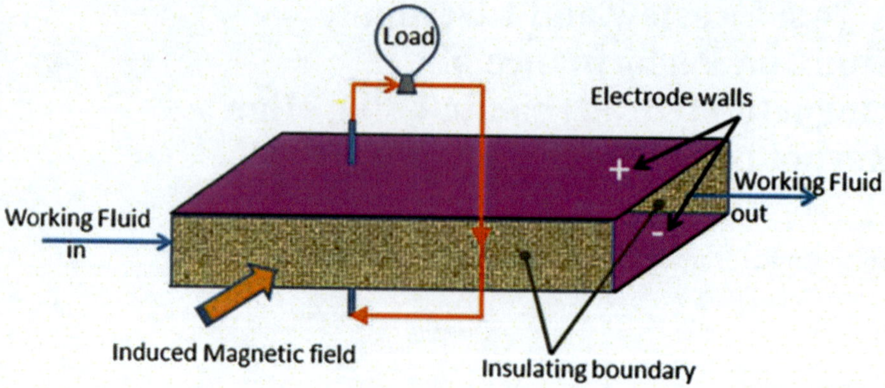

Fig. 1 Basic block diagram of MHD power generation

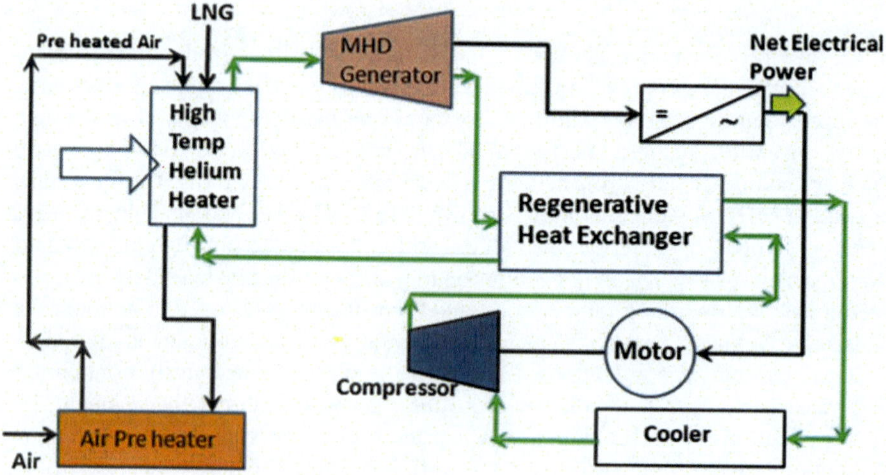

Fig. 2 Energy Re-circulating type MHD single system with LNG as a heat source

generator system by using a low-melting-point Gallium alloy as a working fluid to generate the electric power. The Gallium alloy consists of 68.05% of Gallium, 21.05% of Indium, and 10.00% of Tin. The melting point of Gallium is approximately 19 °C, and high electric conductivity of ~2.3 × 106 S/m [13–15]. Under the low Reynolds numbers, the generators show reliability and achieve maximum power-generation efficiency.

Power generating station combined with the MHD generator will have a high potential of overall efficiency is around 60–65%, which is much enhanced than the ability of conventional thermal power generation station of approximately 30–35%. In the MHD generator, the output electric power per cubic meter channel volume is directly proportional to the square of the strength of the magnetic field and the square of gas velocity and gas conductivity through which the gas flows. The working of the MHD generator usually depends on other issues like generator efficiency, economics, toxic products, etc.

The current research focuses on the review of the importance of MHD power generation, its working principle, the background of development, and the analytical study on the performance of the small-scale MHD power generator using a low-melting-point Gallium alloy. The various phases of MHD generation and their effect on power generation are also highlighted. The analytical result of the study presents the Gibbs energy of different liquid aspects of the liquid metal MHD system.

The organization of the paper is as follows. An introduction to various MHD power generation methods, elements of MHD power generation system, and applications are briefly discussed in Sect. 1. Trends in the MHD power generation system are discussed in Sect. 2. Results and discussions are given in Sect. 3. The conclusion is given in Sect. 4.

2 Trends in MHD Power Generation

This section deals with the various trends of electric power generation methods of MHD and also highlights the working principles with schematic diagrams.

2.1 *Energy Re-circulating LNG/MHD System*

The MHD power generation system by energy re-circulating type with LNG heat source is presented and which is shown in Fig. 3 [15]. The closed-cycle MHD single system does not combine with any other systems and plant efficiency is more than 60% for an enthalpy extraction ratio of 30%. The thermal input to the MHD generator and the corresponding output is less despite heat recovery by the re-generator. The enthalpy extraction ratio is attained with a shock tube facility experimentally. Hence the estimation of efficiency was a realistic reflection [14–16]. Helium is a working medium for re-circulating in High-Temperature Gas-Cooled Reactor (HTGR) this system. This plant efficiency is 47%, as compared with the BWR/steam-turbine system was 35%. The notable increase in efficiency results is more than 25% of nuclear fuel consumption. The increase in operating temperature is very significant in power generation.

2.2 *Gas Turbine/Energy Re-circulating Nuclear System*

Nuclear fission reactor-based power generation system efficiency needs to be increased to decrease overall Carbon dioxide emission per unit power generation.

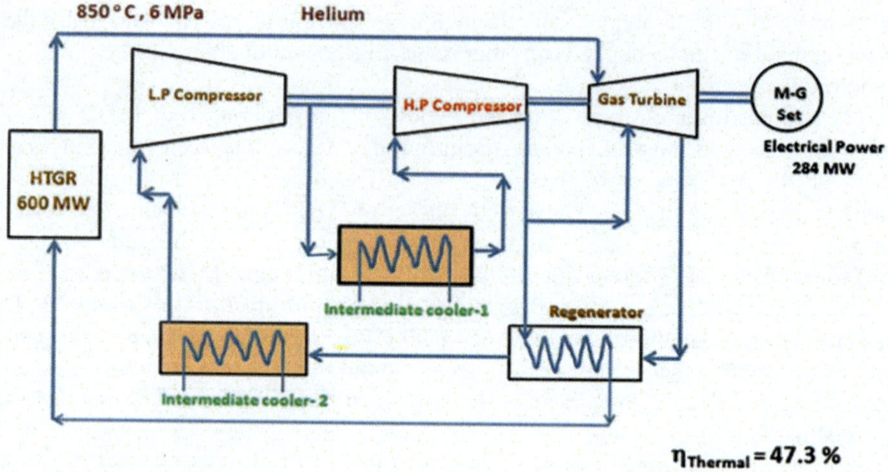

Fig. 3 Energy Re-circulating type gas-turbine single system with a nuclear reactor

The energy re-circulating type gas-turbine single system is recommended with the nuclear reactor. The block diagram of the system is as shown in Fig. 3 [9]. The high-temperature gas-cooled reactor (HTGR) and re-circulating helium are used as working gas in this system. About 47% plant efficiency is anticipated as compared to the BWR/steam-turbine system efficiency of 35%. Approximately a 25% rise in efficiency is achieved by using nuclear fuel. A high working temperature is needed to advance for higher efficiency of the system.

2.3 *Energy Re-circulating Nuclear/MHD System*

The efficiency is relatively low in nuclear/gas-turbine systems, as the maximum operating temperature is 1150 K due to the requirement to develop HTGR is difficult. For improving the efficiency, an MHD generator system is proposed as described in Fig. 4 [9]. As helium (working medium) mixed with xenon is used so that closed-cycle MHD system is connected directly to HTGR. Mixed inert gas (MIG) system was considered for eliminating the complex system of seed injection, mixing and recovery. MIG ionization potential is superior to inert gas as a working medium in seeding with alkali-metal. The electrical conductivity (ionization potential) is not enough at the temperature of the reactor exit 1800 K, hence need to be per-ionized. Simple geometry with fewer electrode connections and simple superconducting magnet structure are elements of a disk-shaped Hall-type MHD generator. Heat exhausted from the generator was collected by a re-generator mounted downstream of the MHD generator to reduce waste heat from the radiation cooler leading to improve plant efficiency. The staged compressor with inter-coolers and radiation cooler are other components of the system [5].

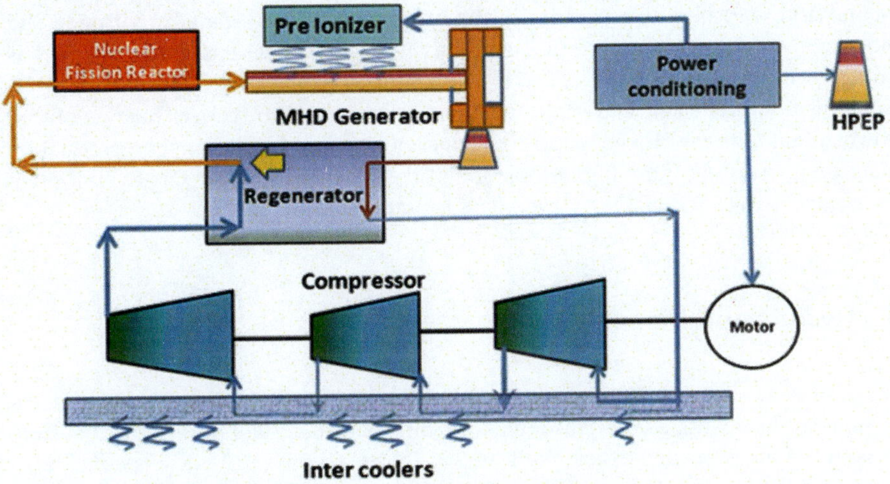

Fig. 4 Energy re-circulating type MHD single system with a nuclear reactor

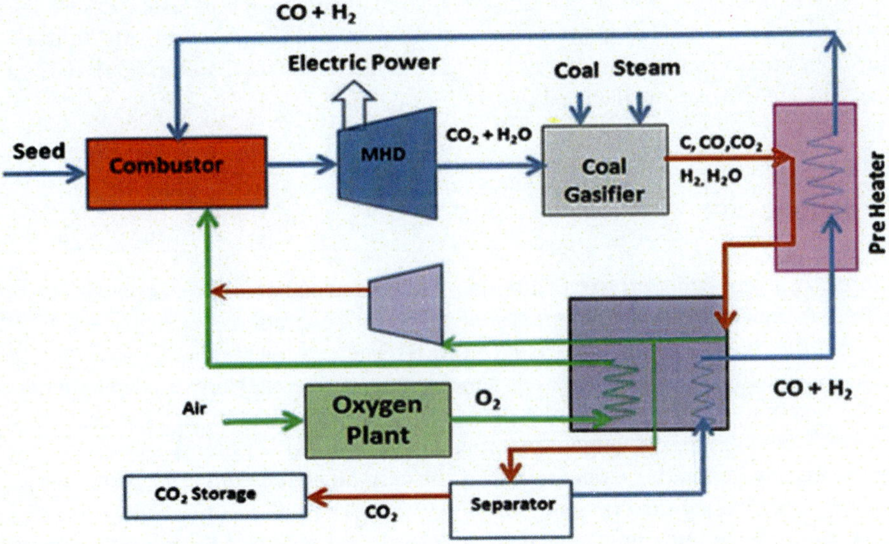

Fig. 5 CO_2 recovery type MHD generator plant

2.4 CO_2 Recovery Type Nuclear/MHD System

Reduction in emissions, particularly of COX, NOX, SOX emissions are the essential requirements to reduce global climate change. The CO_2 recovery type MHD generator plant is shown in Fig. 5 [14]. The coal synthesized gas is a heat source when burnt with pure oxygen [2, 4]. The oxygen separator separates Nitrogen and other elements from air to produce pure oxygen. This MHD generator system was driven by the combustion of CO and H_2 with pure oxygen approximately at 2800 °C. Coal gasifier, preheater, and steam decomposer are used to recover heat as placed downstream to the MHD generator. The energy required for the oxygen production plant be recovered by the high efficiency of the MHD generator. The overall plant efficiency of 50% with CO_2 recovery can be anticipated.

3 Results and Discussions

The performance of MHD power generation using high Reynolds has been investigated in this section. The requirement of different materials in the MHD system is essential in increasing efficiency [26, 27]. The Gallium (Ga) alloy is considered whose operating conditions are most likely for energy generation. In the aspect of its phase diagram is discussed. The micro-structure plays a vital role in deciding its thermal and electromagnetic conducting nature. Hence, the phase diagrams are plotted for the constituents of material considered as shown in Fig. 6a to c describing the microstructure of Gallium (Ga), Indium (In), and Tin (Sn), respectively. Further, the Gibbs energy of the components is given in Fig. 7a to c for Gallium (Ga), Indium (In), and Tin (Sn), respectively (Table 1).

4 Conclusion

The MHD power generation is an auxiliary unit in thermal, and nuclear power plants and it has transformed an efficient system of power generation. Expensive and high temperatures are needed for the plasma type MHD generating system. By using liquid metal as a medium of energy transfer, it increases the efficiency provided and there is the conductivity of element even at lower operating temperatures are not necessary at higher temperatures. The high operating temperature always remains a constraint for the energy generating systems. But if this can be attached to renewable energy sources can enhance the efficiency of the system. The possibility of incorporating MHD power generation in localized usage leads to the eco-friendly solution without any harmful emissions and to reduce the global warming issues.

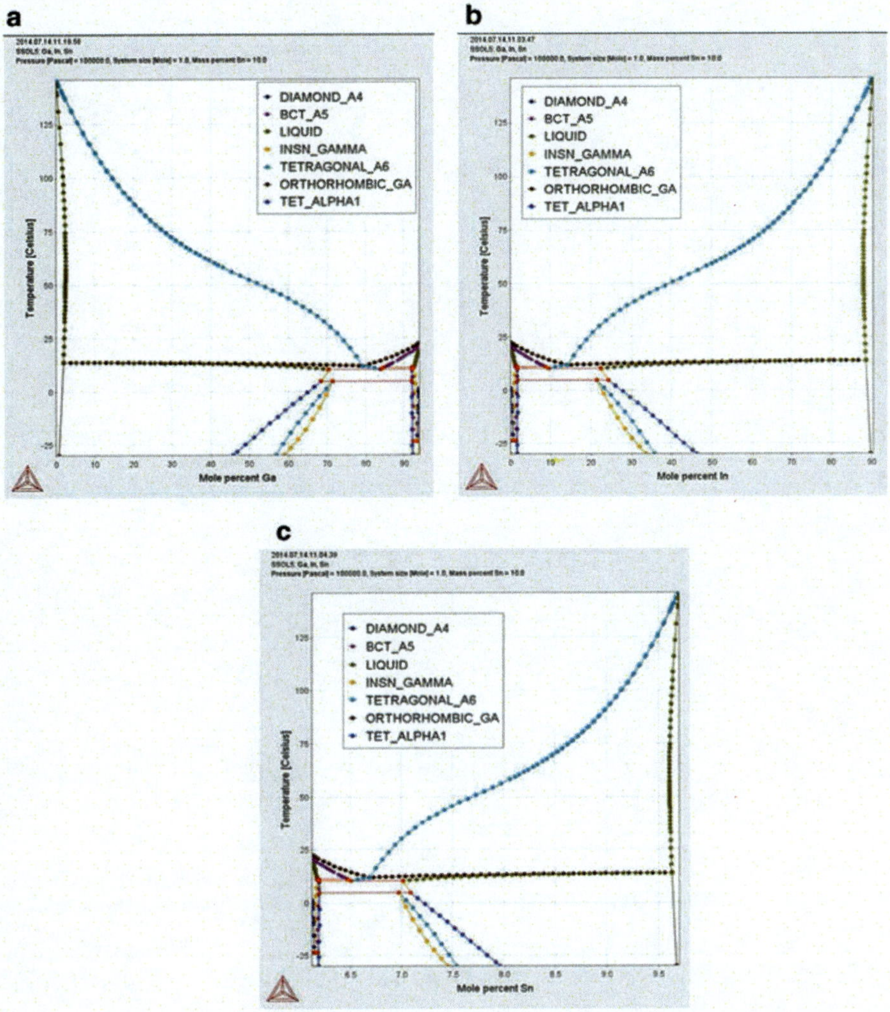

Fig. 6 **a** Phase diagram of Gallium (Ga). **b** Phase diagram of Indium (In). **c** Phase diagram of Tin (Sn)

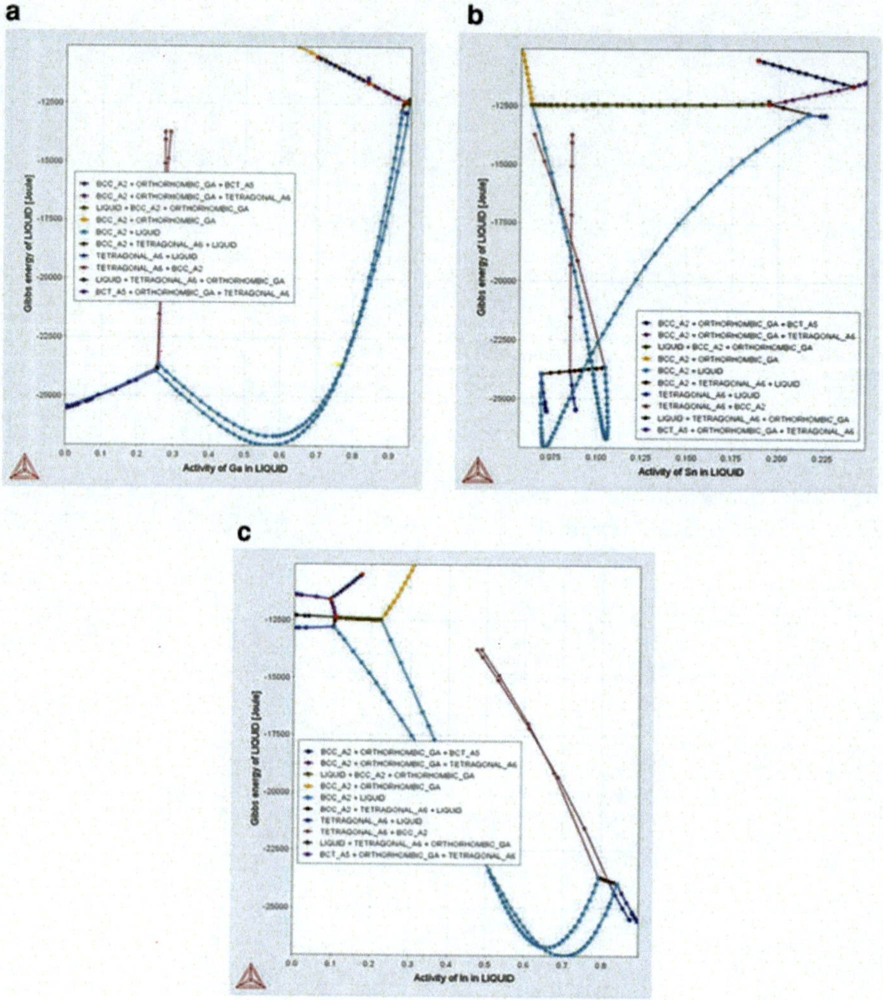

Fig. 7 **a** Gibbs energy of Gallium (Ga). **b** Gibbs energy of Indium (In). **c** Gibbs energy of Tin (Sn)

Table 1 Details of the material at a temperature of 1273 K

Pressure	0.986923267	[Atm]	Moles	1	
Mole fraction Ga	0.783492951		Mass	79.748155	[Gram]
Mole fraction In	0.149328088		Temperature	1273.15	[Kelvin]
Mole fraction Sn	0.067178961		Total GE	−101,502.55	[Joule]
Mass fraction Ga, In, Sn Activity of Ga	0.685, 0.215, 0.1 0.8081137		Enthalpy Component	32,285.166 Activity	[Joule] Potential
Amount of In in LIQUID	0.783492951	[mole]	In	1.15625E-05	−120,334.6
Amount of Sn in LIQUID	0.149328088	[mole]	Ga	0.00012	−95,659.11
Amount of Sn in LIQUID	0.067178961	[mole]	Sn	5.71571E-06	−127,792.7
Chemical potential of Ga	−2255.293102	[J]	GE of System	−101,502.55	[J]
Chemical potential of In	−15,972.26605	[J]	GE of LIQUID	−01,502.55	[J]
Chemical potential of Sn	−26,918.68445	[J]	GE of other phases	−0	[J]

References

1. C. Ambasankaran, Status report on the Indian MHD programme. Indian MHD Power Generation Project, Bhabha Atomic Research Centre
2. P. Haloi, T.K. Gogoi, Exergy modelling of a coal-fired MHD power plant. in *Advances in Applied Mechanical Engineering*, pp. 81–89. Springer, Singapore (2020)
3. P.S. Cicconardi, A. Perna, Performance analysis of integrated systems based on MHD generators. Energy Procedia **45**, 1305–1314 (2014)
4. P. Haloi, T.K. Gogoi, Energy analysis of a coal-fired MHD power plant. IJRTE **8**, 281–285 (2019)
5. S.C. Kaushik, S.S. Verma, A. Chandra, Solar-assisted liquid metal MHD power generation: A state of the art study. Heat Recovery Syst. CHP **15**(7), 675–689 (1995)
6. K. Shimada, S. Kamiyama, Oscillatory pipe flow of a magnetic fluid in a strong magnetic field. J. Magn. Magn. Mater. **122**(1–3), 214–216 (1993)
7. R. Deche, Gas dynamic design optimization of the MHD generator. Int J Energy Res **18**(9), 813–833 (1994)
8. S. Shuchi, K. Shimada, S. Kamiyama, H. Yamaguchi, Hydrodynamic characteristics of steady magnetic fluid flow in a straight tube by taking into account the non-uniform distri- bution of mass concentration. J. Magn. Magn. Mater. **252**, 166–168 (2002)
9. P. Satyamurthy, N. Venkatramani, A.M. Quraishi, A. Mushtaq, Basic design of a prototype liquid metal Magnetohydrodynamic power generator for solar and waste heat. Energy Conversion and Management **40**(9), 913–935(1999)
10. H. Yamaguchi, X.-D. Niu, X.-R. Zhang, Investigation on a low melting-point-Gallium-alloy MHD power generator. Int. J. Energy Res. **35**(3), 209–220 (2011)
11. J.K. Avlyanov, A.S. Zakirov, H.T. Igamberdiev, A. Mavlyanov, A.T. Mamadalimov, P.K. Khabibulaev, Electro- and thermophysical properties of polyaniline. Synth Metals **41**(1–2), 705–709 (1991)
12. P.J. Dellar, Lattice kinetic schemes for Magnetohydrodynamics. J. Comput. Phys. **179**(1), 95–126 (2002)

13. P. Surmann, H. Zeyat, Voltammetric analysis using a self-renewable non-mercury electrode. Anal. Bioanal. Chem. **383**(6), 1009–1013 (2005)
14. R. Ajith Krishnan, B.S. Jinshah, Magnetohydrodynamic Power Generation. Int. J. Sci. Res. Publ. **3**(6), 1–11 (2003)
15. N. Harada, N. Kizuka, T. Okamura, H. Yamasaki, S. Shioda, Improvement of enthalpy extraction over 30% using a disk MHD generator with inlet swirl. Energy Convers. Manage. **36**(5), 355–364 (1995)
16. N. Harada, L.C. Kien, T. Tashiro, Closed cycle MHD generator using He/Xe working plasma. in *33rd Plasmadynamics and Lasers Conference American Institute of Aeronautics and Astronautics*, pp. 1–20. Maui, Hawali (2002)
17. D. Ognerubov, Y. Listratov, V. Sviridov, O. Zikanov, Magnetohydrodynamic heat exchange in next-generation power plants. in *2015 5th International Youth Conference on Energy (IYCE)*, pp. 1–7. Pisa, Italy IEEE (2015)
18. Q. Zhu, Y. li, Marine renewable energy based on the principle of Magneto Hydro Dynamical power generation. in *2019 6th International Conference on Systems and Informatics (ICSAI)*, pp. 1–7, IEEE (2019)
19. Zhengyongshu, Qiuyingzhu, Yongguoli: Literature review of marine renewable energy uti- lization based on Magnetohydrodynamic power generation. In: 2019 6th International Conference on Systems and Informatics (ICSAI), pp. 1594–1599. Shanghai, China IEEE (2019).
20. E. Cosoroaba, B. Fahimi, Magnetohydrodynamics in thermal to electric energy conversion. in *2016 IEEE Conference on Electromagnetic Field Computation (CEFC)*, pp. 1–1. Miami, USA IEEE (2016)
21. D.K. Sarkar, General description of thermal power plants. in *Thermal Power Plant*. pp. 1–31. Elsevier (2017)
22. K.M. Ewis, A New Approach in Differential transformation method with application on MHD flow in non-Darcy medium between porous parallel plates considering hall current. Adv. Water Resour. **143**, 103677 (2020)
23. S.H. Seyedi, B.N. Saray, A.J. Chamkha, Heat and mass transfer investigation of MHD Eyring-Powell flow in a stretching channel with chemical reactions. Physica a **544**(124109), 1–13 (2020)
24. N. Harada, MHD pulsed power generation and applications. in *IEEE Conference Record Abstracts. PPPS-2001 Pulsed Power Plasma Science 2001. 28th IEEE International Conference on Plasma Science and 13th IEEE International Pulsed Power Conference (Cat. No.01CH37255)*, pp. 249. Las Vegas, USA IEEE (2001)
25. O.M. Al-Habahbeh, M. Al-Saqqa, M. Safi, T. Abo Khater, Review of magnetohydrodynamic pump applications. Alexandria Engineering Journal **55**(2), 1347–1358 (2016)
26. X.-D. Niu, H. Yamaguchi, X.-J. Ye, Y. Iwamoto, Characteristics of a MHD power generator using a low-melting-point Gallium alloy. Electr. Eng. **96**(1), 37–43 (2012)
27. H. Yamaguchi, X.-D. Niu, X.-R. Zhang, Investigation on a low-melting-point gallium alloy MHD power generator. Int. J. Energy Res. **35**, 209–220 (2011)

Review of Experimental Study of Carbon Dioxide as Working Fluid Integrated with Phase Change Material in Solar Receiver

Ranjeet Singh and Chandrashekara M.

1 Introduction

Energy scarcity and higher prices of traditional sources of energy are significant threats to human society's economic development. In recent years, several alternatives (solar energy, wind energy, etc.) have been introduced to conventional energy sources. Solar power can provide a multiple of the total current demand for energy. Furthermore, Solar energy is an unreliable form of power, and the energy conversion processes should be used by thermal energy systems. CO_2 has many benefits in heat conveying systems over other working fluids. Owing to its features of zero ozone hole, low global warming, non-toxicity, non-flammability, and inertness, it is an eco-friendly natural fluid and quite healthy to use. The particular collector efficiency will depend on not only the particular procedure of the particular evacuated glass pipe as well because of the motion and heat move nature from the tube's fluid inside. Throughout CO_2 flow through the collector pipe in the suggested Rankine cycle, the CO_2 solution will go through a phase change from a liquid to the supercritical state [1].

The Brayton cycle of CO_2 has attracted tremendous attention recently from investigators creating innovative large efficiency energy cycles. This process was initially tested using nuclear fission as the heat source. Nuclear has driven the CO_2 Brayton cycle and configured the application of various thermal sources to maximize the output of electricity. The CO_2 Brayton cycle provides high efficiencies Compared with other traditional energy cycles (for instance, the air Brayton cycle, the Rankine cycle) at comparatively lower extreme pressure and temperature levels [2]. Also, priority in alternate convertible power systems is increasing to utilize the high reactor outlet temperature successfully. For several decades, the Rankine cycle and gas

R. Singh (✉) · Chandrashekara M.
Department of Mechanical Engineering, NIT Kurukshetra, Kurukshetra, Haryana, India
e-mail: rranjeet1993@gmail.com

Chandrashekara M.
e-mail: Chandru3rvce@gmail.com

S. Vadhera et al. (eds.), *Latest Trends in Renewable Energy Technologies*, Lecture Notes in Electrical Engineering 760, https://doi.org/10.1007/978-981-16-1186-5_32

turbine systems have been used by power plants of enormous size. The Brayton S-CO_2 cycle is an energy conversion system that incorporates the benefit of both the Rankine cycle and the gas turbine system. In other words, in the incompressible area, the fluid is compressed, and the Higher inlet temperature for the turbine can be used with fewer material problems compared to the Rankine cycle. The critical point of CO_2 is 31.98 °C (temperature) and 7.38 Mpa (pressure) [3].

It is accepted that photovoltaic energy can play a vital role in electric power generation given cost concerns, as the thermal source is available and omnipresent. One method for generating Solar-powered energy is the use of the point-focus, power-tower device in which the concentration of solar-thermal energy increases operating temperatures and the associated efficiency. The Brayton cycle of CO_2 has evolved as a potential source for high-performance energy generation. Despite increased involvement in renewable energy, high-efficiency cycles are crucial to maintaining cost-parity despite non-renewable sources [4]. Across developing countries, productivity, whether agricultural, manufacturing or economic, is given considerable importance as a path to affluence. High-grade energy supply and primary electricity are closely related to efficiency. Although demand for the previous continues to increase at a high-speed rate. Climate problems impede huge-scale power generation through the use of widely chosen sources such as hydro. As either a performance, more authorities are converting electricity to sustainable sources of energy. Solar energy is amongst them; the most looked desired [5].

The main function of the CO_2 Brayton process is that after the turbine exhaust, a large amount of heat is regenerated. Therefore, cycle output is heavily dependent on the efficiency of the regenerator and the pressure drop in the heat exchanger's hot and cold channels. Relative to the straight channel PCHE, enhancements in the transfer of heat were observed to be up to four times smaller than the zigzag path structure. Also, the trapezoidal path PCHE's pressure drop was certainly reduced relative to the former. Specific benefits of the S-CO_2 Bryton cycle reflects the lower price relative to helium, higher measured heat transfer performance at constant pressure and atomic weight, and narrower equipment and ease. The main compressor works near the critical point in the S-CO_2 recompression process Brayton while the recompressor acts away from the crucial stage; hence the recompressor function is higher than the main compressor job. The recompressing temperature of the inlet compressor is further kept away from the pseudo-critical region by increasing the temperature of the inlet compressor resulting in increased recompressor work [6, 7].

The growing population and the environmental crisis have dominated the world climate and sustainable development with serious problems. Renewable energy applications (solar and wind power) for generating energy and storing heat are becoming increasingly important, and have gained major criticism. For securing the ozone layer and avoiding greenhouse gases, now there is a strong demand for technology-based on environmentally secure 'ordinary' working fluids, such as water, air, hydrocarbons, ammonia, and CO_2. Improvement of the performance of power plants is key to achieving cost reduction, meeting stringent regulatory requirements, and fighting global warming and climate change. In this scenario, renewable sources of energy that contribute to the energy generation market are among the most effective ways of achieving emission standards. In this context, the concentration of solar power (CSP) is almost relevant among renewable energy sources, as large quantities of energy can be collected at high temperatures at competitive costs. Such facts can minimize the interruption of the process or the transient output of the power plant, but also allow flexibility in the development of electricity dispatch and comprehensive design of the components of power plants that operate under stable state control [8, 9]. Solar energy is the most plentiful resource when considering renewable energy sources, and concentrated solar thermal power technology is domestically defined and provides the chance to combine solar energy production with storage capabilities to address the irregular existence of most resources that are renewable [10].

1.1 Properties of CO_2

1.1.1 Advantage of Carbon Dioxide

a. It is neither costly nor plentiful, neither flammable nor toxic.
b. The temperature is critical close to the ambient temperature, with considerable pressure contributing to acceptable operating conditions.
c. Quite high power density making turbine size ten-times lower than their comparable steam in addition to "switchable" properties of which can be revised with minor strain and temperature adjustments. Higher CO_2 density enables getting more small turbomachinery designs in comparison with the same vapor or air standard power cycles.
d. Air compressor Brayton cycles absorb 1/3 of the energy generated by its turbine, which results in weak cycle performance.
e. Furthermore, in the case of the CO_2 cycle, the compressor can use less than one-seventh of the turbine-generated energy, resulting in higher cycle efficiency [7].

CO_2 is colorless. In low concentrations the particular gas is unsmelling; however, at adequately-high concentrations, it offers a sharp, acid odor. At regular temperature and stress, the density associated with carbon dioxide is about 1. 99 kg/m3, about 1. 68 times that associated with air (Fig. 1). The critical point of CO_2 is temperature 31.1 °C and pressure bar and the triple point is temperature −57.6 °C and pressure 5.18 bar.

1.1.2 Temperature of CO_2 Varying with Specific Heat

In Fig. 2, specific heat increased with increasing temperature.

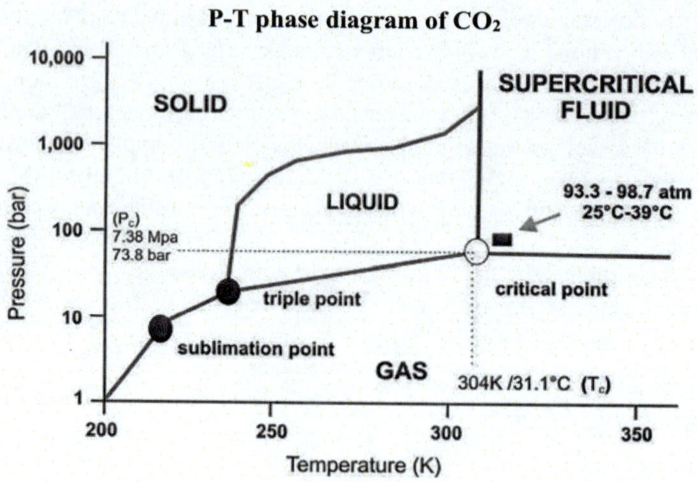

Fig. 1 Schematic diagram of P-T phase diagram of CO_2

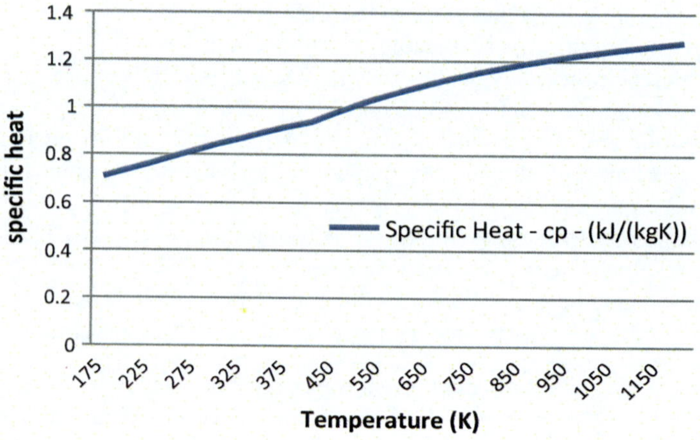

Fig. 2 Schematic diagram of specific heat with temperature

1.1.3 Temperature of CO_2 with Prandtl No. At 1 Bar

In Fig. 3 Prandtl no. of CO_2 decreased with increasing temperature. At 100 bar Prandtl no. first is increased then decreased.

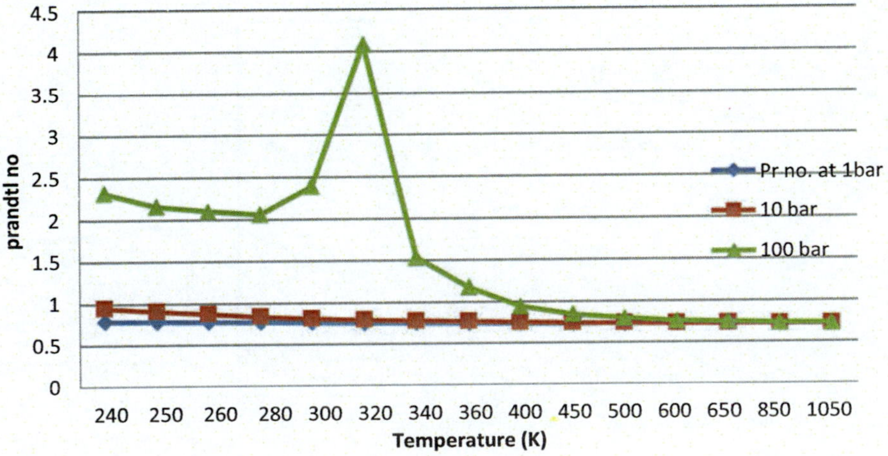

Fig. 3 Schematic diagram of Prandtl no. with temperature

1.1.4 CO_2 Thermophysical Property at 80 Bar

The thermophysical property of CO_2 at 80 bar decreased with increasing temperature (Fig. 4).

2 Experimental Setup

Experimental work aims to generate electricity by turbocharger connected with a micro-generator. The experimental setup consists of a receiver, turbocharger,

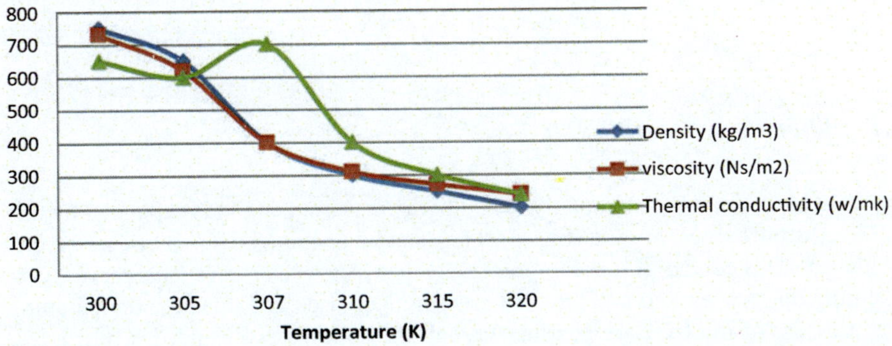

Fig. 4 Schematic diagram of CO_2 thermophysical property with temperature

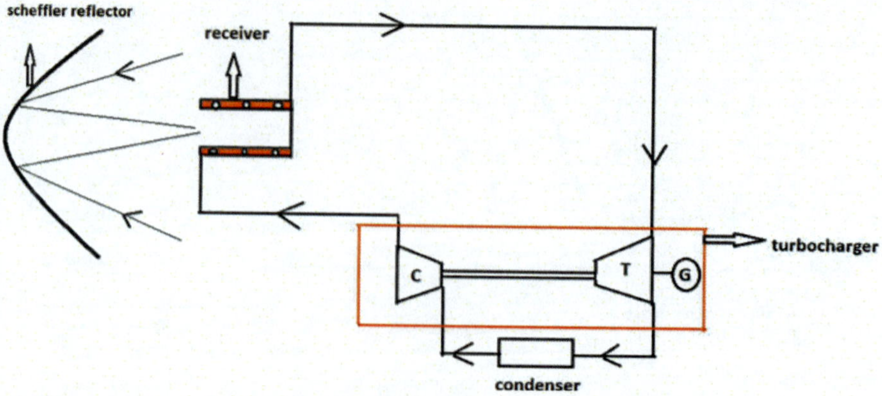

Fig. 5 Brayton cycle integrated with a solar receiver

condenser, Scheffler reflector, phase change material filled inside the receiver, micro-generator connected with a turbocharger, and working fluid use carbon dioxide. First, the receiver is filled by PCM. The PCM is using Acetanilide. The physical properties of PCM like melting point, boiling point, density, and latent heat are 113 °C, 304 °C, 1.219 g/cm^3, and 169.4 (J/kg). The material of the receiver is made of aluminum. The temperature of the receiver is measure by K-type thermocouple. And copper tube wounded around the receiver surface. The sun rays are reflected from the Scheffler reflector and incident on the receiver, which is filled with PCM. During this process, PCM stored solar energy in the form of thermal energy; this process is known as the charging process. Working fluid carbon dioxide gained the heat from the receiver; during this process temperature and pressure of the carbon dioxide increased up to 130 bar and lower pressure 75 bar. The pressure gauge was installed between the turbocharger and condenser. And K-type thermocouple installed between the receiver and condenser. The high pressure and temperature of the CO_2 drive the turbocharger, and this turbocharger connected with the micro-generator generates the electricity (Fig. 5).

2.1 *Thermodynamic Cycle*

In this cycle, CO_2 at high temperature and high pressure (state 1) is expanded by the turbine (turbocharger) at state (Fig. 6). Then state 2 to state 3, CO_2 is cooled by the condenser. This cooled CO_2 is compressed by the Compressor (turbocharger) at state 4, during this process temperature and pressure increased and this high temperature and pressure of the CO_2 are passing through the solar receiver and this process is repeated (Table 1). The efficiency of the CO_2 Brayton cycle is calculated as follows:

Solar power available in cavity receiver

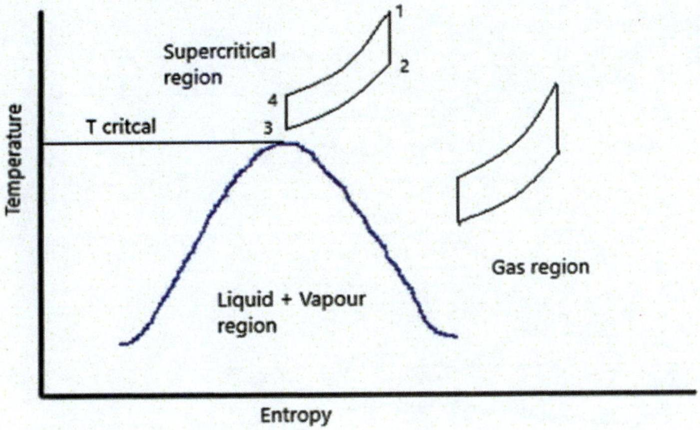

Fig. 6 T-S diagram of CO_2 Brayton cycle

$$Q^* = Q_{\text{loss cond}} + Q_{\text{loss conv}} + Q_{\text{loss rad}} + Q_{\text{net}}$$

The outlet temperature of a receiver

$$Te = Ts - (Ts - Ti)^{-h} rec^{As/mc_p}$$

Heat transfer rate

$$Q_{\text{net}} = h_{\text{rec}} A_s \frac{(Ti - Te)}{\ln\left[\frac{Ts-Te}{Ts-Ti}\right]}$$

Heat transfer coefficient

$$h_{rec} = \frac{kNu}{d} = \frac{0.023k\text{Re}^{0.8}\,\text{Pr}^{0.4}}{d}$$

The efficiency of Brayton cycle

$$\eta = \frac{Wnet}{Qnet}$$

2.1.1 Pressure Ratio Varying with a Mass Flow Rate of CO_2

In Fig. 7, the pressure ratio first is increased with a mass flow rate and then decreased.

Table 1 Input parameter used of the proposed model

Parameter	Value	Reference
Solar intensity	900 w/m^2	[5]
Turbine inlet temperature	500 °C	[5]
Cycle high pressure	130 bar	[1]
Pressure ratio	1.73	[1]
Ambient temperature	35 °C	
Turbine outlet pressure	72–75 bar	[1]
Compressor inlet pressure	30 °C	[5]

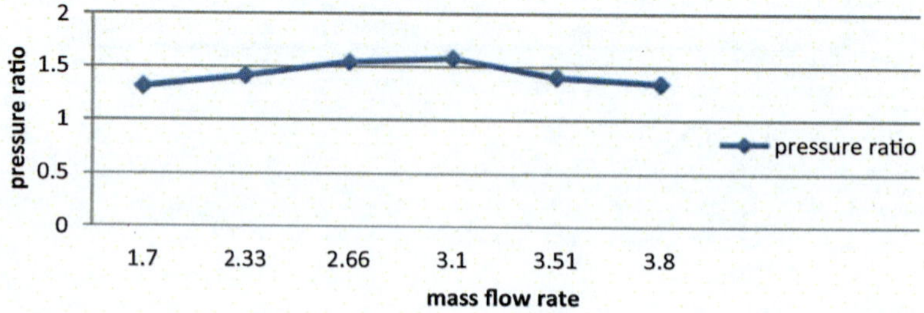

Fig. 7 Schematic diagram of pressure ratio with mass flow rate

2.1.2 Thermal Efficiency Varying with Turbine Inlet Temperature

In Fig. 8 thermal efficiency increased with turbine inlet temperature.

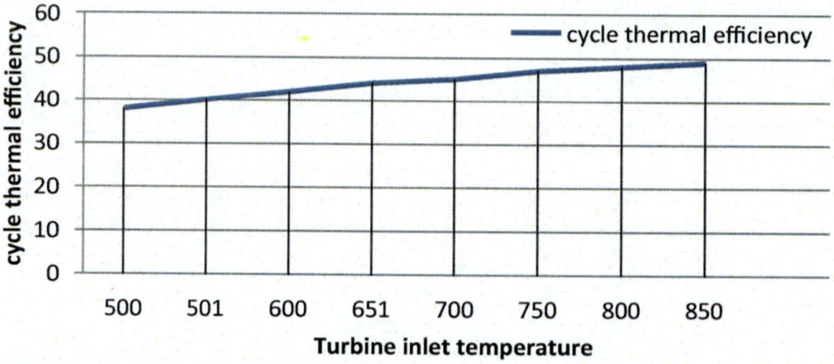

Fig. 8 Schematic diagram of thermal efficiency with turbine inlet temperature

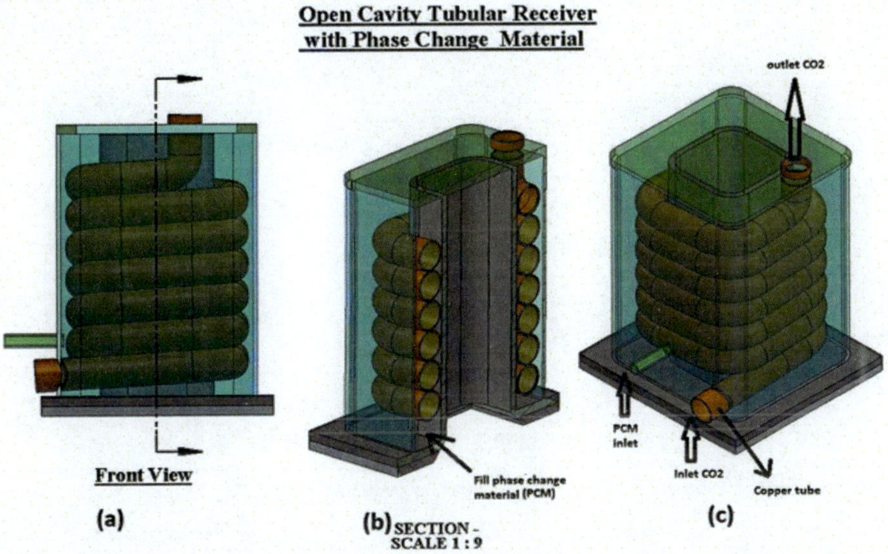

Fig. 9 Schematic diagram of open cavity tubular receiver with phase change material

3 Concept of PCM Integrated with a Solar Receiver

3.1 Solar Receiver

In Fig. 9c, the Solar receiver consists of an aluminum sheet, and in this sheet, a copper tube wounded inside the solar receiver framework and around the particular open cavity coiled solar tubular recipient has installed the compact, high heat conductivity phase to modify thermal storage materials.

In Fig. 9b, typically, the phase changes the substance of high conductivity suggested as a new thermal storage substance, which was attached in an available cavity tubular device across the coiled line, which is in line with insulation for the whole solar receiver. Working fluid is used as carbon dioxide, which is flowing inside the coiled tube. Figure 10b shows the dimension of the solar receiver. The height of the receiver is 300 mm, length 270 mm, width 270 mm, and aperture area 150 × 150 mm^2. The thickness of the copper tube is 1 mm, and the tube diameter is 10 mm. The insulation thickness of the receiver is 20 mm, and the wall thickness of the receiver 1 mm. One thermocouple is provided on the inner surface of the receiver. The outlet of a receiver is connected with the turbine inlet (turbocharger), and the receiver inlet is connected to the compressor outlet (turbocharger) (Tables 2 and 3).

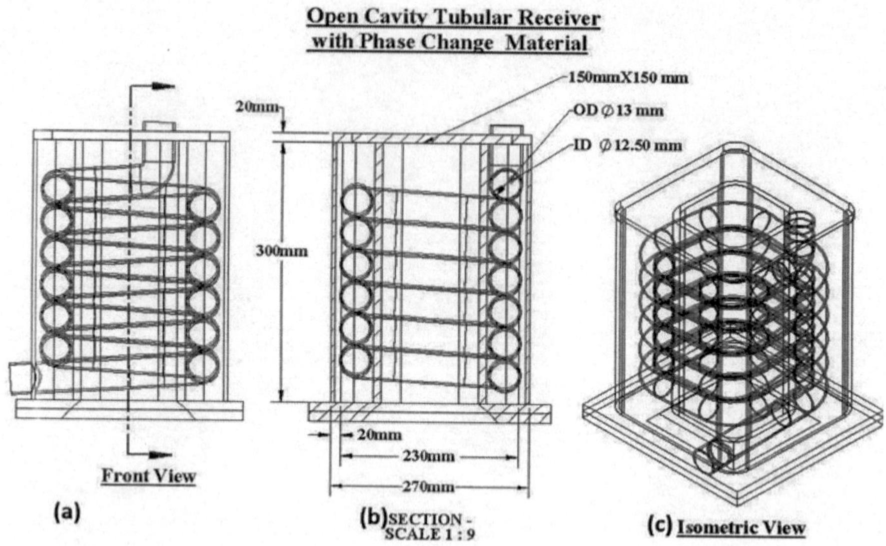

Fig. 10 Schematic diagram of open cavity tubular receiver dimension

Table 2 Open cavity tubular receiver dimension

Constant	Value (mm)
Tube diameter	10
Tube thickness	01
Aperture area (a)	150 * 150 (mm2)
Height	300
Width	270
Insulation thickness	20
Wall thickness of the receiver	01

Table 3 Receiver Material

Part name	Material
Tube	Copper (thermal conductivity −386 W/mK)
Receiver	Aluminum
Phase change material	Acetanilide (melting point −113 °C and latent heat (J/Kg) −169.4)
Insulation	Mineral wool

3.1.1 Phase Change Material

In this experiment, PCM was used as acetanilide. Acetanilide is non-paraffin's type of PCM. PCM has stored energy in the type of latent heat energy. This material is an

Table 4 Property of PCM

Properties	Values
Melting point	113 °C
Latent heat	169.4 (J/Kg)
Boiling point	304 °C
Density	1.219 g/cm^3
Specific gravity	1.214

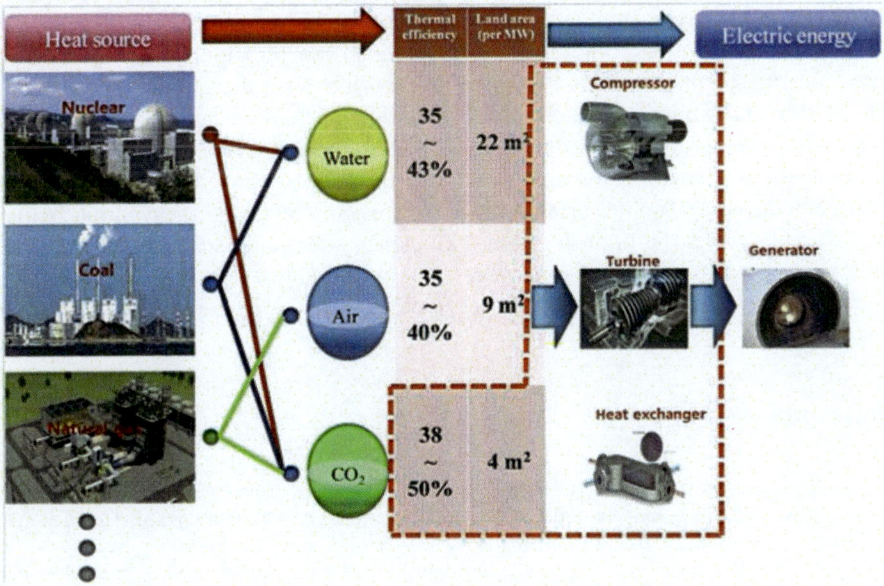

Fig. 11 Schematic diagram of CO_2 Brayton cycle with other power cycle

odorless compound. Acetanilide was soluble in acetone, ethanol, Benzene. It is also soluble in water. The physical and chemical properties are listed in Table 4.

4 Comparison of CO_2 Brayton Cycle with Other Power Cycles

In Fig. 11, the efficiency of the CO_2 power cycle is more than other power cycles and required less space than the steam and airpower cycle. The CO_2 Brayton cycle with recompression intercooling with reheat obtain maximum efficiency (55.3%). Furthermore, in the case of the CO_2 cycle, the compressor can use less than one-seventh of the turbine-generated energy, resulting in higher cycle efficiency.

5 Summary

The CO_2 Brayton cycle can achieve higher efficiency than other power conversion systems. The advantage of the CO_2 cycle is the small size of the overall system. The CO_2 Brayton cycle has a cost advantage because of the high production quantities of turbochargers in the vehicle industry. The main objective of the present work is to used solar energy for power production. Acetanilide is using as a PCM in a solar receiver which is gains thermal energy from the sun rays. The melting point of PCM is 113 °C. In the cycle, the lower limit and upper limit of pressure are 75 bar and 130 bar to assemble the compressor (turbocharger) specification. The receiver has been integrated with phase change material for short-term thermal energy storage (16–31 min) to minimize the effect of solar flux fluctuation and to stabilize the temperature inlet section. Scheffler reflector used to be able to concentrate more sun energy within the cavity receiver. Cavity devices are designed to take in more solar energy which can fell on its surface area and have minimal heat losses onto it. In the CO_2 Brayton cycle, the highest thermal efficiency was obtained in recompression intercooling with reheat. As CO_2 power cycle performance depends on the layout configuration, further study on the layout to design a better performing cycle.

References

1. X.D. Niu, H. Yamaguchi, N. Hashitani, "Experimental study of heat transfer characteristics of supercritical CO_2 fluid in collectors of solar Rankine cycle system", Appl. Therm. Eng. **31**(6–7), 1279–1285 (2011) ISSN 1359–4311.
2. D.K. Sagar, V. Srinivasan, P. Dutta, "Radiative heating of supercritical carbon dioxide flowing through tubes," Appl. Therm. Eng. **109**, Part B, 871–877 (2016) ISSN 1359–4311
3. Y. Ahn, S.J. Bae, M. Kim, S.K. Cho, S. Baik, J.I. Lee, J.E. Cha, "Review of supercritical CO_2 power cycle technology and current status of research and development", Nucl. Eng. Technol. 47(6), 647–661 (2015) ISSN 1738–5733
4. D.I. Brian, T.M. Conboy, J.J. Pasch, A.M. Kruizenga, "Supercritical CO_2 Brayton cycles for solar-thermal energy," Appl. Energy **111**, 957–970 (2013) ISSN 0306–2619
5. K. Pramod, K. Srinivasan, "Carbon dioxide-based power generation in renewable energy systems," Appl. Therm. Eng. **109** Part B, 831–840 (2016) ISSN 1359–4311
6. M. Ajinkya, A.K. Jaiswal, S.D. Khivsara, J.D. Ortega, C. Ho, R. Bapat, P. Dutta, "Modeling and analysis of a printed circuit heat exchanger for supercritical CO_2 power cycle applications", Appl. Therm. Eng. **109** Part B, 861–870 (2016) ISSN 1359–4311
7. M.A. Reyes-Belmonte, A. Sebastián, M. Romero, J. González-Aguilar, "Optimization of a recompression S-CO_2 cycle for an innovative central receiver solar thermal plant", Energy, **112** 17–27 (2016) ISSN 0360–5442
8. H. Yamaguchi, X.R. Zhang, K. Fujima, M. Enomoto, N. Sawada, "Solar energy powered Rankine cycle using supercritical CO_2", Appl. Therm. Eng. **26**17–18, 2345–2354 (2006) ISSN 1359-4311
9. X.R. Zhang, H. Yamaguchi, An experimental study on evacuated tube solar collector using supercritical CO_2, Appl. Therm. Eng. **28**(10) 1225–1233 (2008) ISSN 1359–4311
10. V.T. Cheang, R.A. Hedderwick, C. McGregor, "Benchmarking supercritical carbon dioxide cycles against steam Rankine cycles for Concentrated Solar Power," Solar Energy, **113** 199–211 (2015) ISSN 0038-092X

Modelling and Simulation of Event-Triggered PI Controller for Linear System Using MATLAB

Aniket Karan Chaudhary, Ashavani Kumar, and Shashi Bhushan Singh

1 Introduction

The Event-Triggered controller (ETC) has recently gained lots of attention in the control system. The Event-Triggered mechanisms also occur in many natural circumstances, such as relay systems [1], biological systems [2]. Nowadays most of the control systems are implemented digitally and the digital control system required (shared) wired or wireless networks to connect the different nodes of the systems, embedded electronics circuit and computer to analyse the system and computation. Due to recent development in the field of computer, shared communication network has to create a new type of resources constrained in the wired and wireless embedded control system [3, 4]. The rising requirement for a shared network in the control systems, increases the significance of concern on energy, communication and computation limitations when designing a feedback control loop. Event-based control has some benefits over traditional periodic control when managing those problems, and it introduces some new kinds of theoretical and practical associated problems within this control technique.

Event-triggered control has lots of potentials to reduce these types of constraining, like data communication between the subsystems during the control action task. The event-based control technique lessens the data exchange between the sensor, controller and actuator in the feedback control loop. These components are communicated only when an event had discovered when the control error exceeded a tolerable bound limit [5]. This principle differs basically from the classical time-triggered periodic control feedback loop. In the traditional control technique, the data are communicated continuously at every sampling instance given by a clock. Hence, in

A. K. Chaudhary (✉) · A. Kumar
Department of Physics, NIT, Kurukshetra, Haryana, India

S. B. Singh
Department of Electrical Engineering, NIT, Kurukshetra, Haryana, India
e-mail: sbsingh@nitkkr.ac.in

S. Vadhera et al. (eds.), *Latest Trends in Renewable Energy Technologies*, Lecture Notes in Electrical Engineering 760, https://doi.org/10.1007/978-981-16-1186-5_33

this type of control scheme, the communication takes place when a small control error occurs, when no information feedback loop is required to satisfy the performance requirements.

The Event-triggered control has several advantages over resource utilisation when compared with the classical time-triggered control. The classical time-based control system utilises more resources and energy. The Time triggered control feedback mechanism is based on the periodically sampling process in the Time triggered. The signals are updated at each sampling instant in the loop.

Time-triggered control is also called a periodic control system or classical control system. Some closed feedback systems are still required periodic control. In this control technique, the feedback loop update signal at every certain time period [6, 7]. In the time-triggered control mechanism, the sensor sampled the plant's state periodically. And the sampled data sent to the controller, and also, these data are periodically transmitted over the feedback loop [8, 9]. So this mechanism utilises more resources.

The PID controllers are the special interest because of their effectiveness and flexibility to address the problems of the process industry. In the process industry and application, constant references and disturbances have to observe [10–13]. Digital communication shared network system open new possibility for this effective and simple controller and also opened new challenges concerning its energy efficiency and design [14, 3].

This paper presents a design method to get an event-based PI controller for the continuous linear time-invariant plant. Based on the Lyapunov theorem condition to make sure that the closed-loop system is asymptotic stable (under the event trigger approach) under the specified set of admissible initial state [15, 16, 7, 17–19]. This procedure allows designing the controller's parameters and event trigger illness to achieve a linear quadratic performance criterion [17, 12, 18]. A numerical example illustrates the application of this control method.

This paper has been organised in the given sequence as follows. In Sect. 1 introduction of the system is provided. Section 2 will consist of a problem statement. Section 3 briefs about the Event-Triggered control strategy. Section 4 illustrated a numerical example based on MATLAB Modelling and Simulation, and the conclusion is given in Sect. 5.

2 Problem Statement

The state-space representation of the continuous-time linear plant are described below:

$$\begin{aligned} \tfrac{d}{dt} x_{\mathscr{P}}(t) &= A_{\mathscr{P}} x_{\mathscr{P}}(t) + B_{\mathscr{P}} U(t) \\ y(t) &= C_{\mathscr{P}} x_{\mathscr{P}}(t) \end{aligned} \tag{1}$$

where $x_{\mathscr{P}}(t) \in \mathbb{R}$ is the state vector of this plant; the real constant matrices $A_{\mathscr{P}}, B_{\mathscr{P}}, C_{\mathscr{P}}$ are appropriate dimensions, $y(t) \in \mathbb{R}$ is the plant output, $U(t) \in \mathbb{R}$ is the plant's input vector, and the plant is controllable and observable. The state-space representation of this continuous PI controller described below:

$$\begin{aligned} \dot{x}_C(t) &= -y(t) \\ U(t) &= K_I x(t) - K_P y(t) \end{aligned} \tag{2}$$

where $\dot{x}_C(t) \in \mathbb{R}$ is the controller's state; $U(t) \in \mathbb{R}$ is the controlled output of the controller and, $K_I \in \mathbb{R}$ and $K_P \in \mathbb{R}$ are the integral gains and proportional gain of the controller, respectively.

We assume that the controller and plant at the different nodes in the network and they're connected via a shared network which creates a closed-loop.

An event triggering technique sample the state of the plant and refresh the control signal when the error breached the bound limit, that is the control action task will remain unchanged between successive sampling instants t_k and t_{k+1}, where $k \in \mathbb{R}$. In the time interval $[t_k, t_{k+1})$ the dynamics of the system can be defined below:

$$\begin{aligned} \tfrac{d}{dt} x_{\mathscr{P}}(t) &= A_{\mathscr{P}} x_{\mathscr{P}}(t) + B_{\mathscr{P}} U(t) \\ y(t) &= C_{\mathscr{P}} x_{\mathscr{P}}(t) \end{aligned}$$

$$\dot{x}_C(t) = -y(t_k) \quad \forall t \in \left[t_k, t_{k+1}\right)$$

$$U(t) = K_I x_C(t) - K_P y(t_k) \quad \forall t \in \left[t_k, t_{k+1}\right) \tag{3}$$

We assume that the plant's (3) states continuously measures but it will be sent the control signal to the controller only when the event occurs which are at another node in the network [15]. For the stability, we can we-write the Eq. (4) From the Eqs. (1) and (2).

$$\begin{aligned} \tfrac{d}{dt} x_{\mathscr{P}}(t) &= A_{\mathscr{P}} x_{\mathscr{P}}(t) + B_{\mathscr{P}}(K_I x_C(t) - K_{\mathscr{P}} C_{\mathscr{P}} x_{\mathscr{P}}(t)) \\ \dot{x}_C(t) &= -C_{\mathscr{P}} x_{\mathscr{P}}(t) \end{aligned} \tag{4}$$

3 Event-Triggered Problem Formulation for State Sampling

We focus on the particular case of the linear time-invariant system. One way of proving an event-triggered control system is to use the input to state stability (ISS) of the perturbed system.

Consider a continuous linear plant and the dynamic of plant:

$$\begin{aligned} \tfrac{d}{dt} x_{\mathcal{P}}(t) &= A_{\mathcal{P}} x_{\mathcal{P}}(t) + B_{\mathcal{P}} U(t) \\ y_{\mathcal{P}} &= C_{\mathcal{P}} x_{\mathcal{P}}(t) \end{aligned} \tag{5}$$

where $A_{\mathcal{P}} \in \mathbb{R}$ is the state matrix of demission of $n \times n$ and $B_{\mathcal{P}} \in \mathbb{R}$ is the input matrix of appropriate dimension, $x_{\mathcal{P}}(t) \in \mathbb{R}^n$ is the state vector where n is the number of states, $U(t) \in \mathbb{R}^m$ is an input vector, and the number of inputs is m to the plant. $C_{\mathcal{P}} \in \mathbb{R}$ is the output matrix of the appropriate dimension.

A linear state feedback control law, $U(t) = Kx(t)$ designed to asymptomatically stabilise the ideal close loop system:

$$\frac{d}{dt} x(t) = (A + BK)x(t) \tag{6}$$

where $K \in \mathbb{R}$ is a controller gain matrix of dimension $m \times n$. The control input is computed at the last sampling instant tk.

$$u(t) = Kx(t_k) \tag{7}$$

The sensor measures the error between two successive update instants.

$$e(t) = x(t_k) - x(t) \tag{8}$$

where, $e(t) \in \mathbb{R}^n$ and the close-loop system,

$$\frac{d}{dt} x(t) = (A + BK)x(t) + BKe(t) \tag{9}$$

where $e(t)$ is considered as a disturbance input to the closed-loop system. The controller has chosen such that the $(A + BK)$ is Hurwitz. The event t_k, $k \in N$ are determined by the triggering condition.

The Lyapunov input to output stability theorem (ISS). The closed-loop system is ISS with respect to the measurement error $e(t)$ if:

$$\begin{aligned} t_o &= 0 \\ \|e(t)\| &\geq \rho \|x(t)\| \end{aligned} \tag{10}$$

where $\|\cdot\|$ represent the Euclidean norm. And $\rho = \frac{a\gamma}{b}$ is the eigenvalue of Q, $b = 2\|PBK\|$ and $\gamma \epsilon (0, 1]$

4 Stability Criteria Using Lyapunov Criterion

Let a quadratic Lyapunov function:

$$V = x^T P x \tag{11}$$

where P is a positive definite symmetric matrix of an appropriate dimension, a Lyapunov function is an approach to characterising the closed-loop system's performance and can be defined as:

$$(A + BK)^T P + P(A + BK) = -Q \tag{12}$$

For some symmetric positive definite matrix Q. Which exists if $(A + BK)$ is Hurwitz. Then the derivatives of the Lyapunov quadratic equation:

$$\begin{gathered} \dot{V} = \dot{x}^T P x + x P \dot{x} \\ \dot{V} = \left(x^T (A + BK)^T + e^T (BK)^T P x\right) + x^T P((A + BK)x + BKe) \\ \dot{V} = x^T \left((A + BK)^T P + P(A + BK)\right)x + e^T \left((BK)^T P\right)x + x^T (PBK)e \end{gathered} \tag{13}$$

The right side bounded by:

$$\dot{V} \leq -x^T Q x + 2\|PBK\| \|e\| |x| \tag{14}$$

Using that a symmetric positive matrix Q satisfies:

$$\lambda_{min}(Q)|x|^2 \leq x^T Q x \leq \lambda_{max}(Q)|x|^2 \tag{15}$$

where $\lambda_{min}(Q)$ and $\lambda_{min}(Q)$ represents the minimum eigenvalue and maximum eigenvalue of matrix Q and $\dot{V}$ can be bounded by the following:

$$\begin{aligned} \dot{V} &\leq -a|x|^2 + b|x||e| \\ \dot{V} &= -a|x|(x - a|e|) \end{aligned} \tag{16}$$

where, $a = \lambda_{min}(Q) > 0$ and $b = 2\|PBK\|$.

The function $\dot{V}$ is negative if error is $\left(|x| - \frac{b}{a}|e|\right)$ is negatives and $|x| \neq 0$, Thus if the restarted to satisfy:

$$\begin{aligned} b|e| &\leq a\gamma|x| \\ |e| &= \tfrac{a\gamma}{b}|x| \end{aligned} \tag{17}$$

With $0 < \gamma < 1$, a decrease of Lyapunov function:

$$\dot{V} \leq -(1-\gamma)a|x|^2 \tag{18}$$

It is guaranteed the parameter γ can be used to influence the decrease of the Lyapunov function. The extremes case if $\gamma \rightarrow 0$ would lead to continuous event triggering, which produces the same behaviours as the continuous time-triggered controller, and the other extreme case if $\gamma \rightarrow 1$ would allow the largest event-triggered induced error, but would result in the slow decay of the Lyapunov function and therefore a slow convergence of the state to the origin. The enforce bound $|e| = \frac{a\gamma}{b}|x|$ the event triggering threshold should be selected as $\rho = \frac{a\gamma}{b}$.

5 Event-Triggered Strategy and MATLAB Modelling

In this section, a simulation-based event-trigger strategy algorithm based on the quadratics Lyapunov stability criterion is provided.

The cited in Fig. 1 shows the MATLAB simulation model of the system based on event-based control strategy, which is mentioned in Fig. 2. We assume that, $\delta(t) = x_{\mathscr{P}}(t_k) - x_{\mathscr{P}}(t), x_{\mathscr{P}}(t)$ is the state of the closed-loop system. (shown in [20]). The triggering approach introduced below guarantee that the event-triggered-based control system is stable. The cited Fig. 2 shows the flowchart logic of this event-triggered control strategy.

Where $\|\cdot\|$ represent a Euclidean norm.

$\|\delta(t)\|$ represent the error norm which is calculated from the system state at every sampling instance and ρ_0 is a scalar quantity which selected from the system's dynamics cited in [20].

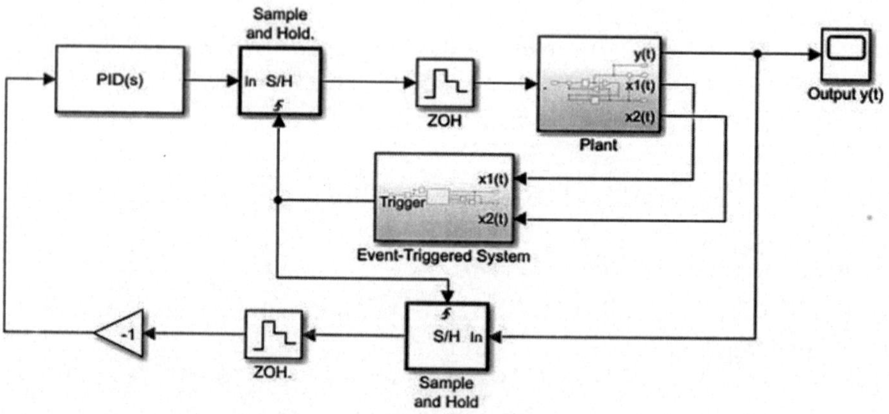

Fig. 1 MATLAB simulation model of event-triggered control

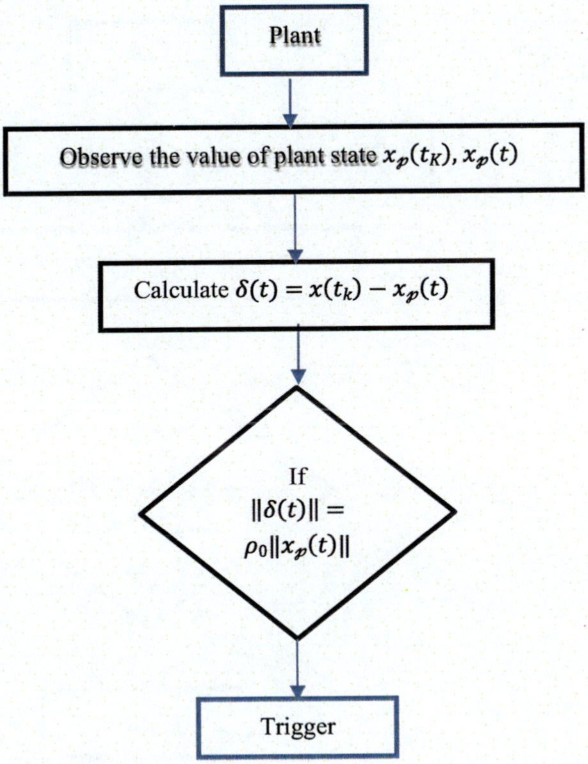

Fig. 2 The flowchart of ETC

This paper uses the following event-triggered strategy, an event-trigger algorithm cited in Fig. 3. Error $\delta(t)$ is measured by the difference between the current sampling and previous sampling of plant state, and it is an array of dimension $(n + 1) \times 1$.

$x(t)$ shown the next sampling state of the plant state and $x_{\mathcal{P}}(t)$ shown the current sampling state. The trigger signal generated only the condition satisfied, which is shown in the algorithm.

Where the Q_x and Q_δ are symmetric and positive definite matrices. The dimension of these matrices is $(n + 1) \times (n + 1)$. Q_x and Q_δ act as a weighted matrix [21, 22]. This matrix plays a similar role as σ in [20]. We choose $Q_\delta = \mu I$ and $Q_x = \sigma I$ as shown in paper [20] and $\sigma_0 = \sigma/\mu$. The large σ_0 it means less sampling activity is expected. $Q_x = Q'_x > 0 \in R$, $Q_\delta = Q'_\delta > 0 \in R$, satisfied the linear matrix inequalities equation given in [23].

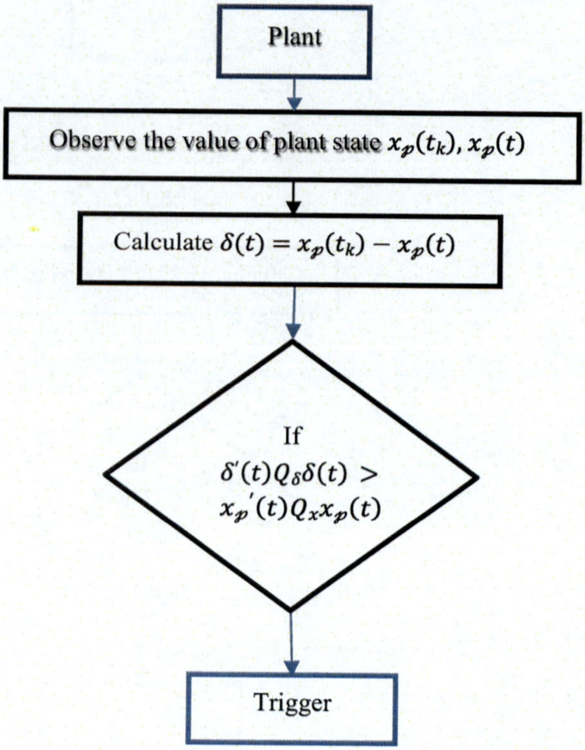

Fig. 3 Flowchart of ETC strategy based on Lyapunov stability

6 Numerical Example

Consider the following unstable plants

$$\frac{d}{dt}x_{\mathscr{P}}(t) = \begin{bmatrix} 0 & 2 \\ 5 & 0 \end{bmatrix} x_{\mathscr{P}}(t) + \begin{bmatrix} 1 \\ 0 \end{bmatrix} U(t)$$
$$y_{\mathscr{P}}(t) = \begin{bmatrix} 1 & 1 \end{bmatrix} x_{\mathscr{P}}(t) \tag{19}$$

The gain of the proportional and integral controllers is respectively $K_I = 22$ and $K_P = 18$ so that the system is stabilised with a PI controller and the closed-loop poles in -14.0657, -0.9111 and -7.0232. The system is controllable and observable (Figs. 4 and 5).

The symmetric positive definite matric Q_x and Q_δ are respectively. Using the quadratic Lyapunov function $V = x'Px$, where P is a symmetric positive definite matrix of appropriate dimension and $P \in \mathscr{R}$, we obtain.

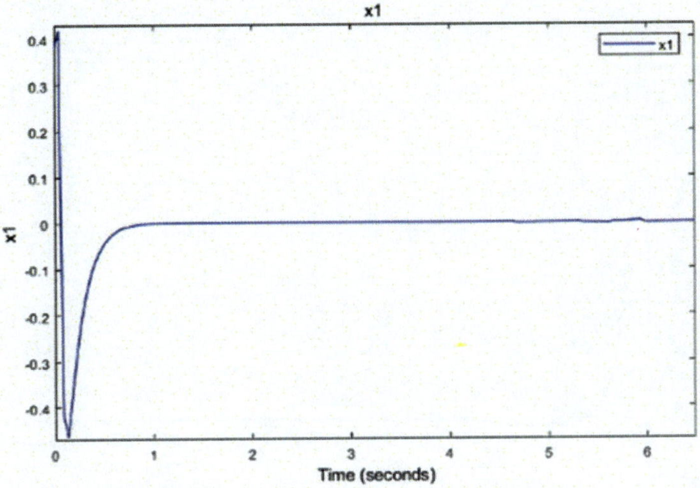

Fig. 4 State of system $x_1(t)$

$$Q_x = \begin{bmatrix} 0.2360 & -0.1015 & -0.05127 \\ -0.1015 & 0.4291 & 0.01756 \\ -0.05127 & 0.01756 & 1.643 \end{bmatrix}$$

$$Q_\delta = \begin{bmatrix} 11.33 & 11.32 & -10.72 \\ 11.32 & 11.33 & -10.72 \\ -10.72 & -10.72 & 12.49 \end{bmatrix}$$

Figure 6 shows the MATLAB simulation results of the system's output $Y(t)$ with initial condition $x(0) = [.\ 38\ .\ 38\ 0]^T$. Figures 4 and 5 show the state of the closed-loop system $x_1(t)$ and $x_2(t)$ respectively and Fig. 7 shows the sampling activity of the event-triggered control system. It shows that Fewer triggers event is needed in the interval of [0, 6.5], and the sampling interval is 45 ms.

7 Conclusion

This paper studied to design an event-triggered-based continuous PI controller using MATLAB simulation. The actuator signal is refreshed when the error goes beyond the threshold limit. The feedback loop works only when the condition is violated. It took less sampling activity.

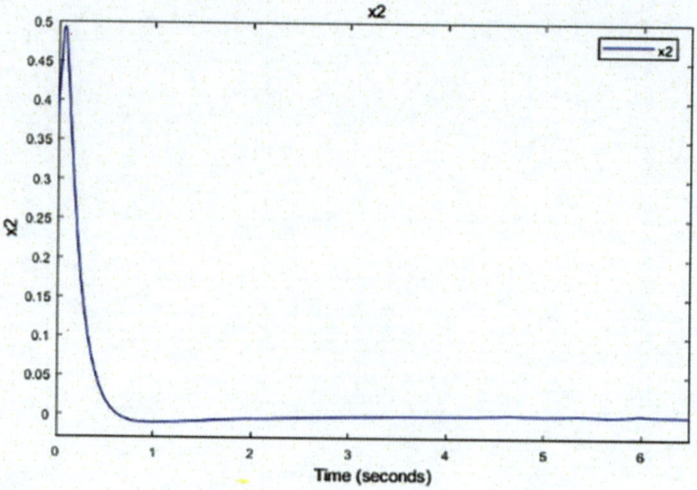

Fig. 5 State of system $x_2(t)$

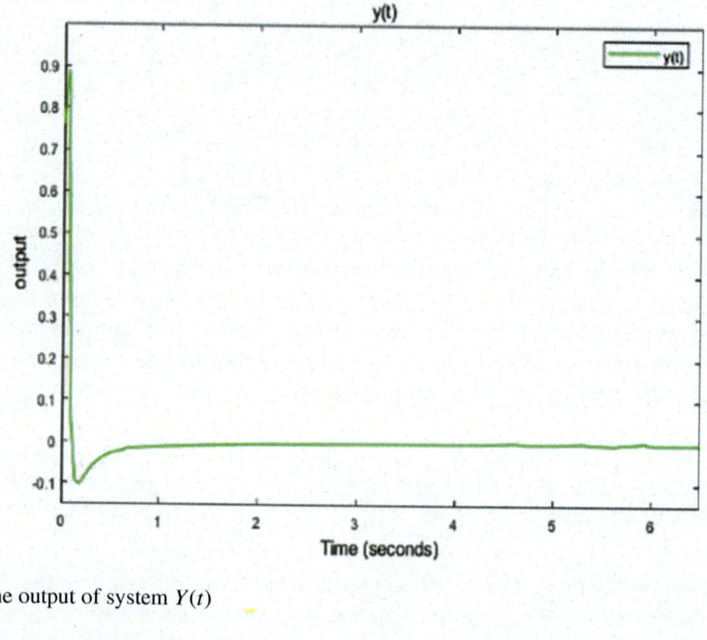

Fig. 6 The output of system $Y(t)$

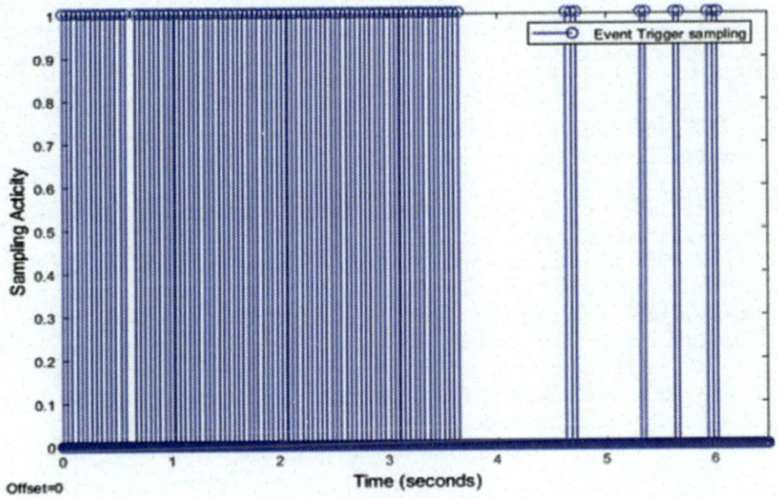

Fig. 7 Sampling of ETC

References

1. S. J. Dodds, "Adaptive, high precision, satellite attitude control for microprocessor implementation." in Automatica, 17:4, 1981, pp. 563–573.
2. J. Keener & J. Sneyd, "Mathematical Physiology II: Systems Physiology" in Springer, Jan. 2009
3. Sylvain Durand and Nicolas Marchand, "Further Results on Event-Based PID Controller" in Proceedings of the European Control Conference 2009 • Budapest, Hungary, August 23–26, 2009
4. L. Zhang, H. Gao, O. Kaynak, Network-induced constraints in networked control systems–a survey. IEEE Trans. Industr. Inf. **9**(1), 403–4016 (2013)
5. W. Heemels, K. Johansson, P. Tabuada, An introduction to event-triggered and self-triggered control, in *51st IEEE Conference on Decision and Control* (CDC), (2012), pp. 3270–3285
6. G. Franklin, J. Powel, A. Emami-Naeini, *Feedback Control of Dynamical Systems*. Prentice-Hall, (2010)
7. J. Lunze, D. Lehmann, A state-feedback approach to event-based control. Automatica **46**(1), 211–215 (2010)
8. K.J. Astrom, B.M. Bernhardsson, Comparison of periodic and event-based sampling for first-order stochastic systems in *Proceedings of the 14th IFAC World Congress*, Beijing, China, pp. 301–306 (1999)
9. B. Wang, M. Fu, "Comparison of periodic and event-based sampling for linear state estimation" in *Proceedings of the 19th World Congress The International Federation of Automatic Control* Cape Town, South Africa, (2014)
10. K.E. Arzen, A simple event-based PID controller, in *Proceedings of the 14th IFAC World Congress*, Beijing, China, 18:423–428 (1999)
11. J.M. Gomes Da Silva Jr., W.F. Lages, D.G. Sbarbaro, Event-triggered PI control design, in *19th IFAC World Congress* (IFAC WC, Cape Town, South Africa, 2014) pp. 6947–6952
12. U. Tiberi, J. Araujo, K.H. Johansson, "On event-based PI control of first- order processes" in *2nd IFAC Conference on Advances in PID Control*, PID'12, Brescia, Italy, (2012), pp. 448–453

13. U. Tiberi, J. Araujo, K.H. Johansson, "On event-based PI control of first- order processes," in 2nd IFAC Conference on Advances in PID Control, PID'12, Brescia, Italy, Mar 2012, pp. 448–453.
14. S. Durand1, J.F. Guerrero-Castellanos, "Event-Based Digital PID Control" in IEEE (2010)
15. M. Beschi, S. Dormido, J. S´anchez, and A. Visioli, Event-based PI controller with exponential thresholds, in *19th IFAC World Congress* (IFAC WC, Cape Town, South Africa, 2014), pp. 5766–5771
16. C. Fiter, L. Hetel, W. Perruquetti, J.P. Richard, A state-dependent sampling for linear state feedback. Automatica **48**, 1860–1867 (2012)
17. A. Sahoo, H. Xu, S. Jagannathan, Approximate optimal control of affine nonlinear continuous-time systems using event-sampled neurodynamic programming, in *IEEE Transactions on Neural Networks and Learning Systems*, vol. 28, no. 3, (2017) pp. 639–652
18. K.G. Vamvoudakis, H. Ferraz, "Model-free event-triggered control for continuous-time linear systems with optimal performance" in Automatica, vol. 87, pp. 412–420 (2018)
19. M. Velasco, P. Mart, E. Bini, "On Lyapunov Sampling for event-driven controllers" in *Proceedings of the 48th IEEE Conference on Decision and Control* (2009)
20. P. Tabuada, "Event-triggered real-time scheduling of stabilizing control tasks," IEEE Transactions on Automatic Control, vol. 52, no. 9, pp. 1680–1685 (2007)
21. G. Kiener, D. Lehmann, K. Johansson, Actuator saturation and anti-windup compensation in event-triggered control. Discrete Event Dynamic Systems **24**(2), 173–197 (2014)
22. L.G. Moreira, L.B. Groff, J.M. Gomes da Silva Jr., S. Tarbouriech, "Event-triggered PI control for continuous plants with input saturation" in, *AmeriCan Control Conference (ACC) Boston Marriott Copley Place July 6–8, 2016* (MA, USA, Boston, 2016)
23. S. Tarbouriech, G. Garcia, J.M. Gomes Da Silva Jr., I. Queinnec, "Stability and Stabilization of Linear Systems with Saturating Actuators" in Springer (2011)
24. L. Fesquet, B. Torr´esani (eds), An event-based PID controller with low computational cost, in *8th International Conference on Sampling Theory and Applications* (SampTA'09), Marseille, France, Special session on Sampling and Industrial Applications (2009)

Dual Mode WECS-Based Two-Stage Hierarchical Control of Hybrid Microgrid

Vipin Kumar Dhiman and Shivam

1 Introduction

Due to the high pressure on non-renewable energy sources by nature fundamentalists, the use of renewable energy sources has increased remarkably over the years. Wind energy is one of the widely used renewable energy sources due to its availability. The complete wind energy conversion system (WECS) is one of the fastest-growing energy resources, doubled in the last 10 years. Various WECSs have been proposed by the researchers working under different capacities with different types of generators like double fed induction generator (DFIG), permanent magnet synchronous generator (PMSG) and more [1, 2]. In most cases, WECS comprises a bidirectional converter to link distributed generators (DGs) to AC side main grid and loads. Due to these interconnections, the decreases of system inertia have encountered the problem of maintaining system frequency and voltage for a microgrid. The microgrid can work under grid-connected mode or islanding mode.

In islanding mode, the load demand is met by the WECS DGs with one of the DG should be working as a controlled voltage source [3]. The main concern during this mode is to avoid circulation current while maintaining system frequency and load voltage. During grid-connected mode, the main grid regulates the frequency and voltage because of its large size impact on the system, which removes the extra burden from the microgrid. Also, the bifurcation of active and reactive powers, stable operation and high power quality must be met during the mode of operation.

Depending upon the complications mentioned above, this paper purposed a microgrid two-stage hierarchical control structure for both grid-connected and islanding modes. The primary stage depends upon droop control based on line impedance

V. K. Dhiman (✉) · Shivam
Department of Electrical Engineering, NIT Kurukshetra, Kurukshetra, Haryana, India
e-mail: vipinkd03@gmail.com

Shivam
e-mail: shivam55ram@gmail.com

S. Vadhera et al. (eds.), *Latest Trends in Renewable Energy Technologies*, Lecture Notes in Electrical Engineering 760, https://doi.org/10.1007/978-981-16-1186-5_34

and load characteristics which helps in active and reactive power bifurcation, but its unanimous implementation may lead to deviations in voltage and frequency. So to overcome it, the secondary stage compensates for the deviations in voltage and frequency [4]. Two steps are introduced to enhance the first stage of droop control, with voltage biasing is done using a low bandwidth link in the first step, then voltage magnitude deviations are compensated.

In this paper, a dual-mode WECS is proposed using a two-stage hierarchical control structure. The primary stage performs P-V/Q-f droop control, based upon the active and reactive power-sharing [5]. The secondary stage controls the voltage and frequency fluctuations. The microgrid operates in grid-connected mode with a smooth transition to islanding mode for ensuring synchronization of voltage and frequency within the specified limits. The microgrid included two types of WECS, one is battery-backed to ensure maximum power transfer from wind and storage of surplus power, and the other one is braced by adaptive operation without any storage element.

The performance of the proposed system is investigated under different loading conditions with mode transition. The complete paper is organized as follows: A brief overview of the proposed system and its control is given in Sect. 1. Microgrid and WECS de- tailed topology analysis is discussed in Sect. 2. The proposed dual-stage controller design strategy is covered in Sect. 3. Section 4 presents MATLAB/Simulink environment analysis and validation of the work by different graphs. Finally, in Sect. 5 conclusion is discussed.

2 Configuration of Microgrid and Control Structure

2.1 WECS-Based Hybrid Microgrid

The microgrid structure is given in Fig. 1 for which the control strategy is proposed which is having two types of wind energy conversion system (WECS) connected to the common microgrid bus. The main AC grid can be connected or disconnected by the means of a static switch. Both types of WECS are having permanent magnet synchronous generator (PMSG) for having higher efficiency, solidity and an acclaimed reputation in offshore wind farms.

2.2 WECS-1 Battery Backed

Figure 2 shows the overall formation of the microgrid. The paper does not offer a pitch-plate design of WECS so it's not shown. After the conversion of mechanical to electrical power by WECS-1, it is given to an uncontrolled rectifier via DC-link and boost converter. The maximum power is achieved by the means of wind turbine

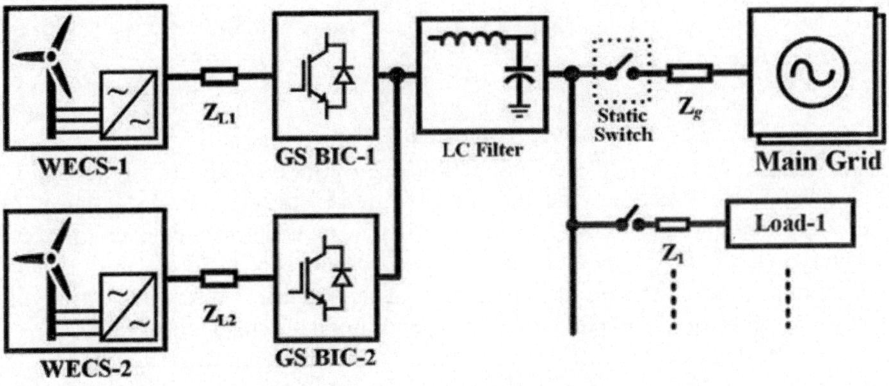

Fig. 1 Configuration of proposed hybrid microgrid

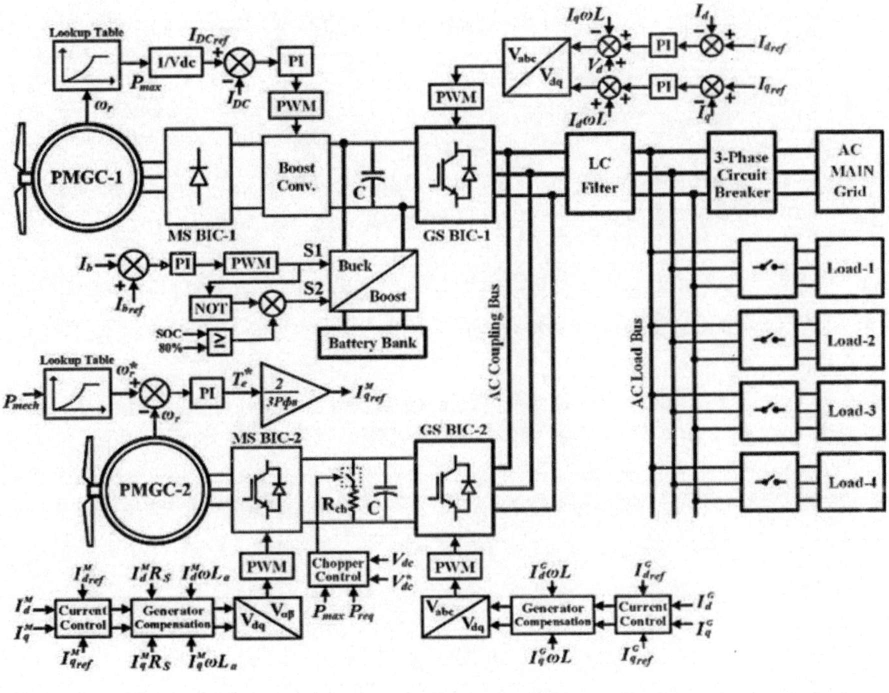

Fig. 2 Structure and control of proposed hybrid microgrid

characteristics and lookup table, further, the buck-boost converter tries to regulate the DC-link voltage by regulating bidirectional battery current. The need for battery becomes unavoidable in islanding mode to regulate DC link voltage and AC side frequency. The battery reference current is defined as:

$$I_{\text{batt}_{\text{ref}}} = k_{bp}(800 - V_b) + k_{bi}\int(800 - V_b)dt \tag{1}$$

where proportional-integral of battery gains are k_{bp} - k_{bi} with V_b as battery voltage for the nominal battery of 800 V and 300 kWh.

The main objective of grid side bidirectional converter (GS BIC-1) is to maintain persistent frequency and voltage for microgrid loads, throughout grid-connected mode and islanding mode. For grid-connected mode, active power is given to loads or batteries according to requirement. So two current control loops are adopted for dual-loop control. The primary loop depends upon current references based on coordination control. The secondary loop depends upon the state of charge (SOC) of the battery. Both loops are described in Sect. 3. After converting the current reference in d-q from, a further voltage regulating loop is used. The output voltage after this loop are described after neglecting the effect of filter capacitance:

$$V_{d\,\text{out}} = V_d - \omega Liq + kpc_d(I_{\text{dref}} - I_d) + kic_d\int(I_{\text{dref}} - I_d)dt \tag{2}$$

$$V_{\text{qout}} = \omega Liq + kpc_q\left(I_{\text{qref}} - I_d\right) + kic_q\int\left(I_{\text{qref}} - I_q\right)dt \tag{3}$$

where ωL is the settlement portion for filter inductance, k_{pc_d}, k_{ic_d}, k_{pc_q} and k_{ic_q} are P-I gains of d-q current regulators respectively [6].

2.3 WECS-2: Adaptive Operated

As shown by Fig. 2, the second type WECS utilizes controlled converters at both grid-side and machine-side. The machine-side BIC is used to extract max power by wind turbine. It uses MPPT algorithm by analyzing rotor speed ω_r to achieve maximum active current *Iq*. The reference current for this converter in dq-frame are:

$$I_{d\ \text{ref}}^{M} = 0 \tag{4}$$

$$I_{q\ \text{ref}}^{M} = \frac{2T_e^*}{3P\phi_B} \tag{5}$$

$$T_e^* = Kp_t\left(\omega_r^* - \omega_r\right) + Ki_t\int\left(\omega_r^* - \omega_r\right)dt \tag{6}$$

where T_e^* is pre-set electrical torque, is P pole pairs, ϕ_B is generator magnetic flux, ω_r^* is reference rotor speed, K_{p_t} and K_{i_t} are the P-I gains for torque regulator.

To avoid power imbalance in islanding mode, WECS-2 utilises the adaptive operation as in this mode real power required could be less than the generated power, which is done by proper active-reactive current references. The active-reactive current reference is set as:

$$I_{d\,\mathrm{ref}}^{G} = \min\begin{cases} Kp_{DCL}\left(V_{DC}^{*} - V_{DC}\right) + Ki_{DCL}\int\left(V_{DC}^{*} - V_{DC}\right)dt \\ \frac{2\mathrm{P_{req}}}{3V_d} \end{cases} \quad (7)$$

$$I_{q\,\mathrm{ref}}^{G} = Kp_r\left(Q_{\mathrm{req}} - Q_{\mathrm{avl}}\right) + Ki\int_r^d\left(Q_{\mathrm{req}} - Q_{\mathrm{avl}}\right)dt \quad (8)$$

$$Q_{\mathrm{avl}} = \frac{3}{2}V_d^I{}_{\mathrm{qavl}}I \quad (9)$$

where V_d is d-axis voltage, $\mathrm{P_{req}}$ is real power needed by the microgrid, I_{qavl} is available reactive current, Q_{req} and Q_{avl} are the required and available reactive power, and K_{p_r} – K_{i_r} are the proportional-integral gains. It is to be noted that Q_{req} must be restricted to Q_{avl} as a threshold boundary.

To achieve optimum DC link voltage, the modulation signal for chopper m_{ch} and maximum chopper power P_{ch} is given as:

$$m_{ch} = \frac{Kp_{ch}\left({V_{DC}^{*}}^{2} - V_{DC}^{2}\right) + Ki_{ch}\int\left({V_{DC}^{*}}^{2} - V_{DC}^{2}\right)dt + \sqrt{\left(P\mathrm{max} - P_{\mathrm{req}}\right)^2 - Q_{\mathrm{req}}^2}}{P_{ch}} \quad (10)$$

$$P_{ch} = \frac{V_{DC}^2}{R_{ch}} \quad (11)$$

where $K_{p_{ch}}$ and $K_{i_{ch}}$ are P-I gain of chopper regulator [7].

3 Proposed Two-Stage Hierarchical Control System

3.1 Primary Stage

To achieve the synchronous generator-like behaviour, P-V/Q-f droop control is utilised as the primary control stage. At this stage, real power adjustment is linked with frequency and reactive power is with the profile of voltage. The consideration of a small system with low voltage microgrids is to be taken for having the line impedance rate (R/X_L) significantly large. So, the line-impedance characteristics consider as mostly resistive with real and reactive powers are calculated as:

$$P = \frac{V.E - V^2}{Z} \quad (12)$$

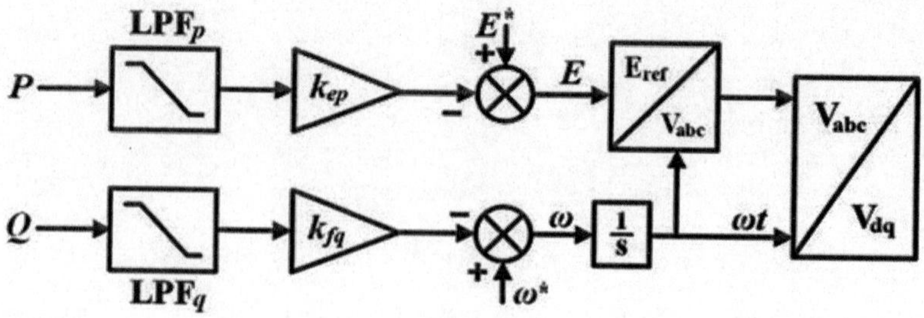

Fig. 3 Primary stage droop control

$$Q = -\frac{V.E}{Z}\varphi \tag{13}$$

where V is grid voltage, E is output voltage of inverter Z and φ are line impedance with its phase angle. The real power depends upon variation in voltage while reactive power is on phase angle. So, the droop equations for resistive lines expressed as:

$$E^* - E = k_{ep} P_{\text{mesr}} \tag{14}$$

$$\omega^* - \omega = k_{fq} Q_{\text{mesr}} \tag{15}$$

where ω^* and E^* are predefined angular frequency and voltage, k_{ep} and k_{fq} are active-re- active droop coefficients power, and defined by Eqs. (16) and (17), P_{mesr} and Q_{mesr} are measured real and reactive power resulting by Eqs. (18) and (19) [6] (Fig. 3).

$$k_{ep} = \frac{V_h - V_l}{P_h} \tag{16}$$

$$k_{fq} = \frac{\omega_h - \omega_l}{Q_h} \tag{17}$$

$$P_{\text{mesr}} = \frac{\omega_c}{\omega_c + s} P \tag{18}$$

$$Q_{\text{mesr}} = \frac{\omega_c}{\omega_c + s} Q \tag{19}$$

where ω_h, ω_l, V_h and V_l are the highest and lowest permissible sets of frequency and voltage. P_h and Q_h are the highest permissible real and reactive power with the cut-off frequency of low pass filter is ω_c.

The expression of pre-set current for grid side converter with power transfer can be given as:

$$I_{\text{pre}} = 2P_{Trans}/3V_{od} \tag{21}$$

$$P_{\text{Trans}} = \frac{3}{2} V_{od} I_{od} + \frac{3}{2} V_{oq} I_{oq} \tag{22}$$

where d-q frame reference output voltages and current are V_{od}, V_{oq}, I_{od} and I_{oq} for grid side converter [8].

3.2 *Secondary Stage*

During load fluctuation and non-uniform generation, the change in voltage and frequency should be bound in the specified limits. At this stage, to die out these undesirable changes, voltage and frequency are compared with pre-set values. The differences found are injected back or removed using PI controllers to the initial stage. The compensation is done with load voltage in d-q reference frame with a comparison of primary stage droop control. The resultant d-q axis newly established of voltages are given as:

$$V_d^{**} = V_d^* + K_{p_{d\,\text{sec}}}\left(V_d^* - V_{dL}\right) + K_{i_{d\,\text{sec}}} \int \left(V_d^* - V_{dL}\right) dt \tag{22}$$

$$V_q^{**} = V_q^* + K_{p_{q\,\text{sec}}}\left(V_q^* - V_{qL}\right) + K_{i_{d\,\text{sec}}} \int \left(V_q^* - V_{qL}\right) dt \tag{23}$$

where $K_{p_{d\,\text{sec}}}$ and $K_{i_{d\,\text{sec}}}$ are d-axis secondary P-I gains, $K_{p_{q\,\text{sec}}}$ and $K_{i_{q\,\text{sec}}}$ are q-axis secondary PI gains, V_{dL} and V_{qL} are real-time d-q reference frame load voltages (Fig. 4).

For the advancement of the secondary stage, two active and reactive terms (δP and δQ) for load damping are added to the above equations [9]. The purpose is to safeguard the integral control loop which was becoming more influenced due to continuous load variations. After taking α as active load-damping and β as the reactive load-damping, the advanced d-q reference frame voltages are established as:

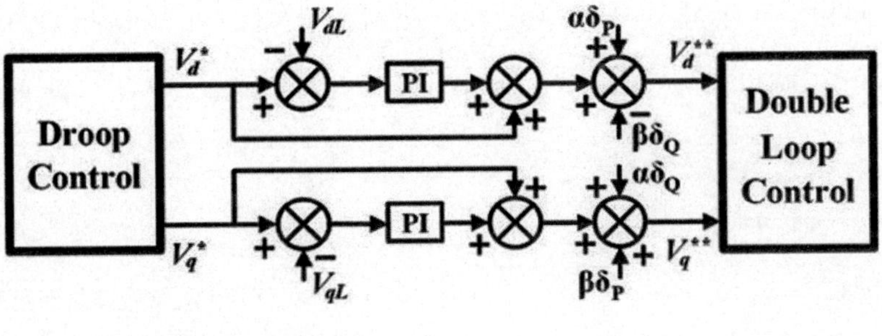

Fig. 4 Proposed secondary-stage droop control

$$V_d^{**} = V_d^* + K_{P_{dsec}}(V_d^* - V_{dL}) + K_{i_{dsec}} \int (V_d^* - V_{dL})dt + \alpha\delta_P - \beta\delta_Q \quad (24)$$

$$V_q^{**} = V_q^* + K_{pq\text{sec}}(V_q^* - V_{qL}) + K_i d_{\text{sec}} \int (V_q^* - V_{qL})dt + \beta\delta_P + \alpha\delta_Q \quad (25)$$

$$\delta_P = \frac{dP_{\text{Load}}}{dt}, \text{ if } \frac{dP_{\text{Load}}}{dt} > \frac{dP_{WD}}{dt}; \text{ or else } 0 \quad (26)$$

$$\delta_Q = \frac{dQ_{\text{Load}}}{dt}, \text{ if } \frac{dQ_{\text{Load}}}{dt} > \frac{dQ_{WD}}{dt}; \text{ or else } 0 \quad (27)$$

where P_{Load} and Q_{Load} are load real and reactive powers, P_{WD} and Q_{WD} are real and reactive powers of WECS. It is to be noted that values of δ_P and δ_Q are taken as zero when power output from wind is lessor than load power.

4 Results and Analysis

In this section, MATLAB/Simulink-based model on the proposed dual-stage hierarchical control of hybrid microgrid with two types of WECS is constructed, simulated and analysed. Based on the related theory formation in the above sections, two modes are considered as grid-connected mode and islanding mode with their transition. Table 1 de- scribes the transition period while Table 2 describes the system parameters. During the whole operation and transition, both WECS, battery-bank and AC loads are uninterruptedly associated while the AC main grid is detached from the system at 2 s by three-phase circuit breaker as static switch henceforth the system drives to islanding mode.

4.1 Grid-Connected Mode

For an initial 0–2 s, the AC main grid is linked to the system henceforth system remained in grid-connected mode, while an AC load 2 of 50 + j20 kVA is added at 1 s to verify the loading effect on the system. For the WECS-1, the DC link voltage

Table 1 Operational periods

Case	Operation mode	Time (s)	Modification
A.	Grid-connected mode (up to initial 2 s)	1	Load-2 added
		2	Gird disconnection
B.	Islanding mode (after initial 2 s)	3	Load-2 removed
		3.5	Load-3 removed

Table 2 System parameters

Sl. No.	Parameter names	Ratings
1.	WECS-1	360 kVA
2.	WECS-2	300 kVA
3.	Battery capacity of WECS-1	300 kWh
4.	DC link capacitance in WECS-2	3.5 mF
5.	AC main grid voltage	480 V
6.	System frequency	50 Hz
7.	Feeder Impedance	0.1 + j0.018 Ω
8.	Loads: 1, 2, 3 and 4 respectively	100, 50 + j50, 50, 50 + j20 kVA

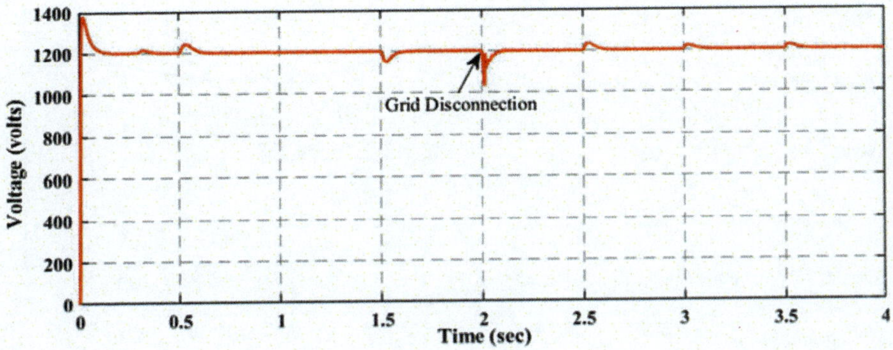

Fig. 5 DC-link voltage at WESC-1

is maintained at the level of 1200 V as shown by Fig. 5. Initially, after the transient period over the first stage of droop control and charge controller of battery helps to maintain the DC- link voltage at the desired level with minimal variations [10]. The AC load-side frequency is shown by Fig. 6, it has a small variation during mode transition at 2 s and load addition at 3 s. The battery SOC is shown in Fig. 7, the battery is charging by storing extra power available. Figure 8 shows the AC side voltage and current at the main grid and loads, at 2 s when AC main grid is separated from the system, then the grid voltage and the current becomes zero p.u. as required and for loads, the voltage remains almost 1 p.u. while load current is changing according to the requirements, supported by taking power from both WECS. Figure 9 shows the active and reactive powers provided by the AC main grid and consumed by AC loads. During this mode, real and reactive power for the loads is mainly provided by the AC main grid and becomes zero after this mode at 2 s. Figure 9 shows the active and reactive power provided by both WECS, for WECS-1 during grid-connected mode real power is provided to the battery while WECS-2 provides power to load throughout the operation because its system does not have energy storage devices.

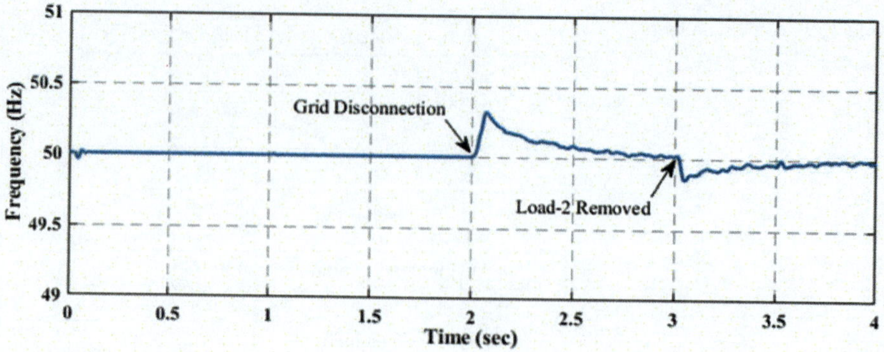

Fig. 6 Load frequency of the system

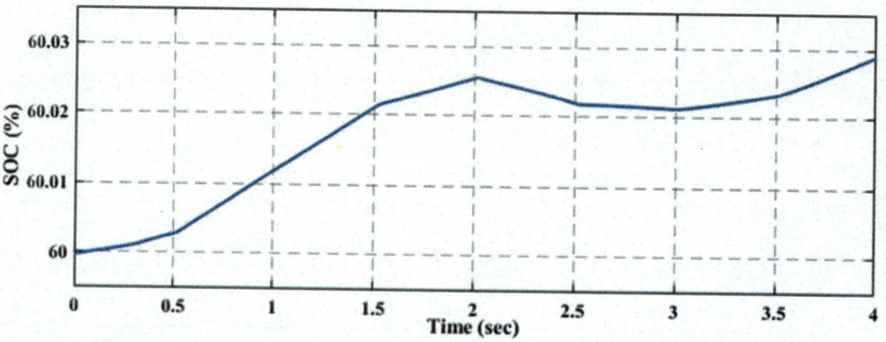

Fig. 7 Battery SOC

During this mode, the reactive power is mainly provided by the AC main grid for maintaining proper operation.

For considering the change effect of loads on the system, load 2 is added to the system at 1 s, the real and reactive power requirement increases as shown in Fig. 9. At this change magnitude of current at the grid and load is increased because of load change while voltages are maintained at the same level. At this instant other DC bus voltage, system frequency and SOC of the battery remain the same.

4.2 *Islanding Mode*

The mode transition from grid-connected mode to islanding is performed at 2 s by removing the AC main grid from the system using a 3-phase circuit breaker as a static switch. At this instant, system performance is critically decisive as the system can go into an unstable mode. At this instant, all the loads are connected to the system which

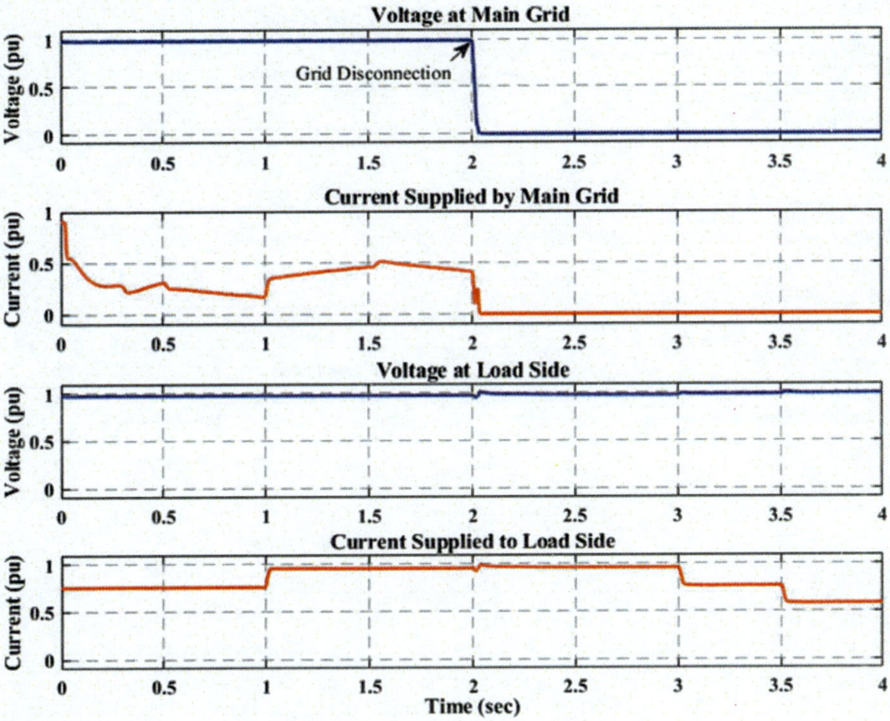

Fig. 8 Voltages and currents at the main grid and load

gets power from the two WECS. The proposed control strategy helps to achieve the above mode transition reliably.

At the starting of this islanding mode, the DC bus voltage for WECS-2 has a dip but recovers instantly to 1200 V as shown by Fig. 5, further have small variations which are in limits [6]. Figure 6 indicates the system frequency which increases a bit from its desired value of 50 Hz, droop control further maintains system frequency within operational limits. The battery SOC indicates discharging but after a small instant its again starts charging by taking surplus power from WECS-2 as shown by Fig. 7. The voltage and current at the main grid become zero and a small variation is observed in the load voltage which decays out further as shown by Fig. 8 and the load current is maintained according to the load demand supported by grabbing power from both WECS. The real and reactive powers provided by the AC main grid becomes zero throughout this mode shown by Fig. 9. The required real and reactive powers for loads are supplied from both WECS as mentioned above and shown by Fig. 10.

Further to observe load variations, at 3 s load-2 and at 3.5 s load-3 is removed from the system, the current supplied to load indicates a dip in its value as shown by Fig. 8. At the same time power supplied from WECS increased because power requirement decreased at load. A small dip in the system frequency is observed at

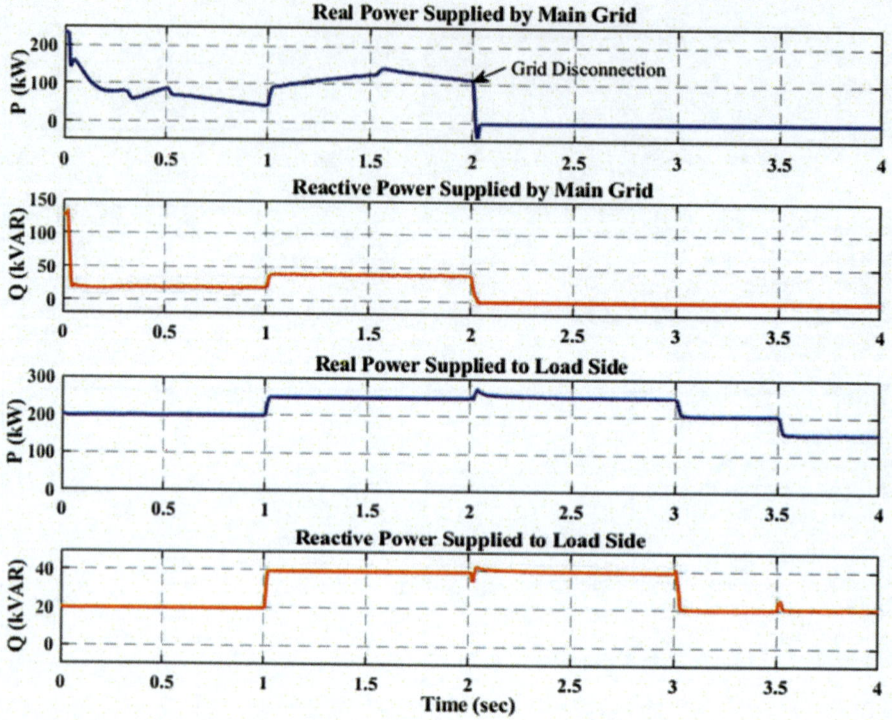

Fig. 9 Real and reactive powers at the main grid and load

3 s due to the load removed which is further maintained at the desired level. During mode transition and load variation throughout the whole simulation, load voltage remains almost constant (Table 3).

5 Conclusion

The modelling, examination and simulation of the dual-stage hierarchical control strategy for coordinating two types of WECS with AC main grid are seamlessly performed in the MATLAB/Simulink environment. The battery-backed WECS-1 extracts maximum power from the wind during both grid-connected and islanding mode of operation. The battery-free WECS-2 supported by adaptive control helps to avoid power imbalance and improves system flexibility. The structure of two-stage hierarchical control consists PV/Q-f droop control as the primary stage whose task to share real and reactive power was found satisfactory. The secondary stage improves transients stability by compensating the deviations in load voltage and current, further to support overshoot damping during large load change two damping terms were added in each d-q frame. Moreover, the dual-stage control strategy also

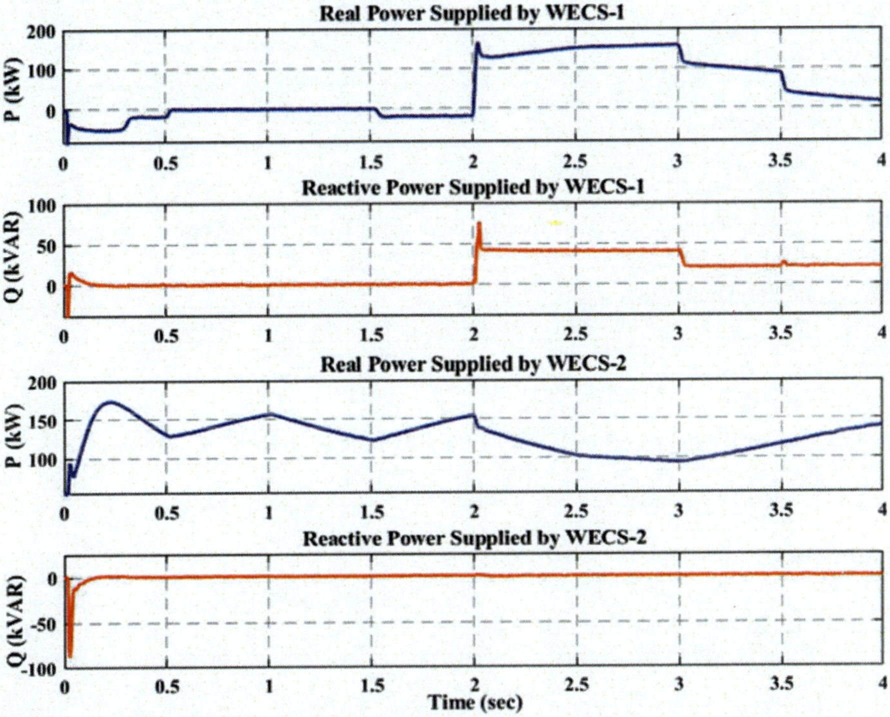

Fig. 10 Real and reactive power supplied by two WECS

Table 3 Control parameters

Sl. No.	Parameter names	Ratings
1.	WECS-1 grid side converter voltage regulator PI gains for d-q axis respectively	0.86, 4 89; 0.65, 4.6
2.	WECS-1 grid side converter current regulator PI gains for d-q axis respectively	3.3, 40; 1.8, 66
3.	WECS-1 current regulator PI gains for DC boost converter	0.56, 13.8
4.	WECS-1 current regulator PI gains for DC buck boost converter	4.5, 60
5.	WECS-2 grid side converter current regulator PI gains for d-q axis respectively	5.4, 82; 10.6, 63
6.	WECS-2 grid side DC regulator PI gains	18, 78
7.	WECS-2 chopper voltage controller PI gains	0.2, 6.9

(continued)

Table 3 (continued)

Sl. No.	Parameter names	Ratings
8.	WECS-2 machine side converter current regulator PI gains for d-q axis respectively	4, 12; 4, 14
9.	Current compensator PI gains for unplanned outrage	5, 52
10.	Pre-synchronized controller PI gains for voltage, frequency and phase angle respectively	0.2, 4; 4.2, 2; 12, 7.5
11.	Active load damping adaptive gain, α	0.65
12.	Reactive load damping adaptive gain, β	0.25

regulates DC bus voltage for WECS-1, system frequency and power bifurcation and transition of grid-connected mode to islanding mode even though load variations.

References

1. K.B. Tawfiq, A.S. Mansour, H.S. Ramadan, M. Becherif, E.E. El-Kholy, Wind energy conversion system topologies and converters: comparative review. Energy Proc. **162**, 38–47 (2019)
2. S. Muhammad, S. Imran, R. Shafiqur, K. Shamim, M.A. Luai, Assessment of wind energy potential using wind energy conversion system. J. Clean. Prod. **216**, 346–360 (2019)
3. X. Tang, X. Hu, N. Li, W. Deng, G. Zhang, A novel frequency and voltage control method for islanded microgrid based on multienergy storages. IEEE Trans. Smart Grid **7**, 410–419 (2016)
4. L. Che, M. Shahidehpour, A. Alabdulwahab, Y. Al-Turki, Hierarchical coordination of a community microgrid with AC and DC microgrids. IEEE Trans. Smart Grid **6**(6), 3042–3051 (2015)
5. M. Rezkallah, A. Chandra, B. Singh, S. Singh, Microgrid: configurations, control and applications. IEEE Trans. Smart Grid **10**(2), 1290–1302 (2019)
6. V.K. Dhiman, Shivam, Enhanced two-stage coordination control of AC/DC hybrid microgrid, in *2020 IEEE 9th Power India International Conference (PIICON)*, SONEPAT, India, 2020, pp. 1–6
7. U.K. Kalla, H. Kaushik, B. Singh, S. Kumar, Adaptive control of voltage source converter based scheme for power quality improved grid-interactive solar PV–battery system. IEEE Trans. Ind. Appl. **56**(1), 787–799 (2020). https://doi.org/10.1109/tia.2019.2947397
8. P. Omid, K. Kimmo, Hierarchical control structure in microgrids with distributed generation: Island and grid-connected mode. Renew. Sustain. Energy Rev. **44**, 797–813 (2015)
9. J.M. Guerrero, J.C. Vasquez, J. Matas, L.G. de Vicuna, M. Castilla, Hierarchical control of droop-controlled AC and DC microgrids—a general approach toward stand- ardization. IEEE Trans. Industr. Electron. **58**(1), 158–172 (2011)
10. IEEE Standard for the Specification of Microgrid Controllers," in *IEEE Std 2030.7–2017*, (2018), pp. 1-43. Accessed 23 April 2018

Modelling and Simulation of Grid Connected Solar Photovoltaic System with LCL Passive Filter

Pankaj Dhal and Shashi Bhushan Singh

1 Introduction

India is located on the Equatorial Earth's sunbelt, receiving abundant sunlight. Conventional resources for energy could expire in the future, alongside the world's fast-growing economy. Energy consumption and quality have become two things of great concern today. The development of renewable energies could help preserve social sustainability, which also prevents a possible energy crisis [1]. In order to take care of most of the energy generation, renewable energy projects have been considered as one of the most relevant options for the future. There are many upcoming and varied levels of maturity technology available [1].

The original 20 GW capability goal was met by the Indian Government for 2022, four years in advance. In India, closely 42 solar parks have been built to make land accessible to solar plant advertisers [2]. The nation installed 3GW of solar capacity in 2015–2016, 6GW in 2016–2017 and more than 10GW in 2017–2018, while the retail price of solar energy dropped to 18% below that of its coal-fired co-payers. However, its solar capacity has grown from 26650 MW to over 20GW as at 31 January 2018. In India, over 82,580 MW of renewable power energy, with roughly 31,150 MW output, was deployed at different stages of construction by the end of September 2019 [2].

SPV system's output power characteristic depends on the solar radiation level and temperature. In addition, the temperature and solar radiation on the photovoltaic cell depend on the continued changing atmospheric conditions. Thus, we will monitor the point of availability of maximum power at different levels of irradiation in the SPV

P. Dhal (✉)
SREE, NIT, Kurukshetra, India
e-mail: Pankaj.dhal9466@gmail.com

S. B. Singh
Department of Electrical Engineering, NIT Kurukshetra, Kurukshetra, India
e-mail: sbsingh@nitkkr.ac.in

S. Vadhera et al. (eds.), *Latest Trends in Renewable Energy Technologies*, Lecture Notes in Electrical Engineering 760, https://doi.org/10.1007/978-981-16-1186-5_35

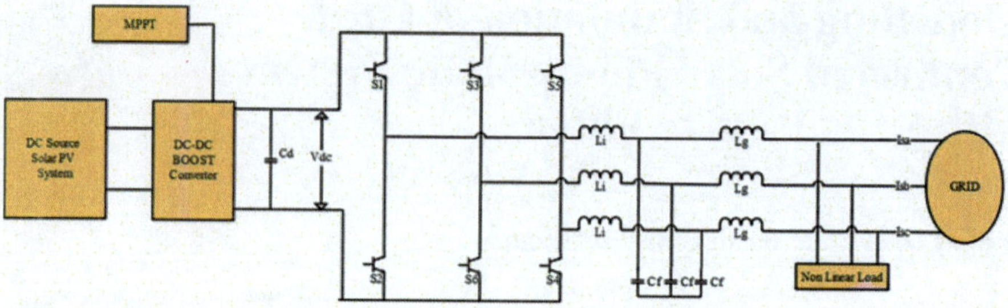

Fig. 1 Block diagram of grid connected solar photo-voltaic system

unit. Development in power electronics devices further stimulates the development of SPV-based energy systems. More-over in a high-frequency transient switching system, power electronic devices can be integrated with the SPV system but the systems have some drawbacks due to the presence of harmonics [3].

In the power distribution systems, the existence of current and voltage harmonics increases energy loss on line and lowers power factor and may cause resonance in the circuit. Converter harmonics can interact with stable and sensitive domestic electrical equipment which is unwanted. Harmonics may trigger possible physical harm or unnecessary trip relays in motor drives or electricity networks. Harmonic interference is a major concern for power system experts when losses escalate, unnecessary heating of spinning machines and communication systems conflict dramatically with typical AC power line and noise production, contributing to incorrect device activity [4].

In this paper, the LCL passive filter joined between grid and Solar PV array and simulated with filter design consideration. The synchronous reference frame (SRF) theory is utilized for system control with an inherent grid synchronization capability. The MPPT method based on perturbation and observation (P & O) is refined and executed by using a boost converter. The entire evaluation is carried out with the peak power tracking from the PV source. In addition, the grid side converter is used to provide current harmonics and reactive power requirements locally. One simple and common approach for minimizing harmonics of the input current is the use of multi-pulse connections with multiple windings based on transformers. Passive filters were implemented in the last decade to reduce the harmonics pumped into the mains [5] (Fig. 1).

2 System Description

2.1 PV Array

The solar photo-voltaic system contains PV array which is a solid-state semiconductor device and it generates electrical current from solar insolation. The PV array is made up of by using a group of solar panel that works in the principle of solar photovoltaic effect. Three to 5-W power can be produced from a single solar cell and then we can increase power by connecting a number of such cells in series [1, 2]. Under the open-circuit voltage and the short-circuit current, the o/p current of the PV array is

$$I_d = I_{sc}\left[1 - e^{\frac{(V_d - V_{oc})q}{N_s nkt}}\right] \tag{1}$$

where.

I_d	Output current of PV array.
I_{sc}	Short circuit current.
V_d	Diode voltage.
V_{oc}	Open circuit voltage.
N_s	Number of photovoltaic cell in series.
T	Temperature of solar cell.

2.2 *MPPT with DC/DC Converter*

We mainly required peak power from our PV array for all operations in the system. By changing the duty ratio of the boost converter we can extract peak power from the PV array by using different MPPT techniques. There are many MPPT methods are used but here we use the most frequently used method that is perturbation and observation (P & O) due to its ease of use and simplicity. The system measures the peak power point by watching, comparing and perturbing the power which is generated from the PV array [4, 5].

Voltage and current of Photovoltaic cells are needed for the power feedback type MPPT method, which has a relationship between the output voltage and the o/p power of the PV array. In the P & O methods, we can get maximum power at which point the power to voltage derivation is zero, this derivative is +(ze) at the left side and become –(ve) at the right side of peak powerpoint. In power voltage characteristics at MPP

$$\frac{\partial P}{\partial V} = 0 \tag{2}$$

The basic track of P & O algorithms is given in Fig. 2 we use a DC-DC boost converter here, which runs in continuous driving mode. The duty cycle of the boost converter is given as

$$\frac{V_O}{V_{PV}} = \frac{1}{1 - D} \tag{3}$$

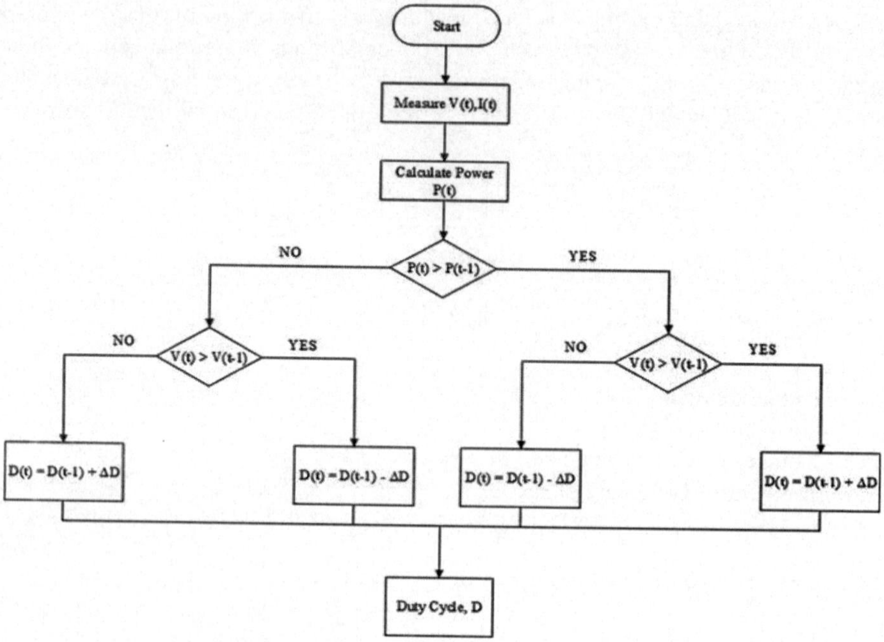

Fig. 2 Flow diagram of P & O MPPT method

where V_{PV} is the input voltage, V_O is the o/p voltage and D is the duty cycle of the DC-DC boost converter. The flow chart of P & O algorithm is given below in Fig. 2.

2.3 *Three Phase Inverter*

Here we use a Full-bridge three-phase voltage source converter, which is made up of six switches. The main operation of the VSI is to transform the Dc voltage given by the DC-DC boost converter into an AC voltage of 230 V RMS having a frequency of 60HZ. Three alternative pulses are developed by Hysteresis Current Controller(HCC) which is supplied pulses to the corresponding inverter gate. In several intervals, the control signal of pulses ON for some section of the time and then again off for the remaining section of the time. The O/P signal from the three-phase full-bridge VSI is pass through a passive filter to get desirable output voltage which is given to the load and grid [6, 7].

2.4 Filters

The design of the LCL passive filter is classified into two parts. In the 1st part inductance (Li) at the inverter, the side is designed and in the 2nd part inductance (L_g) at the grid side with capacitance (C_f) is designed. The inductance which is on the grid side is having a relation with the inductance is in the inverter side and the ratio between them depends upon the attenuation of the ripple current. For the LCL filter design here we used a simpler equation and according to the design principle and the equations are given below.

$$L_I = \frac{V_s}{2\sqrt{6} f_s i_{ripple}} \tag{4}$$

$$C_f = \frac{0.05}{\omega_n Z_{base}} \tag{5}$$

$$Z_{base} = \frac{V_{sLL}^2}{P_n} \tag{6}$$

$$L_g = L_i \tag{7}$$

where L_g is inductance in grid side, L_i inductance in inverter side, f_s switching frequency of the inverter, C_f capacitance of LCL filter, ω_n operating frequency, i_{ripple} is the peak value of rated output current, P_n rated power of the inverter.

The inductance value is very costly, bulky and high for the above cases. Here we take inductance value 1.4 mH for both grid side and inverter side inductance and also take 5 μf for capacitance.

3 Control Strategy Implementation

3.1 SRF Theory for Synchronization

SRFT theory is mainly utilized for the independent control in both quadrature and direct axis. Mainly it converts equivalent d-q components from three-phase AC supply for our control propose. As per this theory, the a-b-c variables of current signals are converted into d-q components and then filtered. For generating gate pulses for inverter switches again we convert the current variable from d-q component into a-b-c frame and given it to the hysteresis current controller.

In this type first, we identified the source current variable that is i_a, i_b, i_c then transformed into (α-β-0) component which is a two-phase frame from the a-b-c component which is a three-phase frame. This transformation is given in Eq. 8

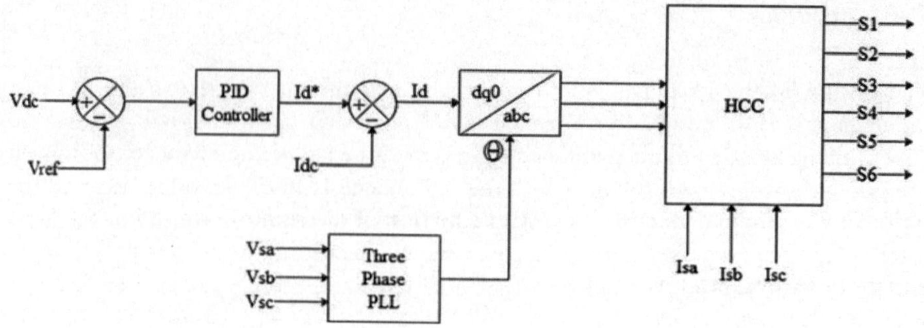

Fig. 3 Main control scheme

$$\begin{bmatrix} i_a \\ i_q \\ i_0 \end{bmatrix} = \sqrt{\frac{2}{3}} \begin{bmatrix} 0 & -\frac{1}{2} & -\frac{1}{2} \\ 0 & \sqrt{\frac{3}{2}} & -\sqrt{\frac{3}{2}} \\ \frac{1}{2} & \frac{1}{\sqrt{2}} & \frac{1}{\sqrt{2}} \end{bmatrix} \begin{bmatrix} i_a \\ i_b \\ i_c \end{bmatrix} \tag{8}$$

Again we obtain d-q current variable by convert α-β variable into three-phase synchronous frame and this transformation is given in Eq. 9

$$\begin{bmatrix} i_d \\ i_q \end{bmatrix} = \begin{bmatrix} \cos\theta & \sin\theta \\ -\sin\theta & \cos\theta \end{bmatrix} \begin{bmatrix} i_\alpha \\ i_\beta \end{bmatrix} \tag{9}$$

3.2 Main Control Scheme

Main Control scheme is described in Fig. 3. The dc output from the boast converter is compared with the reference dc voltage Vref*. The voltage from boast converter contains ripples. These ripples needs to be separated to have admirable power quality at ac side. The error is then reduced by PID controller with values $K_{pd} = 2$ and $K_{id} = 1.5$ [8].

The output of PID controller i.e. Id* is then compared with dc current output of boast converter and the resultant is considered as Id i.e. d phase current to be fed to dq0 to a-b-c converter (Inverse Park's Transform) which generates 3 phase currents i.e. Ia*, Ib* and Ic* which is then compared with reference three-phase current in hysteresis current controller which generates triggering signals for respective MOSFET's in the inverter circuit.

4 Results and discussion

The photovoltaic system coupled to the grid using PV arrays is modelled and simulated in MATLAB SIMULATION. The simulation is performed under non-linear load. Figure 4 displays the result of grid voltage which is 3 phase sinusoidal with 230 V rms.

Figure 5 shows the DC side result which generates 500 V constant DC voltage by a group of PV array and is supplied to the grid and non-linear load though boost converter and three-phase inverter.

Figure 6 shows the results of the system on the AC side with RL non-linear load. With this figure, it is cleared that the ac side parameter like load voltage and load current have sinusoidal wave shapes due to the compensation of reactive power and load harmonic done by the LCL passive filter.

Figure 7 shows the results of AC side grid voltage and grid current which is quite randomistic sinusoidal figure due to non-linear connection at PCC. It injects harmonic and reactive power to the grid so we get this type of wave shape.

Figure 8 shows the result of both active and reactive power on the AC side. Power

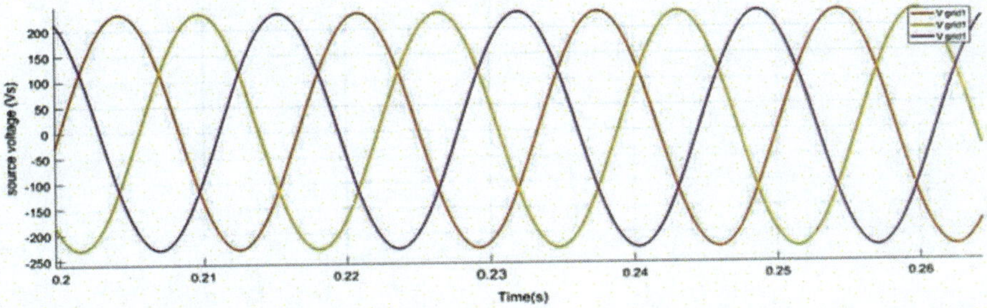

Fig. 4 Grid voltage

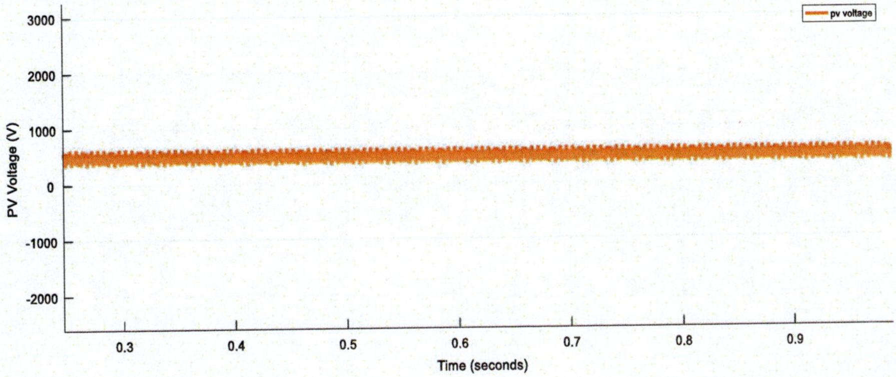

Fig. 5 PV voltage

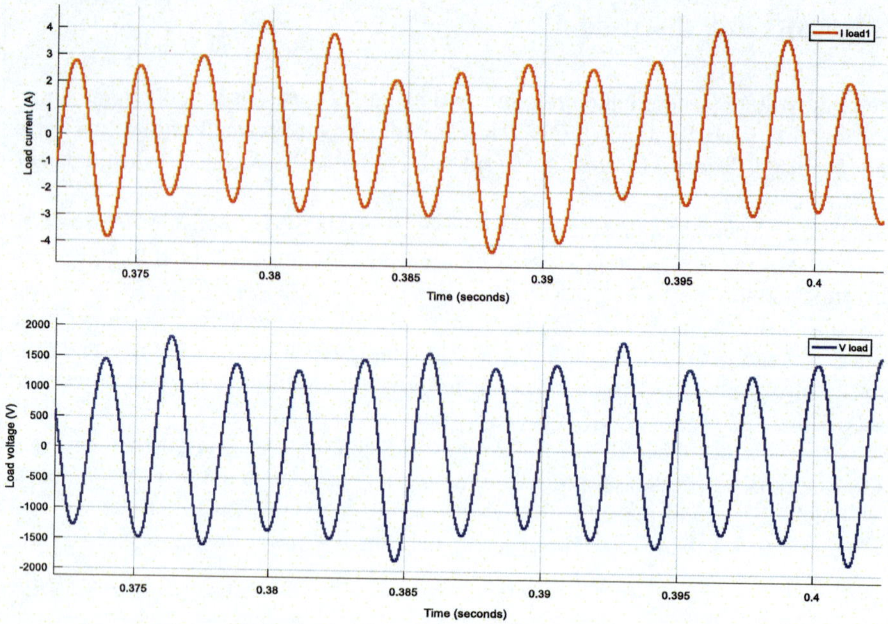

Fig. 6 AC side Results of current and voltage with RL load

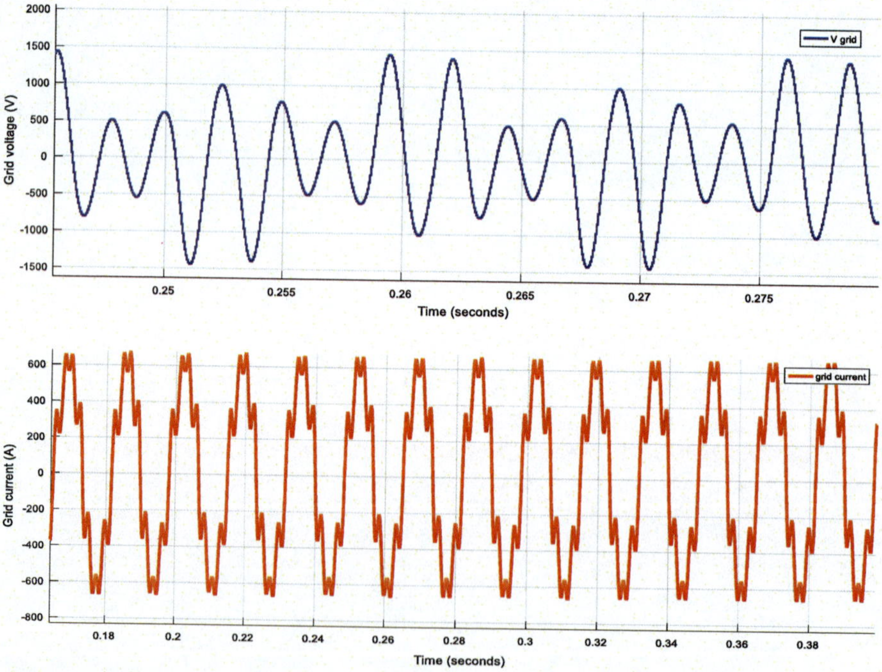

Fig. 7 AC side grid voltage and current

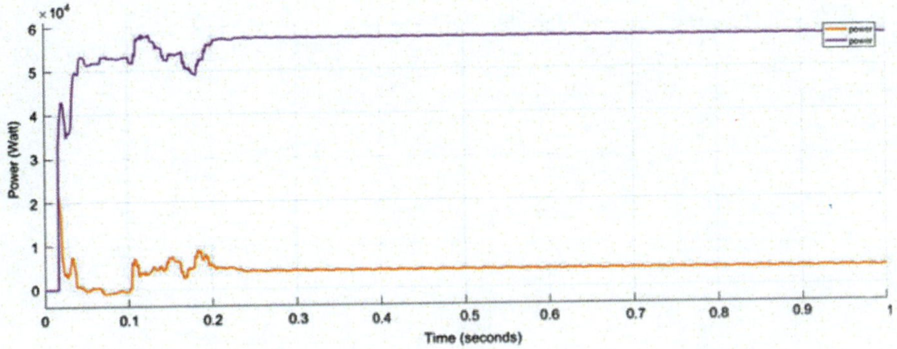

Fig. 8 Grid power with RL load

transferred to the load from the solar photovoltaic system at unity power factor. Power is nearly constant after initial transients hence keeping the system stable.

5 Conclusions

By using the SRF theory algorithm we effectively simulate the solar photovoltaic system with an LCL passive filter. The required results are obtained for the evacuation of PV resources and by using a passive filter we compensate reactive power for the grid which is used under the unity power factor. By using this method, we boost the power factor of the system. By using non-linear load and irradiance of the light we analyze the photovoltaic network linked to the grid. The simulation results have been shown to verify the different parameters from PV source to non-linear load and grid.

References

1. J. Singh, S. Ganguly, Study and design of grid connected photovoltaic system, thesis report electrical and instrumentation dept.,Thapar University, Patiala
2. A.Nottrott, J.Kleissl, B.Washom, Energy dispatch schedule optimization and cost benefit analysis for grid-connected, photovoltaic-battery storage systems. Int. J. Renew. Energy- Elsevier **55**, 230–240 (2013)
3. C. Schelly, E.P. Louie, J.M. Pearce, Examining interconnection and net metering policy for distributed generation in the United States. Renew. Energy Focus- Elsevier **22–23**, 10–19 (2017)
4. S. Sumathi, L.A. Kumar, P. Surekha, *Solar PV and Wind Energy Conversion Systems- An Introduction to Theory, Modelling with MATLAB/SIMULINK, and the Role of Soft Computing Techniques* (Springer International Publishing, Switzerland, 2015).
5. P.S. Bimbra, *Power Electronics*, 6th edn. (2018)
6. I. Bhattacharya, Y. Deng, S.Y. Foo, Active filters for harmonics elimination in solar photovoltaic grid-connected and stand-alone systems, *IEEE 2nd Asia Symposium on Quality Electronic Design, Penang, Malaysia, 3–4 Aug. 2010*

7. P.C. Sen, *Power Electronics*, 46th reprint edition, 2017.
8. A.K. Verma, B. Singh, D.T. Sahani, Grid interfaced solar photovoltaic power generating system with power quality improvement at AC mains, in *IEEE 3rd International Conferences Sustainable Energy Technology, Kathmandu, Nepal, Oct. 2012*

Comparison of Illumination at Different Workplaces Using Optical Sensor

Shashi Kant Vij and Sandeep Gupta

1 Introduction

Spectral evaluation of an optical spectrum is a well-known method in biology, physics, and chemistry [1, 2]. In the chemical investigation, the fluorescence spectrum is commonly used to identify the composition of a sample solution and to evaluate its concentrations [3, 4]. Fluorescence signals are also examined for the supervising of photosynthesis [5, 6]. Based on such measurements, it might be possible to monitor plant conditions and use this information for online control of illumination conditions in such a way that photosynthesis is maximized, without causing plant stress.

A mobile phone has been used both as an illumination source and an image detector for quantitative optical investigation [7]. The paper [3] gives an overview of micro spectrometers operating in the visible and infrared spectral range. A full detail model of the point spread function (PSF) is presented for the complete optical system of aberration-free light field cameras with low to moderate numerical apertures [8–10].

Therefore, this paper has undertaken the different case studies on sample classrooms or workplace, where the measurement of illumination in Lux. is carried out with the help of an optical sensor; based on which different results are obtained, compared and analyzed with the ideal standards (250 lx for easy office work or classes) as provided by [11, 12] with the help of tables and graphical representation.

S. K. Vij (✉) · S. Gupta
Department of Electrical Engineering, JECRC University, Jaipur, India
e-mail: shashikantvij@yahoo.com

S. Gupta
e-mail: jecsandeep@gmail.com

S. Vadhera et al. (eds.), *Latest Trends in Renewable Energy Technologies*, Lecture Notes in Electrical Engineering 760, https://doi.org/10.1007/978-981-16-1186-5_36

Table 1 Illumination in lux at 10:30 AM and 11:30 AM for 2nd floor nearest to light source

Date	Max. position of sensor	Horizontal position of sensor	Date	Max. position of sensor	Horizontal position of sensor
14.2.18	4090	654	6.3.18	2780	470
	2303	403		2230	400
15.2.18	3490	535	7.3.18	3150	1130
	2285	400		2685	617
16.2.18	4717	620	8.3.18	4017	626
	2547	527		2658	517
17.2.18	4123	615	9.3.18	4112	594
	2490	322		2781	526
19.2.18	3310	480	10.3.18	4317	629
	2170	410		2862	510
20.2.18	2190	322	12.3.18	2792	482
	1978	301		2018	298
5.3.18	4434	639	13.3.18	4003	531
	2357	577		2331	492

2 An Illumination Study in the Various Circumstances

2.1 Case 1

Illumination received (readings were taken for the maximum light-receiving angle of sensor and at horizontal level) at a place in II floor class which is nearest to the light source for different calendar days at Jaipur Rajasthan at 10:30 AM and 11:30 AM when all the inner sources of light were all illuminated as shown in Table 1 and Fig. 1.

2.2 Case 2

Illumination received (readings were taken for the maximum light-receiving angle of sensor and at horizontal level) at a place in II floor class which is farthest from the light source for different calendar days at Jaipur Rajasthan at 10:30 AM and 11:30 AM when all the inner sources of light were all illuminated. These results are shown in Table 2 and Fig. 2.

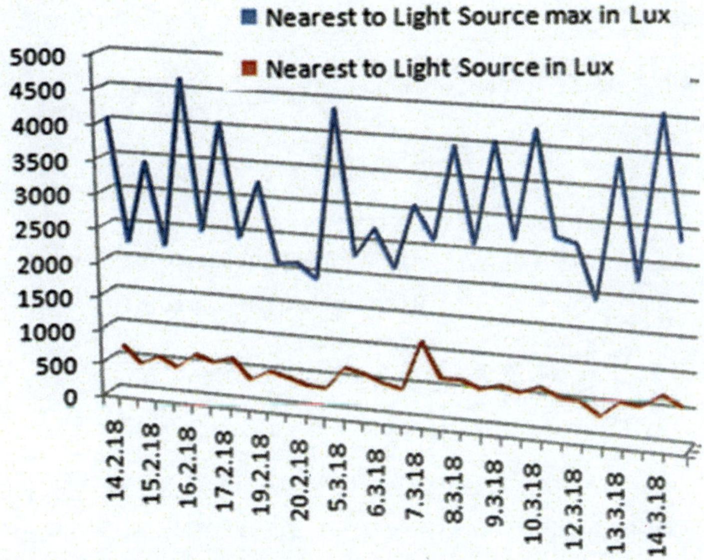

Fig. 1 Illumination at 2nd floor when a sensor is nearest to the light source

Table 2 Illumination in lux at 10:30 AM and 11:30 AM for 2nd floor farthest to light source

Date	Max. position of sensor	Horizontal position of sensor	Date	Max. position of sensor	Horizontal position of sensor
14.2.18	86	56	6.3.18	85	58
	76	56		80	55
15.2.18	88	58	7.3.18	89	61
	75	57		85	58
16.2.18	88	56	8.3.18	87	57
	81	58		82	56
17.2.18	89	58	9.3.18	87	58
	80	56		82	56
19.2.18	81	56	10.3.18	89	59
	79	56		81	55
20.2.18	85	58	12.3.18	92	56
	81	56		82	56
5.3.18	87	57	13.3.18	86	56
	82	56		81	53

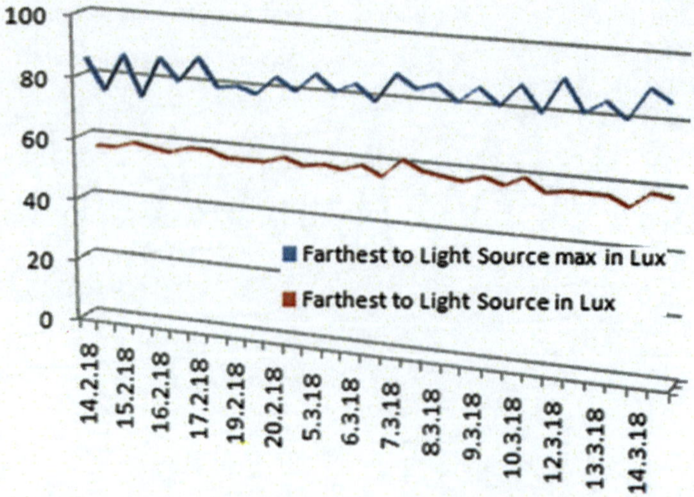

Fig. 2 Illumination at 2nd floor when a sensor is farthest to the light source

2.3 *Case 3*

Illumination received (readings were taken for the maximum light-receiving angle of sensor and at horizontal level) at a place in I floor class which is nearest to the light source for different calendar days at Jaipur Rajasthan at 10:30 AM and 11:30 AM when all the inner sources of light were all illuminated. The outputs are shown in Table 3 and Fig. 3.

2.4 *Case 4*

Illumination received (readings were taken for the maximum light-receiving angle of sensor and at horizontal level) at a place in I floor class which is farthest from the light source for different calendar days at Jaipur Rajasthan at 10:30 AM and 11:30 AM when all the inner sources of light were all illuminated. Table 4 and Fig. 4 are represented the outcomes.

2.5 *Case 5*

Illumination received (readings were taken for the maximum light-receiving angle of sensor and at horizontal level) at a place in ground floor class which is nearest to the light source for different calendar days at Jaipur Rajasthan at 10:30 AM and

Table 3 Illumination in lux at 10:30 AM and 11:30 AM for 1st floor nearest to light source

Date	Max. position of sensor	Horizontal position of sensor	Date	Max. position of sensor	Horizontal position of sensor
14.2.18	3950	480	6.3.18	1839	483
	1415	145		1604	322
15.2.18	2377	350	7.3.18	1470	475
	1267	160		1098	392
16.2.18	5661	947	8.3.18	3556	507
	2000	400		2211	492
17.2.18	4085	490	9.3.18	3867	568
	1565	358		2438	474
19.2.18	1936	586	10.3.18	3919	576
	1158	156		2315	470
20.2.18	2050	650	12.3.18	3792	482
	1210	205		2180	465
5.3.18	5434	837	13.3.18	3308	460
	2085	654		2417	450

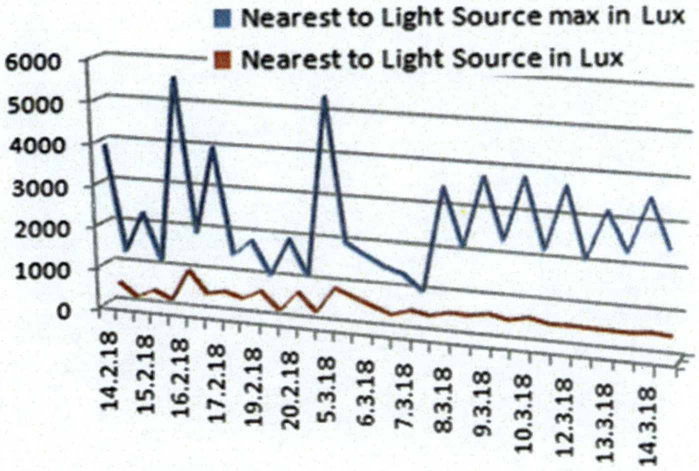

Fig. 3 Illumination at 1st floor when sensor is nearest to the light source

11:30 AM when all the inner sources of light were all illuminated as shown in Fig. 5 and Table 5.

Table 4 Illumination in lux at 10:30 AM and 11:30 AM for 1st floor farthest to light source

Date	Max. position of sensor	Horizontal position of sensor	Date	Max. position of sensor	Horizontal position of sensor
14.2.18	220	122	6.3.18	265	135
	280	140		204	125
15.2.18	270	146	7.3.18	320	120
	275	142		310	125
16.2.18	189	145	8.3.18	322	137
	170	120		301	139
17.2.18	280	146	9.3.18	327	140
	331	136		303	142
19.2.18	147	129	10.3.18	320	138
	178	127		300	137
20.2.18	261	97	12.3.18	292	125
	270	111		285	129
5.3.18	353	140	13.3.18	312	139
	318	146		305	141

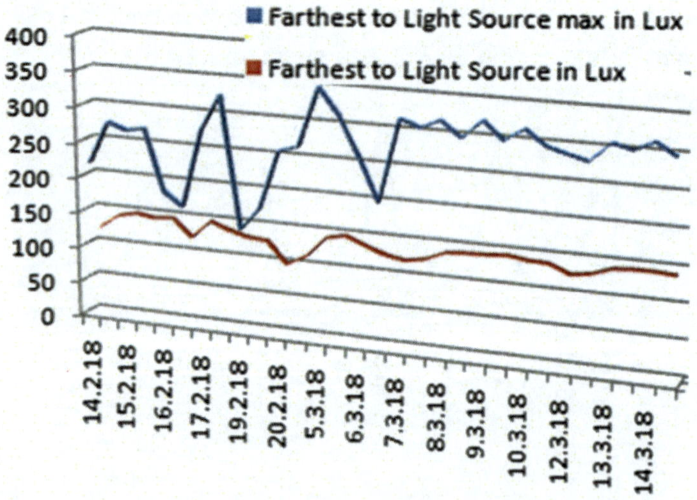

Fig. 4 Illumination at 1st floor when a sensor is farthest to the light source

2.6 *Case 6*

Illumination received (readings were taken for maximum light-receiving angle of sensor and at horizontal level) at a place in ground floor class which is farthest from the light source for different calendar days at Jaipur Rajasthan at 10:30 AM and

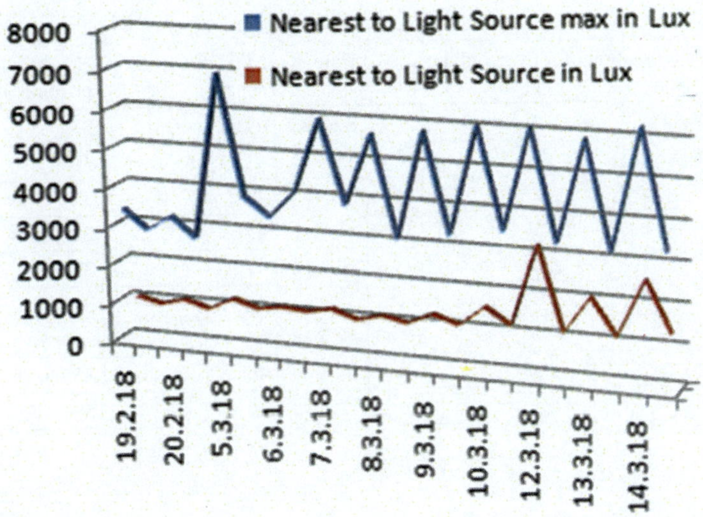

Fig. 5 Illumination at the ground floor when a sensor is nearest to the light source

Table 5 Illumination in lux at 10:30 AM and 11:30 AM for ground floor nearest to light source

Date	Max. position of sensor	Horizontal position of sensor	Date	Max. position of sensor	Horizontal position of sensor
19.2.18	3480	1007	8.3.18	5788	1076
	2976	857		3255	945
20.2.18	3380	1030	9.3.18	5956	1234
	2859	840		3445	1011
5.3.18	7062	1157	10.3.18	6172	1531
	4014	957		3647	1124
6.3.18	3548	1059	12.3.18	6215	3186
	4222	1002		3448	1075
7.3.18	6091	1128	13.3.18	6027	2010
	4002	908		3338	1085

11:30 AM when all the inner sources of light were all illuminated. These results are shown in Table 6 and Fig. 6.

2.7 *Case 7*

Comparative analysis of illumination in Lux for two readings at 10:30 AM and 11:30 AM on same days at a place where the maximum light angle of sensor is there

Table 6 Illumination in lux at 10:30 AM and 11:30 AM for ground floor farthest to light source

Date	Max. position of sensor	Horizontal position of sensor	Date	Max. position of sensor	Horizontal position of sensor
19.2.18	48	16	8.3.18	47	33
	35	14		41	30
20.2.18	39	23	9.3.18	49	34
	34	19		45	32
5.3.18	57	41	10.3.18	47	33
	47	32		44	31
6.3.18	49	27	12.3.18	53	36
	51	35		46	35
7.3.18	48	35	13.3.18	45	38
	43	31		42	34

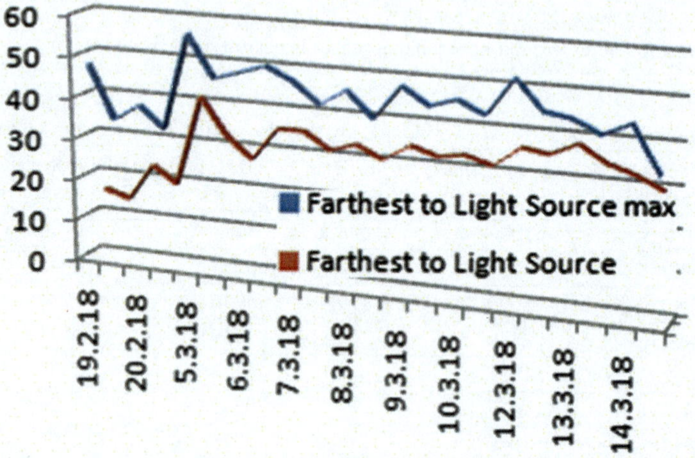

Fig. 6 Illumination at the ground floor when the sensor is farthest to the light source

for ground floor, first floor and second floor. The comparison is clearly shown in Fig. 7.

2.8 *Case 8*

Illumination received at a place of study is also compared with the temperature on that particular day as shown in Table 7 and Fig. 8.

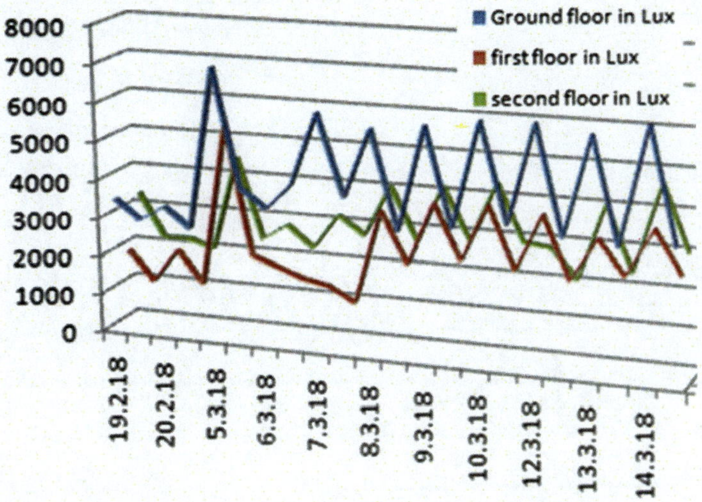

Fig. 7 Illumination at all floors when the sensor is nearest to the light source

Table 7 Maximum illumination in lux received based on the temperatures on a particular day for all floors

Temperature İn Deg. C	Ground Floor	I Floor	II Floor	Temperature İn Deg. C	Ground Floor	I Floor	II Floor
24	3480	1936	3310	23	5956	3867	4112
	2976	1158	2170		3445	2438	2781
26	3380	2050	2190	27	6172	3919	4317
	2859	1210	1978		3647	2315	2862
24	7062	5434	4434	25	6215	3792	2792
	4014	2085	2357		3448	2180	2018
25	3548	1839	2780	26	6027	3308	4003
	4222	1604	2230		3338	2417	2331
26	6091	1470	3150	28	6383	3666	4637
	4002	1098	2685		3443	2551	2930
23	5788	3556	4017				
	3255	2211	2658				

3 Conclusions and Remedies

From the observation data and graphical representations there on we can reach the following conclusions:

1. It is observed that irrespective of the floor which is under study, whenever an optical sensor is placed at a favorable angle; it detects the large magnitude

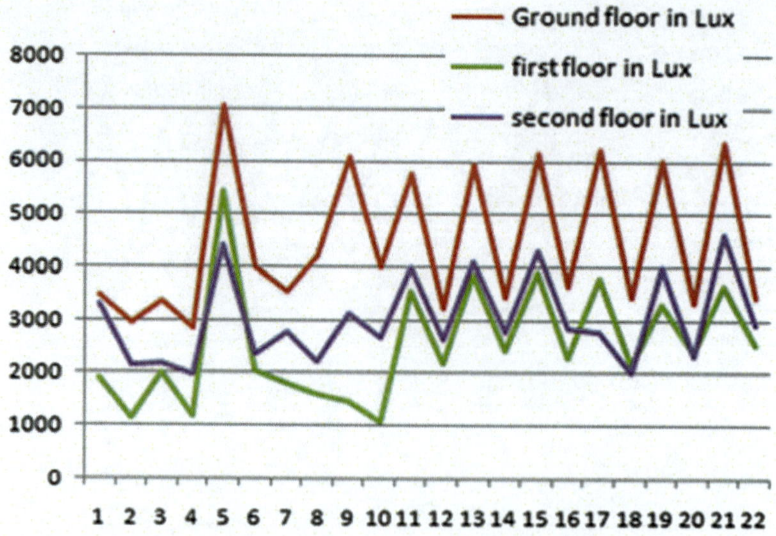

Fig. 8 Illumination at all floors to check the effect of temperature on different days

of illumination in comparison to that when held in a horizontal position. The horizontal position reading is important because at this very position most of the office task is performed or reading/writing is done.

2. It is found that there is insufficient light to undergo academic activity in the farthest area of the classrooms or office areas.
3. It was observed that rising temperature does not have a significant impact on the illumination as the temperature range in our case is not much.
4. It was observed that the ratio of window area to the total area did not have a significant impact on the illumination; as the total areas were different.
5. Here we can conclude that the illumination is very much affected by the architecture of the surroundings understudy, time of taking the readings, the climate on a particular day, calendar days under observation and trajectory of the sun. So, there must be a good provision of light sources within the working areas or classrooms and should not much depend on nature much.

References

1. R. Edward Fischer, et al. Optical system design. Vol. 1. No. 2 (McGraw Hill, New York, 2000)
2. http://www.illumenate.com/lightlevels.htm
3. R.F. Wolffenbuttel, State-of-the-art in integrated optical microspectrometers. IEEE Trans. Instrum. Meas. **53**(1), 197–202 (2004)
4. F. Pérennès, P.C. Beard, T.N. Mills, Analysis of a low-finesse Fabry-Perot sensing interferometer illuminated by a multimode optical fiber. Appl. Opt. **38**(34), 7026–7034 (1999)
5. Z. Malacara, M. Servin, Interferogram analysis for optical testing (CRC press, 2016)

6. S. Gupta, A. Sharma, Global Scenario of Solar Photovoltaic (SPV) materials, in *International Conference on Advanced Computational and Communication Paradigms (ICACCP)*, Lecture Notes in Electrical Engineering (LNEE), Springer, Vol. 475, pp. 126–133 (2018)
7. Z. Iqbal, M. Eriksson, Classification and quantitative optical analysis of liquid and solid samples using a mobile phone as illumination source and detector. Sens. Actuators B: Chem. **185**, 354–362 (2013)
8. M. Ardebili, A. Erdmann, Point spread and illumination analysis of lightfield cameras, in *2018 25th IEEE International Conference on Image Processing (ICIP)*, pp. 659–663 (2018)
9. E. Van Horn, et al, Devices, systems, and methods for optical validation. U.S. Patent Application No. 15/388,082
10. S. Gupta, K. Kumar et al, An IoT based approach to minimize air pollution. Int. J. Recent Technol. Eng. (IJRTE) 7(6S5), 596–599 (2019)
11. National Optical Astronomy Observatory, NOAO (in US)
12. http://www.engineeringtoolbox.com/light-level-rooms-d_708.html